Frontiers in Bioprocessing

Editors

Subhas K. Sikdar, Ph.D.
Group Leader
Transport Processes Group
Center for Chemical Engineering
National Institute of Standards and Technology
Boulder, Colorado

Milan Bier, Ph.D.
Professor of Chemical Engineering
Director
Center for Separation Science
University of Arizona
Tucson, Arizona

Paul Todd, Ph.D.
Biophysicist
Center for Chemical Engineering
National Institute of Standards and Technology
Boulder, Colorado

CRC Press, Inc.
Boca Raton, Florida

035798 76
CHEMISTRY

Library of Congress Cataloging-in-Publication Data

Frontiers in bioprocessing/editors, Subhas K. Sikdar, Milan Bier,
  Paul Todd.
       p.   cm.
     Plenary lectures and poster presentations presented at a
  conference held June 28—July 2, 1987, Boulder, Colo.
     Bibliography: p.
     Includes index.
     ISBN 0-8493-5839-6
     1. Biotechnology—Technique—Congresses.   2. Separation
  (Technology)—Congresses.   I. Sikdar, Subhas K.   II. Bier, Milan.
  III. Todd, Paul, 1936-
  TP248.24.F76   1989
  660′.6—dc19

                                                             89-30528
                                                             CIP

Direct all inquiries to CRC Press, Inc., 2000 Corporate Blvd., N.W., Boca Raton, Florida, 33431.

© 1990 by CRC Press, Inc.

International Standard Book Number 0-8493-5839-6

Library of Congress Card Number 89-30528
Printed in the United States

To the Memory

of a Little Girl, Manjori Sikdar

PREFACE

There is a great demand for generic research in process measurement and control and in bioseparations as they pertain to the rapidly maturing biotechnology industry. The National Institute of Standards and Technology (NIST) and the National Aeronautics and Space Administration (NASA) sponsor basic physical research in process technology and foster the diffusion of technology into the commercial sector. NIST, an agency of the U.S. Department of Commerce, maintains one of the nation's leading physical research capabilities, while NASA is responsible for access to low-gravity environments, an important resource for the investigation of gravity-dependent fluid phenomena relevant to bioprocessing. It seemed fitting that researchers representing the goals of these two agencies should, with the concurrence and support of the agencies, convene a conference dealing with subjects at the leading edge of bioprocessing research and development. Thus was born the Frontiers in Bioprocessing conference, which took place in Boulder, Colorado from June 28 to July 2, 1987. We wish to thank the sponsors of the Frontiers in Bioprocessing conference: the U.S. Department of Commerce, NASA, and the corporate sponsors (Schering-Plough, Coors, Synergen, and Synthetech).

The rapid dissemination of bioprocessing research results that characterized the conference was only made possible by the generous participation of representatives of leading companies, universities, and governments, who freely shared their common problems and scientific progress. The pages that follow contain the plenary lectures and several poster presentations that were given at the conference in the spirit of exploring the frontiers and creating a spirit that leads beyond them. The problems and challenges of modern bioprocessing are aptly defined in the keynote paper by Professor Humphrey. Each of these problems is addressed by contributors in six categories: process integration, fermentation control, sensor development, free-fluid bioseparations, chromatography, and emerging technologies such as integrated systems, protein engineering, and mineral processing. Each of these contributions was reviewed by volunteer scientists to whom we are extremely grateful for their patience, willing contribution of time, and, in many cases, promptness. As a token of our gratitude to them, we have listed their names and affiliations below. Finally, we wish to thank and acknowledge those who contributed in a major way to the organization of the conference and the publication of the proceedings: Kent Zimmerman and his staff at the University of Colorado Conference Center; Jesse Hord, director of the NIST's Center for Chemical Engineering, and his staff, including the local organizing committee; Dr. V. N. Schrodt; Dr. R. Perkins; Mrs. Linda Banks, secretary; and the editorial staff of CRC Press, Inc.

Subhas K. Sikdar
Milan Bier
Paul Todd
Boulder, Colorado
May 1988

REVIEWERS OF PAPERS WRITTEN FOR THE CONFERENCE

THE EDITORS

Subhas K. Sikdar, Ph.D., is Group Leader, Transport Processes Group, Center for Chemical Engineering, National Institute of Standards and Technology, Boulder, Colorado. Dr. Sikdar graduated from Calcutta University, Calcutta, India with a B.Sc. in chemistry (1964) and a B.Tech. and an M.Tech. in chemical engineering and chemical technology (1966 and 1967, respectively). He received his M.S. and Ph.D. degrees in chemical engineering in 1971 and 1975, respectively, from the University of Arizona, Tucson. Prior to his present position, Dr. Sikdar worked as a Senior Research Engineer for Occidental Research Corporation in California and as Unit Manager of Process Technology at the General Electric Research & Development Center in Schenectady, New York.

Dr. Sikdar is a member of the American Institute of Chemical Engineers, the American Chemical Society, and the American Society for Testing and Materials, where he is the chairman of the subcommittee on unit processes and their control. In his present job, he directs research on membrane separation, heat and mass transfer, bioreactions, bioseparations, and microgravity processes. He has published 27 journal papers and has registered 14 U.S. patents.

Milan Bier, Ph.D., is Professor in Chemical Engineering and the founder and Director of the Center for Separation Science at the University of Arizona, Tucson. The Center is internationally recognized for its leadership in the study of electrophoretic transport processes.

Dr. Bier completed his undergraduate training in chemistry at the University of Geneva, Switzerland and obtained his Ph.D. in chemistry in 1950 at Fordham University, New York. He was Adjunct Professor at the University of Arizona and Research Biophysicist at the Veterans Administration Hospital in Tucson, Arizona from 1962 to 1977. Since 1977 he has been associated solely with the University of Arizona's College of Engineering.

Dr. Bier is a member of the American Association for the Advancement of Science, the American Society of Biological Chemists, and a past president of the Electrophoresis Society. He has edited four books, and he serves on the editorial boards of three journals. Dr. Bier has over 150 publications and 14 patents. He has served on advisory boards and held numerous consultancies in industry and government, and he has held long-term grants from the National Aeronautics and Space Administration and other government agencies and private foundations.

Paul Todd, Ph.D., is a Biophysicist in the Center for Chemical Engineering, National Institute of Standards and Technology and Bioseparations Program Leader in the Transport Processes Group, Boulder, Colorado; he also is affiliated with the Department of Chemical Engineering, University of Colorado, Boulder.

Dr. Todd graduated in 1959 from Bowdoin College, Brunswick, Maine with a B.A. degree in physics (cum laude) and from the Massachusetts Institute of Technology with an S.B. in 1959, and he obtained his M.S. degree at the University of Rochester and a Ph.D. in biophysics at the University of California, Berkeley. From 1966 to 1986 he served as Assistant Professor, Associate Professor, and Professor of Biophysics at Pennsylvania State University, University Park and for 5 years was Chairman of the Graduate Program in Genetics. He was appointed Director of the Philadelphia Bioprocessing and Pharmaceutical Research Center in 1984 and served in that capacity for 3 years before joining the National Institute of Standards and Technology.

Dr. Todd is a member of the Society for Analytical Cytology, the Electrophoresis Society, the American Association for the Advancement of Science, Sigma Xi, the New York Academy of Sciences, and the American Society for Photobiology. He serves on the editorial boards of five scientific journals and has served on several national and regional boards, committees, and panels, including the Space Applications Board of the National Academy of Sciences/National Research Council. He has received international research grants from the Eleanor Roosevelt Foundation, the International Union against Cancer, the Fogarty International Center of the National Institutes of Health, and the Universities Space Research Association. His research has been supported by the U. S. Atomic Energy Commission, the National Institutes of Health, the Pennsylvania Research Corporation, and the National Aeronautics and Space Administration.

Dr. Todd has co-edited one technical book and has published over 100 scientific articles. His current research interests are cell separation and cell culture and the application of electrokinetics and low-gravity methods to bioprocessing research.

CONTRIBUTORS

M. M. Ataai, Ph.D.
Department of Chemical and
  Petroleum Engineering
University of Pittsburgh
Pittsburgh, Pennsylvania

William E. Bentley, M.S.E.
Department of Chemical Engineering
University of Colorado
Boulder, Colorado

M. Bier, Ph.D.
Center for Separation Science
University of Arizona
Tucson, Arizona

J. Douglas Birdwell, Ph.D
Department of Electrical Engineering
University of Tennessee
Knoxville, Tennessee

F. E. Brinckman, Ph.D.
Institute for Materials Science
  and Engineering
National Institute of Standards
  and Technology
Gaithersburg, Maryland

Peter E. Brodelius, Ph.D.
Institute of Biotechnology
Swiss Federal Institute of Technology
Zurich, Switzerland

Donald Elliott Brooks, Ph.D.
Departments of Pathology and Chemistry
University of British Columbia
Vancouver, British Columbia
Canada

James B. Callis, Ph.D.
Center for Process Analytical Chemistry
Department of Chemistry
University of Washington
Seattle, Washington

K. Chan
Information Industries, Inc.
Kansas City, Missouri

M. H. Coan, Ph.D.
Protein Purification Research
  and Process Development
Cutter Biological
Miles Laboratories, Inc.
Berkeley, California

S. Cozzette, B.S.
I-STAT Corporation
Princeton, New Jersey

Lois Davidson, B.S.
Department of Chemistry
University of Texas
Austin, Texas

G. Davis, D.Phil.
I-STAT Corporation
Princeton, New Jersey

N. Egen, Ph.D.
Center for Separation Science
University of Arizona
Tucson, Arizona

William Elsasser
Bioprocessing and Pharmaceutical
  Research Center
Philadelphia, Pennsylvania

Joseph Feder, Ph.D.
Invitron Corporation
Saint Louis, Missouri

Ronald L. Fournier, Ph.D.
Department of Chemical Engineering
University of Toledo
Toledo, Ohio

T. K. Ghose, Ph.D
Biochemical Engineering Research Center
Indian Institute of Technology
New Delhi, India

P. Ghosh, Ph.D
Biochemical Engineering Research Center
Indian Institute of Technology
New Delhi, India

Loni Gibbs
Dupont Chemical Company
Freeport, Texas

Jan-Gunnar Gustafsson, M.S.
Biochemical and Biological R & D
Pharmacia LKB Biotechnology AB
Uppsala, Sweden

Per Hedman, M.S.
Biochemical and Biological R & D
Pharmacia LKB Biotechnology AB
Uppsala, Sweden

Linda L. Henk, B.S.
Department of Agricultural and
  Chemical Engineering
Colorado State University
Fort Collins, Colorado

Susan Hershenson, Ph.D.
Process and Product Development
Cetus Corporation
Emeryville, California

Arthur Humphrey, Ph.D.
Center for Molecular Bioscience
  and Biotechnology
Lehigh University
Bethlehem, Pennsylvania

R. C. Humphreys, B.S.
Center for Separation Science
University of Arizona
Tucson, Arizona

J. Jachowicz
Chemical Engineering Department
Polytechnic University
Brooklyn, New York

J. W. Jeong
Department of Chemical and
  Petroleum Engineering
University of Pittsburgh
Pittsburgh, Pennsylvania

Göte Johansson, Ph.D.
Department of Biochemistry
University of Lund
Lund, Sweden

Ramesh Kagalanu, M.S.Ch.E.
Department of Chemical Engineering
University of Toledo
Toledo, Ohio

M. N. Karim, Ph.D.
Department of Agricultural and
  Chemical Engineering
Colorado State University
Fort Collins, Colorado

Isao Karube, Ph.D
Research Center for Advanced
  Science and Technology
University of Tokyo
Tokyo, Japan

John A. Kehoe, B.S.
Development and Technical
  Services Administration
Eli Lilly and Company
Indianapolis, Indiana

G. Barrie Kitto, Ph.D.
Department of Chemistry
University of Texas
Austin, Texas

W. J. Ko, M.S.
Center for Separation Science
University of Arizona
Tucson, Arizona

Dhinakar S. Kompala, Ph.D.
Department of Chemical Engineering
University of Colorado
Boulder, Colorado

R. Kosecki, Ph.D.
Department of Biotechnology
Schering-Plough Research
Bloomfield, New Jersey

Jeffrey Kurdyla
Smith Kline and French Laboratories
King of Prussia, Pennsylvania

K. H. Kwan
Center for Separation Science
University of Arizona
Tucson, Arizona

J. Labdon, Ph.D.
Department of Biotechnology
Schering-Plough Research
Bloomfield, New Jersey

I. R. Lauks, Ph.D
I-STAT Corporation
Princeton, New Jersey

Howard L. Levine, Ph.D.
Pilot Plant Operations
Xoma Corporation
Berkeley, California

N. H. Lin
Planning Department
City of North Charleston
North Charleston, South Carolina

James C. Linden, Ph.D.
Departments of Agricultural and
  Chemical Engineering and Microbiology
Colorado State University
Fort Collins, Colorado

Jorge L. Lopez, Ph.D.
Research and Development
Sepracor, Inc.
Marlborough, Massachusetts

Christine Markeland-Johansson, M.S.
Biochemical and Biological R & D
Pharmacia LKB Biotechnology AB
Uppsala, Sweden

F. S. Markland, Ph.D.
Department of Biochemistry
University of Southern California
  School of Medicine
Los Angeles, California

Stephen L. Matson, Ph.D.
Research and Development
Sepracor, Inc.
Marlborough, Massachusetts

G. Mitra, Ph.D.
Protein Purification Research
  and Process Development
Cutter Biological
Miles Laboratories, Inc.
Berkeley, California

Harold G. Monbouquette, Ph.D.
Department of Chemical Engineering
University of California, Los Angeles
Los Angeles, California

Charles F. Moore, Ph.D.
Department of Chemical Engineering
University of Tennessee
Knoxville, Tennessee

T. L. Nagabhushan, Ph.D.
Department of Biotechnology
Schering-Plough Research
Bloomfield, New Jersey

Samer F. Naser, M.S.E.
Department of Chemical Engineering
University of Toledo
Toledo, Ohio

Michael G. Norton, Ph.D.
Biotechnology and Separations Division
Warren Spring Laboratory
Stevenage, Hertfordshire
United Kingdom

David F. Ollis, Ph.D.
Department of Chemical Engineering
North Carolina State University
Raleigh, North Carolina

G. J. Olson, Ph.D.
Institute for Materials Science
  and Engineering
National Institute of Standards
  and Technology
Gaithersburg, Maryland

S. Piznik, Ph.D
I-STAT Corporation
Princeton, New Jersey

B. Pramanik, Ph.D.
Department of Molecular Spectroscopy
Schering-Plough Research
Bloomfield, New Jersey

Percy H. Rhodes, M.S.
Space Science Laboratory
Marshall Space Flight Center
Huntsville, Alabama

Peter W. Runstadler, Jr., Ph.D.
Verax Corporation
Lebanon, New Hampshire

D. W. Sammons, Ph.D.
Center for Separation Science
University of Arizona
Tucson, Arizona

Burton E. Sarnoff, M.S.
Princeton Applied Research Corporation
Princeton, New Jersey

Ze'ev Shaked, Ph.D.
Manufacturing and Development
Triton Biosciences
Alameda, California

M. L. Shuler, Ph.D.
School of Chemical Engineering
Cornell University
Ithaca, New York

N. J. Smit, Ph.D.
I-STAT Corporation
Princeton, New Jersey

Robert S. Snyder, D.Sc.
Space Science Laboratory
Marshall Space Flight Center
Huntsville, Alabama

Gregory Stephanopoulos, Ph.D.
Department of Chemical Engineering
Massachusetts Institute of Technology
Cambridge, Massachusetts

W. J. Strohm
Chemical Engineering Department
Polytechnic University
Brooklyn, New York

Eugene Sulkowski, Ph.D.
Department of Molecular and
 Cellular Biology
Roswell Park Memorial Institute
Buffalo, New York

Eiichi Tamiya, Ph.D.
Research Center for Advanced
 Science and Technology
University of Tokyo
Tokyo, Japan

Paul Todd, Ph.D.
Center for Chemical Engineering
National Institute of Standards
 and Technology
Boulder, Colorado

P. P. Trotta, Ph.D.
Department of Biotechnology
Schering-Plough Research
Bloomfield, New Jersey

Kevin M. Ulmer, Ph.D.
SEQ, Ltd.
Cohasset, Massachusetts

Joseph J. Vallino, M.S.
Department of Chemical Engineering
Massachusetts Institute of Technology
Cambridge, Massachusetts

Tse-Wei Wang, Ph.D.
Chemical Engineering Department
University of Tennessee
Knoxville, Tennessee

John G. Watt, M.R.C.V.S.
Bioseparation Associates Limited
Livingston, West Lothian
Scotland

Kristina Wiberg
Biochemical and Biological R & D
Pharmacia LKB Biotechnology AB
Uppsala, Sweden

H. J. Wieck, Ph.D.
I-STAT Corporation
Princeton, New Jersey

Michael W. Young
Verax Corporation
Lebanon, New Hampshire

TABLE OF CONTENTS

Chapter 1

PROBLEMS AND CHALLENGES IN THE PRODUCTION AND PROCESSING OF BIOLOGICALLY ACTIVE MATERIALS

Arthur Humphrey

TABLE OF CONTENTS

I. INTRODUCTION

Biotechnology is no longer a fantasy. It offers numerous opportunities to improve the quality of life on many fronts. It presently represents a more than $2 billion/year industry in the U.S. alone. Most economists predict it will be a $40 billion/year industry by the turn of the century. The U.S. has been, and continues to be, the world leader in genetic biotechnology research.[1-4] Maintenance of that leadership and timely commercialization of the fruits of biotechnology research are vitally important to the future economic well-being and international technological stature of the U.S. The key to commercialization of this research is a strong knowledge base on which to design bioprocessing systems for the manufacture of bioproducts.[5] Unfortunately, little attention has been given to date to downstream bioprocessing research. This text will focus on the development of new and improved processing techniques for the manufacture of products of this "new biology". I have been asked by the editors to "set the stage" for the chapters to follow by highlighting the problems and challenges in the production and processing of biologically active materials.

A. WHAT IS THE PAYOFF?

Virtually every economist and every biotechnologist has a personal assessment of the potential of biotechnology. Busche and Hardy[6] have reported on a consensus survey of 75 experts in the field on biotechnology opportunities (see Figure 1). Their survey suggests that the greatest potential is in health care, followed by agriculture and, then, chemicals. A number of reports on biotechnology potentials have been summarized in a National Research Council (NRC) research briefing report entitled *Chemical and Process Engineering for Biotechnology*.[1] This panel has estimated that potential worldwide markets for biologically derived products will range from $40 to $100 billion annually by the year 2000. The high and low ranges of these estimates for each area are given in Table 1. This ranking indicates pharmaceuticals will be first, followed by chemicals, then agriculture. Clearly, the consensus is that health care will be the most important application of biotechnology. There is disagreement over whether chemical or agricultural products have the next greatest potential for biotechnology applications.

Pollack[13] suggests that the potential for biotechnology products will be at least $10 billion/year by 2000. He notes that the present market capitalization of the seven largest new genetic engineering companies (see Table 2) is nearly $6 billion and for all 200 new genetic engineering companies it approaches $10 billion. He points out that the four pharmaceuticals produced to date by genetic engineering techniques exceeded $150 million in sales last year. He suggests that the potential for genetically engineered drugs in 1988 could approach $0.5 billion (see Table 3). When one adds the sales of diagnostic bioproducts, income from breeding better plants and animals, and high-value-added chemicals produced by bioprocesses, it is not difficult to believe that by the year 2000 this industry will have a value in the tens of billions of dollars.

B. CHEMICALS VIA BIOTECHNOLOGY?

The debate over the potential for biotechnology in the chemical industry is very understandable. It stems from the fact that in terms of absolute market sales, commodity chemicals derivable from biotechnology processes have far and away the largest sales potential of any of the bioproducts, including medicinals[6] (see Table 4). However, most forecasters do not take into account the fact that most bioproducts are produced in dilute aqueous solutions. It takes a lot of energy to "squeeze the water out" of the bioproducts. Note that the selling price of bioproducts can be related to the final concentration of product achieved in the fermentation broth[2] (see Figure 2) and the annual production rate[5] (see Figure 3). In the case of large-volume chemicals, bioprocesses have to compete with efficient high-product-concentration chemical processes. Biotechnology probably will not make significant headway

3

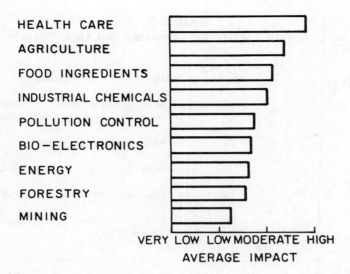

FIGURE 1. Expected worldwide impact of biotechnology. (From Busche, R. M. and Hardy, R. W. F., *Biotechnol. Bioeng. Symp.*, 15, 651, ©1985. Reprinted by permission of John Wiley & Sons, Inc.)

TABLE 1
Estimates for Annual Sales of Biotechnology
Products by the Year 2000[11]

Category	Millions of Dollars	
	Low estimate	High estimate
Medical products	7,000	45,000
Chemical products	5,000	25,000
Agricultural products	3,000	9,000
Food/feed	3,000	4,000
Associated equipment and engineering systems	10,000	24,000
Sum of combined low and high estimates	28,000	107,000

in the production of commodity chemicals because of end-product inhibition effects on cellular biology and the difficulty genetic engineering has had solving this end-product inhibition problem. Because of these problems, biotechnology will impact only on those high-value chemical products having a value greater than $10/kg. Therefore, one should not consider biotechnology research on commodity chemicals unless it is directed at process improvement, i.e., nonaqueous bioprocessing, elimination of end-product inhibition, or development of cheaper substrates. Future biotechnology research in the chemical area should be focused on high-value-added chemicals produced by tissue culture, where small yield improvements can result in significantly increased process economies. However, more on this later.

II. SPECIFIC OPPORTUNITIES

The specific opportunities for commercializing the products and devices emerging from biotechnology are both tantalizing and highly diverse. Among the more exciting prospects are those described in the following sections.

TABLE 2
Leading Genetic Engineering Companies and Their Products[13]

Company	Market capitalization (millions)	Products
Genentech	$3200	Human growth hormone; tissue plasminogen activator; gamma interferon; tumor necrosis factor
Cetus	776	Interleukin-2; beta interferon
Amgen	628	Erythropoieten; granulocyte colony stimulating factor
Genetics Institute	398	Granulocyte colony stimulating factor; factor VIII:C
Chiron	348	Hepatitis B vaccine; superoxide dismutase; epidermal growth factor
Biogen	243	Alpha interferon; gamma interferon
Immunex	180	Interleukin-2; granulocyte colony stimulating factor
California Biotechnology	155	Atrial natriuretic factor

TABLE 3
First Genetically Engineered Pharmaceuticals[13]

Pharmaceutical	Company	Sales, $/year (potential)
Human insulin	Eli Lilly	100,000,000
Human growth hormone	Genentech	44,000,000
Alpha interferon	Biogen/Schering Plough	5,000,000[a]
Hepatitis B vaccine	Chiron/Merck	10,000,000[a]
Tissue plasminogen activator	Genentech	(200,000,000)
Interleukin-2	Cetus/Immunex	(10,000,000)

[a] Estimated values.

A. HUMAN AND ANIMAL HEALTH CARE

A revolutionary new family of diagnostic products based on enzymes, monoclonal antibodies, and other genetically engineered proteins promises to provide quick and highly accurate detection of immunity to or infection by viral and bacterial diseases, susceptibility to autoimmune diseases, the presence of genetic defects, or the existence of neoplasms. Other significant opportunities include novel prophylactic products: vaccines for the prevention of viral, bacterial, and protozoal diseases such as hepatitis, typhus, and malaria; new therapeutic biologicals for the treatment of cardiovascular and cerebrovascular disease, neurological diseases, rheumatoid arthritis, diabetes, and cancer; and peptide hormonal substances that minimize dwarfism, stimulate red blood cell production, increase milk production of dairy cattle, stimulate growth, and enhance feed utilization by cattle and other farm animals.

B. HUMAN AND ANIMAL NUTRITION

Opportunities include the utilization of low-cost carbon sources for new microbiological and enzymatic syntheses of amino acids, sugars, and edible fats and oils, and the use of low-value feedstocks in the manufacture of nutritionally balanced single-cell protein for

TABLE 4
Bioproduct Markets

Current world sales ($ million U.S.)	Product
$14,180	Organic solvents and acids
1,700	Amino acids
1,625	Antibiotics
667	Vitamins
440	Industrial enzymes
380	Steroids and alkaloids
260	Polypeptides and hormones
160	Nucleotides and nucleosides
155	Medicinal enzymes
100	Polysaccharide gums

Modified from Busche, R. M. and Hardy, R. W. F., *Biotechnol. Bioeng. Symp.*, 15, 651, 1985.

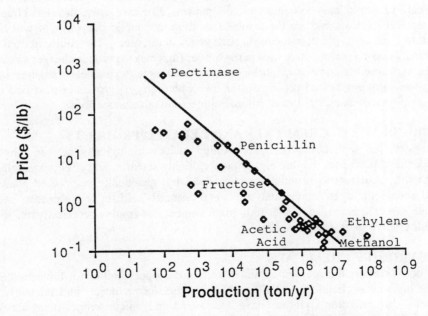

FIGURE 2. Price vs. production volume for selected commodity and specialty chemicals and biochemicals. (Data courtesy of Dr. C. L. Cooney, Massachusetts Institute of Technology.)

human or animal consumption. This area is particularly challenging for developing nations with rapidly expanding populations and with special needs for nourishment of their peoples.[7,8]

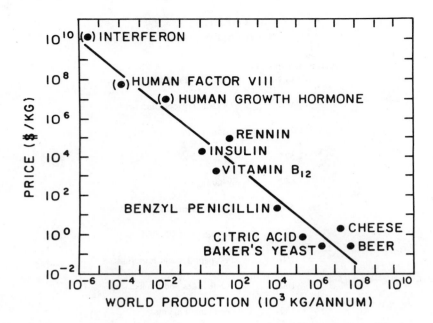

FIGURE 3.   World production and prices per kilogram of some products of biotechnology.
Supporting data are given in Table 7. (From Dunnill, P., *Biochem. Soc. Symp.*, 48, 9,
1983. With permission.)

Another hope for biotechnology or genetic engineering applications in the food area is
to transfer the genes coding for various food flavors, colors, and fragrances from plant and
animal cells to simple, easy-to-culture microorganisms. However, since flavor and fragrances
are complex chemicals and are the result of multiple metabolic pathways, it will be very
difficult to transfer, express, and control such genes in microbes. The ability to do this will
undoubtedly occur a considerable time in the future. Enzymes to produce cheaper sweeteners,
e.g., the high-fructose corn syrups, have been obtained through genetically engineered cells.
Also, genetically engineered proteins offer the opportunity of providing modified texture
and other physical characteristics to future engineered convenience foods.

C. AGRICULTURAL CHEMICALS AND RELATED PRODUCTS

Prospects for applications of biotechnology include biologically derived fungicides,
herbicides, and pesticides that are highly potent, highly specific, and environmentally safe;
plant growth regulators to stimulate crop productivity; genetically engineered plants with
reduced sensitivity to environmental stress and reduced fertilizer requirements; and new
techniques for crop propagation and strain improvement that employ genetically manipulated
plant cell embryoids to replace seeds.

D. COMMODITY CHEMICALS

As noted earlier, the specific commodity chemical opportunities for biotechnology lie
with the high-value chemicals. This is especially true for countries with relatively small
economies. Just as some countries have specialized and unique crops, there are related
opportunities for developing specialized products, particularly from plant tissue culture. The
focus can be directed to those opportunities which by their size and volume are not attractive
to giant chemical companies such as DuPont and Monsanto.

Sahai and Knuth[12] have suggested some plant products of commercial interest (see Table
5). At present, some of these products have sufficient volume and value to interest even big
chemical companies. For example, diosgenin has a 200,000 kg/year market demand and a

TABLE 5
Plant Products of Commercial Interest

Food colors, flavors, and fragrances
   Colors: anthocyanins, betacyanin, saffron
   Flavors: strawberry, grape, vanilla, tomato
   Oils: mint, rose, vetiver, jasmine, patchouly, garlic, sandalwood, lemon, onion
   Sweeteners: stevioside, thaumatin, miraculin, monellin
Agricultural chemicals
   Pyrethrins, rotenone, azadirachtin, nerlifolin, salannin
Pharmaceuticals
   Codeine, morphine, scopolamine, atropine, vinblastine, L-dopa, hyoscyamine, diosgenin, digoxin, quinine,
     shikonin, ajmalacine, serpentine

Modified from Sahai, O. and Knuth, M., *Biotechnol. Prog.*, 1, 1, 1985.

selling price of nearly $700/kg. It is presently derived from the *Dioscorea deltoides* plant, where it occurs in concentrations of up to 2% of the dried plant tissue. Big companies worry about the long-term profitability of such products when produced on a large scale. Because of economies in scale, fermentation products such as penicillin sell for around $40/kg, monosodium glutamate for $2/kg, and citric acid for $1.4/kg. Obviously, if diosgenin could be produced by large-scale tissue culture for prices approaching that of penicillin, it would have little attraction to a large chemical manufacturer. Instead of a $140 million/year potential, it might become only an $8 million/year product. Because of this, there probably is a niche for emerging bioentrepreneurs in the high-value-added specialty chemical market which would be marginal for the big chemical companies.

E. ENVIRONMENTAL IMPROVEMENT AND PROTECTION

New microbial and enzymatic techniques for removing or destroying toxic pollutants in municipal and industrial wastes provide a promising horizon for the application of biotechnology to environmental problems. Opportunities exist for collaboration among biotechnologists, toxicologists, and environmental engineers for developing and implementing these techniques.

F. NATURAL RESOURCE UTILIZATION

New microbiological processes for the recovery of metals and nonmetals (e.g., iodine) from low-grade ores and subsurface aquifers are under development. These processes could increase our supply of essential minerals substantially. The possibility of increasing recoveries of petroleum from depleted reservoirs by microbiological means is being explored, as are *in situ* microbial and enzymatic techniques for transforming solid fossil carbon sources (coal, shale) into gaseous and liquid fuels. As the various fossil fuel crises repeat themselves (e.g., acid rain, oil cartels, etc.), biotechnology processes will again be examined and possibly utilized for producing fuels and energy from biomass and wastes.

III. WHAT ARE THE CHALLENGES?

A. THE NEED FOR RESEARCH

In a 1984 report on commercial biotechnology from the Office of Technology Assessment of the U.S. Congress,[2] it was noted that "in 1983 the Federal Government spent $511 million on basic biotechnology research compared to $6.4 million on generic applied research in biotechnology." The report went on to state that "in the next decade the competitive advantage (to the U.S.) in areas related to biotechnology may depend as much on developments in bioprocess engineering as on innovations in genetics, immunology and other areas of basic science."

TABLE 6
U.S. Firms Commercializing
Biotechnology[2]

Total number — January 1984	219
New firms since 1973	118
New firm starts — 1980 through 1982	91

Many congressmen ask, "Why can't the development of this bioprocess knowledge base be left to the private sector?" The answer lies in the character of the biotechnology industry. At the present time, nearly 250 U.S. companies are involved in the commercialization of biotechnology. One half of these companies have been established in the last decade. Roughly 100 are less than 5 years old (see Table 6). Few have yet built a commercial plant to produce their first product. Most are in the product development and testing stage.

Part of the problem is that most bioproducts are high-unit-value materials. For this reason, there is little concern for optimizing the process until the commercial value of the product has been established. As a consequence, many "new" biotechnology companies initially make their products virtually in pails.

Most directors of research of these genetic engineering companies say they haven't had time to worry about scaleup and bioprocess improvement. Their greatest concern is rapid approval of their products. At present, they are focusing on developing better gene vectors, gaining a handle on protein excretion from the cloned cells, and maintaining the activity of their protein products during processing, i.e., understanding protein folding and unfolding. For the most part, gaining a better knowledge base for the design of bioprocesses is rather remote in their thinking. They point out that a new company cannot afford to spend its precious investment dollars on process research until the value of its product(s) has been established. However, to maintain a competitive advantage one must have not only the best product, but also a product made by the most economical process. To achieve this, an adequate knowledge base must exist to enable the company to do the following:

1. Design and scale up bioreactors
2. Design and scale up separation and purification operations
3. Develop biosensors and process-control algorithms enabling effective and economic control of the bioprocesses

B. WHAT ARE THE CONSEQUENCES OF *NOT* DOING THIS?

The present U.S. lead in biotechnology research is being unwittingly lost to the Japanese and Western European countries who are poised to establish commercial positions in biotechnology by practicing effective and forward-looking bioprocess engineering. For virtually no cost they are utilizing our basic research in biotechnology, which is made conveniently available to them through the open literature.

Past case histories illustrate lost opportunities in related fields because of a lack of aggressiveness in U.S. research in process engineering. For example, U.S. companies led the world in manufacture of amino acids prior to the 1960s. Then, an intensive Japanese effort began to improve amino acid production through biological and biochemical engineering techniques. First, classical genetic strain improvement techniques were applied to isolated mutant microorganisms that produced greatly increased quantities of amino acids. Second, biochemical engineers utilized very large bioreactors and developed efficient methods for product recovery. Today, Japan dominates the amino acid business, with annual world sales of $1.7 billion.

In another example, process technology for penicillin manufacture was developed in the U.S. in the 1940s. Protected by patents, several U.S. companies dominated the world

TABLE 7
Supporting Data for Figure 3

Product	Production (tons × 10³/year)	Price ($/lb)
Chemicals		
Ethylene	15,000	0.25
Propylene	7,000	0.24
Toluene	5,000	0.21
Benzene	5,000	0.23
Ethylene	4,600	0.14
Methanol	4,200	0.11
Styrene	3,300	0.38
Xylene	3,200	0.21
Formaldehyde	1,100	0.29
Ethylene oxide	2,600	0.45
Ethylene glycol	2,100	0.33
Butadiene	1,600	0.38
Acetic acid	1,400	0.26
Phenol	1,300	0.36
Acetone	1,100	0.32
Pyopylene oxide	900	0.44
Isopropanol	800	0.32
Adipic acid	600	0.57
Ethanol (synthetic)	600	0.27
Ethanol (ferm.)	700	0.27
Dextrose	430	0.45
Citric acid	300	0.80
Gluconate	75	0.48
Fructose	22	1.10
L-Glutamine	270	1.8
D,L-Methionine	100	2.9
L-Lysine	25	7.3
L-Aspartic acid	0.6	2.7
L-Arginine	0.5	13.6
L-Phenylalanine	0.1	38
L-Tryptophan	0.06	44
Enzymes		
Glucose isomerase	1	25
Glucoamylase	0.35	30
Bacterial amylase	0.35	35
Bacterial protease	0.5	60
Pectinase	0.1	700
Other		
Penicillin	7	20
D-Calcium pantothenate	4.5	6.4

Data courtesy of Dr. C. L. Cooney, Massachusetts Institute of Technology.

penicillin market for many years. Now that the patents have expired and U.S. companies have decided to invest in research on new, higher-return products rather than on penicillin process improvement, European companies with modern process technology account for the preponderance of worldwide sales of natural penicillin.

Japan and Western Europe can be expected to be no less aggressive as the results of the U.S. research in the "new biology" diffuse abroad.

C. WHAT, THEN, ARE THE SPECIFIC PROCESS CHALLENGES?
Frankly stated, our ability to perform and control large-scale tissue culture for the

manufacture of bioproducts is primitive. Our knowledge of techniques for efficient large-scale recovery and purification of bioproducts from the complex mixtures of proteins in which they occur is little more than rudimentary. Successful manufacture of bioproducts requires the availability of highly sophisticated on-line monitoring instrumentation and process control systems. Practical biosensors for monitoring large-scale systems are virtually nonexistent. This is also true of process models for control and optimization of bioprocesses. Let us now look at these problems in greater detail.

1. Large-Scale Tissue Culture

A major intellectual challenge exists in the scaleup of plant and animal cell culture systems. Because of the complexities of some plant and animal genomes and because of problems of prokaryote cells in reading the genetic messages of higher cell genomes and translating those messages to correctly folded proteins, future genetically engineered cell culture systems will necessarily involve both plant and animal cells. As a consequence, a knowledge base for the design of bioreactors for handling plant and animal cells is needed. This includes a knowledge of

1. Effect of mechanical shear on genetically engineered plant and animal cells, in both suspension and microcarrier culture systems
2. Cell-surface interactions for those plant and animal cells that require attachment to surfaces in order to grow (anchorage-dependent cells) and make product
3. Surfaces and materials for anchorage-dependent cells that can be used in large bioreactors
4. Transport phenomena in plant cells forming large cell aggregates (organized cells) or in gel-entrapped cells

2. Separation and Purification of Bioproducts

It is the bioseparation and purification area in which many biotechnologists see the greatest need. For my part, I see three principal challenges to rapid bioprocess commercialization in this area.

The first challenge is the modification of conventional, large-scale industrial separation systems to render them more efficient, reliable, and/or economic when employed for biological product recovery. Examples in this area include

- Column chromatography
- Pressure-driven membrane separation processes
- Liquid-liquid extraction

Adaption of biochemical laboratory separations methods to large-scale bioprocessing is the second challenge. Life scientists of all subdisciplines have, over the past several decades, developed extremely powerful and sophisticated separation tools which today are used predominantly for analytical and small-scale preparative purposes. Significant rewards await those who are able to extend the scale and range of these techniques to meet industrial separation process requirements. Examples are

- Electrophoretic separations
- Affinity separations
- Fractional precipitation

Finally, a third principle challenge to rapid bioprocess commercialization is the investigation of novel separation/purification process concepts. The special properties and stringent process requirements for the safe and economic production of biological products emphasizes the need for discovery and development of truly novel separation/purification process concepts which embody unique combinations of physical, chemical, and biological phenomena outside the area of conventional chemical process technology. A number of such process concepts have recently been described, and they are illustrative of the kind of imaginative synthesis which must be stimulated and encouraged if process innovation of genuine industrial utility is to be accomplished. These process concepts include

- Aqueous two-phase separations
- ''Multiple-field'' fractionation processes
- Electromolecular phase transfer
- Separations by selective enzymatic transformation
- Separations by genetically manipulated intracellular processes
- Separations based on modifications of cell-wall permeability

When one begins to look in detail at these bioseparation process challenges, one is struck by two main features. First, the bioproducts are for the most part fragile protein molecules in which the activity (i.e., three-dimensional structure) must be preserved in the processing. Second, the processes, with few exceptions, involve some surface interaction. An understanding of the surface chemistry involved is fundamental to an understanding of the process.

With respect to the protein nature of bioproducts, there is considerable variation. They can be anything from small polypeptides of the several thousand molecular weight range, as is the case with hormones, to very large enzyme molecules with molecular weights of 100,000 or greater. They occur as highly polar dynamic molecules that readily change their shape and behavior in various pH and osmotic environments. For this reason, each bioproduct protein tends to be unique, with its own shape, size, surface, and tertiary characteristics.

3. Protein Behavior

The key to many bioprocesses is an understanding of protein behavior. Most biomanufacturing systems will require processes to produce, concentrate, and separate a target protein from a myriad of other proteins while maintaining its activity (i.e., three-dimensional structure).

The protein must first be excreted from the biosystem into the reaction solution. This solution is generally aqueous. Consequently, data on protein solubilities and a knowledge base are needed to predict the relation of solubility to protein structure and size, as well as to ionic strength of the solution. There is a need to develop affinity cosolvents for proteins. Virtually nothing is known about this field. Some enzymes are known to function in doped organic solutions. An understanding is needed of the solubility of proteins in organic solvents and of how organic solvents and cosolvents affect their behavior.

Most separation and purification processes involve surface interactions. These interactions are of three basic types:

1. Particle/particle interactions, such as the effect of micro- and macroenvironments on adhesion of animal cells to various supports and on antigen/antibody affinity applications
2. Molecule/particle interactions, such as the effect of local environments on the adsorption of a particular protein to a chromatographic column support or enzyme adsorption on a membrane
3. Molecule/molecule interactions, including environmental effects on intramolecular reactions such as the denaturation/renaturation (folding/unfolding) of proteins or enzyme/inhibitor interactions

It seems obvious, therefore, that there is a need to better understand the behavior of various protein molecules in different physical and chemical environments and to develop chemical theories to explain the reaction of protein molecules with various surfaces, as well as with other proteins and with solute and solvent molecules.

4. Process Monitoring and Control

The successful and economical operation of bioreactors and separation equipment requires process control. This, in turn, depends upon accurate and timely measurement of the critical process variables. Reliable, noninvasive sensors are needed to measure cell concentrations and activity, substrate and product concentrations, plus those of certain growth promoters and inhibitors.

a. Monitoring

Biosensing measurements fall into two categories: (1) devices for measuring parameters on-line and (2) instruments for rapid, sophisticated off-line measurements. On-line biosensors will steadily involve more fiber optics technology to permit the noninvasive spectroscopic examinations of materials within a bioreactor. We need to adapt the present knowledge of nuclear magnetic resonance (NMR), infrared, Raman, and fluorometric spectroscopy to biosystems.

Sensors using tagged moieties such as antibodies will permit continuous off-line monitoring of particular proteins and cell types. More immuno-type sensors will appear.

b. Control

The difficulty with controlling most bioprocess systems is that they are very complex, and we cannot measure many of the cellular parameters. Thus, most bioreactors operate without an adequate control model. They are, for the most part, controlled to match historical operating patterns that over time have proved optimal. As we wait for needed biosensors to be developed, it may be feasible to adapt tendency control techniques to our bioprocesses.

c. Continuous Processing

Most industrial processes involving biotechnology are being carried out in a batch mode. For handling large volumes, the advantages of continuous processing, such as economy and uniformity of product quality, are well known. The translation from batch to continuous processing is not technically trivial. For example, beer is still batch brewed and aged in huge vessels. Past attempts to manufacture beer continuously did not yield acceptable taste. An understanding of the aging process is lacking.

Continuous processing of biological systems will place additional demands on asepsis and biocontainment in terms of equipment design, instrumentation, and operation.

d. Asepsis, Containment, and Detection of Trace Contamination in Large-Scale Bioreactors

Bioreactor systems have an additional element of complexity when compared to chemical reactors: microbial contamination must be prevented. Bioreactor design demands that a single bacterium or virus cannot be allowed to penetrate the reaction system over a period of 200 h, during which time as many as 5 vessel transfers and 50 samplings of the system might be made. One or the other, i.e., containment or contamination prevention, is relatively easy to achieve by itself; however, to achieve both simultaneously is very difficult. Reliable yet economical systems to do this have yet to be developed. A further complication with bioreactors is the need to detect low levels of contamination, i.e., single microorganisms in 10^8 ml of media. This is necessary in order to ensure the purity and safety of the bioproduct. No one has an idea to date how this level of detection can be achieved, but research could vastly improve the gross detection methods that are used today.

IV. SUMMARY

Biotechnology presents many possiblities for improving the quality of life in the future. There are many commercial opportunities in biotechnology in the areas of health care delivery, agriculture, foods, chemicals, environmental protection, and resource utilization. In the short term, opportunities in health care delivery, particularly in diagnostic and therapeutic medicine, will dominate the applications opportunities. In the long run, agriculture and human nutrition will benefit most from biotechnology. The chemical industry will mainly benefit from biotechnology in the high-value chemicals area. Commercialization of biotechnology may depend as much on developments in bioprocess technology as on innovations in genetics and immunology. The intellectual challenge in developing the needed process technology includes the following:

1. The design and scale up of bioreactors
2. Development of separation and purification techniques; in particular, those that will preserve the three-dimensional structure of protein bioproducts
3. Creation of biosensors and process models enabling the optimal control of bioprocesses

If we meet these intellectual challenges quickly, the biotechnology will be commercialized more rapidly.

REFERENCES

1. National Research Council, Chemical and Process Engineering for Biotechnology, Research Briefings 1984, National Research Council, National Academy of Sciences, Washington, D.C., September 1984.
2. Office of Technology Assessment, Commercial Biotechnology: An International Assessment, Office of Technology Assessment, U.S. Congress, U.S. Government Printing Office, Washington, D.C., February 1984.
3. Office of Technology Assessment, Impacts of Applied Genetics: Microorganisms, Plants, and Animals, Office of Technology Assessment, U.S. Congress, U.S. Government Printing Office, Washington, D.C., April 1981.
4. Committee on Technology and International Economics and Trade Issues, The Competitive Status of U.S. Pharmaceutical Industry: The Influence of Technology in Determining International Competitive Advantage, National Research Council, National Academy of Sciences, Washington, D.C., 1983.
5. **Dunnill, P.,** The future of biotechnology, *Biochem. Soc. Symp.,* 48, 9, 1983.
6. **Busche, R. M. and Hardy, R. W. F.,** Biotechnology: potential impact and issues, *Biotechnol. Bioeng. Symp.,* 15, 651, 1985.
7. **Thijssen, A. C. and Roels, J. A.,** The impact of biotechnology on the food industry, in *Impact of Biotechnology in Food Production and Processing,* Knorr, D., Ed., Marcel Dekker, New York, 1985.
8. **Shimilt, W. J., Ed.,** *Chemistry and World Food Supply: The New Futures,* Pergamon Press, Oxford, 1983.
9. **Michaels, A.,** The impact of engineering, *Chem. Eng. Prog.,* 9, October 1983.
10. **Michaels, A.,** Adapting modern biology to industrial practice, *Chem. Eng. Prog.,* December, 19, 1983.
11. **Humphrey, A. E.,** Commercializing biotechnology: challenge to the chemical engineer, *Chem. Eng. Prog.,* December, 7, 1984.
12. **Sahai, O. and Knuth, M.,** Commercializing plant tissue culture processes: economics, problems and prospects, *Biotechnol. Prog.,* 1, 1, 1985.
13. **Pollack, A.,** *New York Times,* Business Day Section, June 10, 1987.

Part 1. Process Integration

In most cases, process integration still consists of the sequential ordering of feedstock supply, production, waste management, formulation, and packaging. In a more recent context, the efficient integration of processes needs to include the development of simultaneous and compatible technologies of reactor maintenance, continuous product removal, gas exchange, prevention of toxic waste accumulation, and downstream processing of extremely dilute product solutions. In this section, researchers from three corporations and a university in a developing nation describe their experiences in the development of new and significant integrated processes.

Chapter 2

DEVELOPMENT OF A REVERSE-PHASE HPLC PROCESS STEP FOR RECOMBINANT β-INTERFERON

Susan Hershenson and Ze'ev Shaked

TABLE OF CONTENTS

I. INTRODUCTION

Recombinant techniques allow production of many potentially useful proteins that otherwise could not be obtained in large quantities. The challenge in bioprocessing is to develop purification methods that yield sufficient quantities of product in a form suitable for the intended use without sacrificing efficiency.

Factors such as high expression levels and ease of fermentation often make bacterial production, most commonly in *Escherichia coli,* the method of choice. Certain problems must be addressed in the purification of recombinant proteins or polypeptides from *E. coli*.[1] If the product is intended for parenteral use in humans, very thorough removal of *E. coli* antigens and pyrogens must be achieved. Also, problems specific to synthesis of the particular protein in the *E. coli* host must be addressed. For example, heterologous proteins are sometimes deposited within the host cell in insoluble inclusion bodies. This can often be used to advantage in the initial stages of the purification,[2] but eventually the protein must be solubilized. Use of strong denaturants and disruption of covalent bonds, particularly disulfide bonds, are sometimes required.[3] Processes must then be developed to refold the protein to the proper, active conformation. Especially for oligomeric proteins, this may be a tricky process, since subunit refolding and interaction can be competing processes requiring different conditions.[4,5] Also, *E. coli* is incapable of certain types of posttranslational processing, such as glycosylation. Lack of glycosylation can affect the solubility of some proteins, as in the case of vesicular stomatitis virus (vsv) glycoprotein,[6] and may lead to altered immunogenicity. Another problem that may require attention is synthesis of alternate forms of the protein of interest during fermentation. Incompletely synthesized or proteolized fragments have been reported for several recombinant proteins.[3,7] Other alternate forms can include oligomers,[3] oxidation states,[8] and N-terminal heterogeneity.[9]

The product that will be discussed here is β-interferon (IFN-β) that has been cloned and expressed in *E. coli*. Native IFN-β is a glycoprotein.[10] The glycosylation does not appear to be required for activity, since the *E. coli* product is fully active in *in vitro* assay systems.[11] The molecule is deposited in refractile bodies during expression in the *E. coli* host and must be solubilized using a denaturant and then refolded. During this process, the native IFN-β sequence, containing three cysteines, can form incorrectly disulfide-bonded conformations which are inactive. Therefore, the protein has been engineered to replace the third cysteine (not involved in disulfide bond formation) with serine.[12,13] The replacement simplifies production of a fully active, correctly refolded molecule and improves the stability of the product.[14,15] However, other issues remain to be addressed during the purification process. *E. coli* antigens and pyrogens must, of course, be removed. Also, certain minor forms of the protein are produced during fermentation and must be removed. In the case of Betaseron, these include incomplete fragments of the protein, an oxidized methionine species, and oligomers.

This chapter describes a reverse-phase high-performance liquid chromatography (RP-HPLC) purification step for recombinant IFN-β, expressed in *E. coli*, that addresses some of these issues. The process can be used to achieve good removal of *E. coli* antigens and endotoxin. It also separates certain minor forms of the protein that are produced during fermentation: low-molecular-weight fragments, oxidized methionine, and a proportion of the oligomers.

II. RP-HPLC PROCESS

Figure 1 shows some of the minor IFN-β species produced during fermentation. They have been separated by SDS-polyacrylamide gel electrophoresis (SDS-PAGE) and detected by Western blotting with a monoclonal antibody to IFN-β. Dimers and oligomers, as well

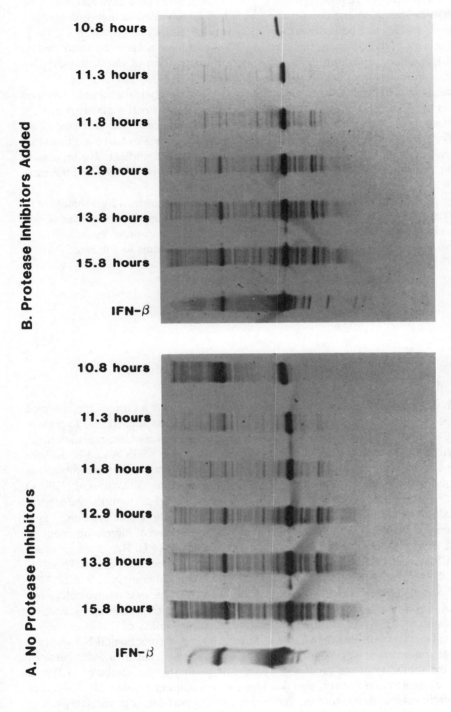

FIGURE 1. IFN-β-ser$_{17}$ in fermentation samples. Samples were separated by SDS-PAGE in 15% polyacrylamide gels. The Western blot was developed with a monoclonal antibody to the recombinant IFN-β-ser$_{17}$. All lanes contain 0.20 μg protein. Low-molecular-weight fragments, dimers, oligomers, and other minor species are detected by this method, whether or not protease inhibitors are added during sample preparation. (A) No protease inhibitors; (B) a mixture of protease inhibitors was added during cell lysis and sample solubilization. The mixture included leupeptin, pepstatin A, E-64, aprotonin, guanidinoethyl-mercaptosuccinic acid (GEMSA) (all 0.5 μg/ml), phenylmethylsulfonyl fluoride (PMSF, 0.2 μg/ml), and EDTA (5 mM).

as low-molecular-weight fragments, begin to appear shortly after induction (approximately 10.8 h) and accumulate throughout the fermentation. The same species in roughly the same amounts are apparent whether or not protease inhibitors are added during cell disruption, implying that the fragments are produced at the time of synthesis. The dimers and oligomers may result from denaturation and cross-linking as IFN-β is deposited in inclusion bodies.

Other minor species can be separated by analytical RP-HPLC, as shown in Figure 2. The most prominent of these is a species labeled peak A, which has been shown to contain an oxidized methionine in position 62, analogous to a minor form of interleukin-2.[8] Oligomers, if present, elute after peak B, the main species.

The species shown in Figure 2 were separated on a biphasic analytical gradient developed by Kunitani et al.[22] The system gave excellent separations on a small scale. However, it would have been difficult to scale up and run as a preparative process. The first step in the development of a preparative reverse-phase procedure, therefore, was to find conditions that would give good separation of minor species using a simple, linear gradient. The separation of peak A (the most prominent of the minor species by analytical RP-HPLC) and the main peak, peak B, was used as the benchmark for a good separation.

A number of column and solvent combinations were screened using a linear gradient of solvent B (10 to 85%) with a fixed slope and flow rate. A single gradient was used to test all combinations of columns and solvents in order to allow comparison of the selectivities of each system. Selectivity (α) has been defined for isocratic systems as follows:

$$\alpha = \frac{K_b'}{K_a'}$$

where

$$K' = \frac{t_r - t_0}{t_0}$$

t_r = retention time, and t_0 = injection peak time. Essentially, it is a ratio of the retention times of two peaks adjusted for the injection peak time.[16] It has been applied here to provide some measure of the separation between two peaks, independent of absolute retention time.

The selectivities of several systems, separation times between peaks A and B, and the retention time of peak B are reported in Table 1. Of the systems shown here, the combination of a C_{18} column and a gradient of acetonitrile with 0.1% heptafluorobutyric acid (HFBA) gives the best selectivity between peaks A and B. Combinations of acetonitrile and isopropanol, reported in some cases to give improved separations of hydrophobic proteins,[17] give increased separation between peaks A and B, but only at the expense of increasing retention times for both species; overall selectivity actually decreases. Use of HFBA, a longer-chain analogue of trifluoroacetic acid (TFA), with C_{18} increases selectivity somewhat over TFA.[18,19]

Although the results are not shown on Table 1, isopropanol in combination with TFA or acetic acid was also tested. During these gradients, backpressure rose dramatically, and the recovery of IFN-β was poor. Apparently, recombinant IFN-β precipitates on the column in this solvent system. This was confirmed by examining the solubility of IFN-β in various concentrations of isopropanol, shown in Table 2. Judging by eye, recombinant IFN-β appears to be soluble at concentrations of isopropanol less than 10% or greater than 50%; between 10 and 50%, the solubility of the protein decreases. Surprisingly, the solubility of IFN-β in acetonitrile, examined as a control, also decreases at concentrations below 50%. In order to keep the protein soluble throughout the entire range of the gradient, it is necessary to use very high concentrations of acid. Although separations of hydrophobic proteins have been reported under such conditions,[20] this was not considered desirable for a preparative process.

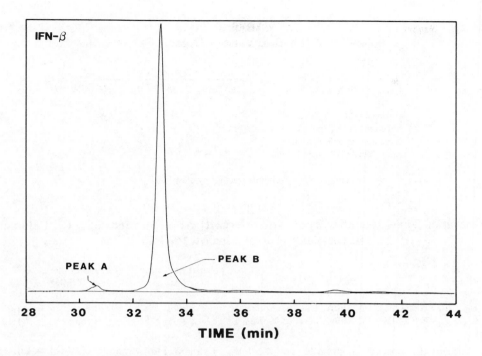

FIGURE 2. Analytical RP-HPLC of IFN-β-ser$_{17}$. The chromatography was done as described in the text. The proportions of the separated species in this sample are 2.95% peak A, 94.8% peak B, and 2.2% other species.

TABLE 1
Separation Time, Retention Time, and Selectivity for Each of the RP-HPLC Systems

Column	Solvent B	Separation time (min)	Retention time of Peak B (min)	α
C$_4$	Acetonitrile/0.1% TFA	1.93	54.8	1.04
C$_4$	Acetonitrile/0.1% HFBA	3.02	53.9	1.06
C$_4$	Acetonitrile:propanol-2 (4:1), 0.1% HFBA	3.20	68.0	1.05
C$_{18}$	Acetonitrile/0.1% TFA	1.38	47.1	1.03
C$_{18}$	Acetonitrile/0.1% HFBA	3.20	56.1	1.07
C$_{18}$	Acetonitrile:propanol-2 (4:1), 0.1% HFBA	3.77	67.0	1.06

During initial testing on a small scale, no problems were encountered using a gradient of acetonitrile (10 to 80%). However, to be economical, preparative reverse-phase processes are often run under "overloaded" chromatography conditions.[21] Here, the reduction in IFN-β solubility at intermediate concentrations of acetonitrile could potentially lead to problems in running the process or to reductions in yield. Moreover, precipitation might be expected to affect the separation adversely, even at the analytical scale. Therefore, the system was retested using a gradient with a starting concentration of 50% acetonitrile. The slope and flow rate were identical to the gradient used previously. The results are summarized in Table 3. Beginning the gradient at 50% acetonitrile increases the absolute separation time between peaks A and B and increases selectivity. The change will also lead to shorter processing time and lower solvent consumption at the production scale.

The conditions selected for the preparative process, then, were C$_{18}$-derivitized silica with a linear gradient of acetonitrile beginning at 50% and 0.1% HFBA. The next step was

TABLE 2
Solubility of IFN-β in Various Organic Solvents

	Percent B[a]				
Solvent B	10	30	50	70	100
Propanol-2/0.1% TFA	+	−	+/−	+	+
Acetonitrile/0.1% TFA	−	−	+	+	+
Acetonitrile/60% formic acid	+	+	+	+	+
Acetonitrile/10% formic acid	−	−	+/−	+	+

[a]   + = soluble, − = insoluble (judged by eye).

TABLE 3
Separation Time, Retention Time, and Selectivity for Separations on a C$_{18}$ Column in Gradients of Acetonitrile/0.1% HFBA

Column	Solvent B	Separation time (min)	Retention time of Peak B (min)	α
C$_{18}$	Acetonitrile/0.1% HFBA, 10—80% Solvent B	3.20	56.1	1.07
C$_{18}$	Acetonitrile/0.1% HFBA, 50—80% Solvent B	3.35	16.0	1.35

to scale up the process to an intermediate level, using a 1-in.-diameter (90-ml volume) column and a load of approximately 50 mg of IFN-β. The results are shown in Figure 3. Close to baseline separation of peaks A and B is achieved under these conditions. Analysis of fractions from each peak shows that almost pure fractions of each material are recovered (Figure 4). Serendipitously, Western blot analysis, shown in Figure 5, reveals that low-molecular-weight fragments are also separated.

Next, the load was scaled up an order of magnitude to a level equivalent to that desired for a production process. The chromatogram is shown in Figure 6. Baseline separation of peaks A and B is not achieved; however, analysis of fractions through the chromatogram reveals that separation of the two species does occur. Results are shown in Table 4. Early fractions are enriched in peak A and other early-eluting peaks. Later fractions contain very little peak A, but are enriched in late-eluting species. Western blots of the IFN-β species separated by SDS-PAGE reveal that separation of low-molecular-weight fragments is preserved at the higher loads (Figure 7).

To examine the power of the separation system to resolve *E. coli* antigens and endotoxin from IFN-β, material at an early stage of purification was loaded. An elution profile is shown in Figure 8. Within the error of the bioassay, all of the IFN-β is recovered during the gradient. Some of the *E. coli* proteins are eliminated during application of the sample to the column. Within the gradient, *E. coli* proteins are most concentrated in the early-eluting fractions, with some increase in the late fractions, as shown in Figure 9. IFN-β is concentrated in the middle fractions. Analytical RP-HPLC and Western blots of IFN-β confirm that the separations of peaks A and B and of low-molecular-weight fragments are maintained (not shown). Depending on the concentration in the starting material, endotoxins can be reduced by up to five orders of magnitude, from as much as 250 μg/mg protein in a very impure starting material to less than 0.8 ng/mg protein in a pool from the column.

When the process is scaled up to production levels, preserving the ratio of load and column dimensions, identical separations are obtained. A process chromatogram is shown in Figure 10; it is indistinguishable from the chromatogram obtained at the 1-in. stage. Western blots and analytical RP-HPLC demonstrate the same separation of *E. coli* contaminants and minor IFN-β species (not shown). The process can, therefore, be implemented in production to obtain reductions in minor IFN-β species that can be difficult to separate

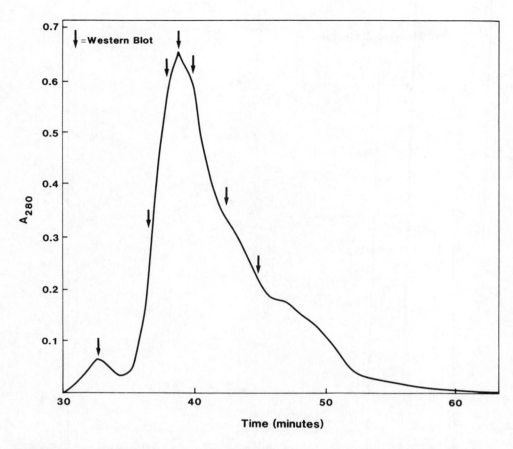

FIGURE 3. Elution of approximately 50 mg IFN-β on a 22-mm C_{18} column. The separation was performed as described in the text. At this load, almost baseline resolution between peaks A and B is attained.

from the major species. The process can also be used to obtain large reductions in *E. coli* proteins and endotoxin contamination. Recovery of IFN-β bioactivity from the column appears to be close to 100%, although some activity must be sacrificed to eliminate fractions high in minor IFN-β species. In combination with other process steps, this procedure can be a simple and efficient method to obtain a high purity product.

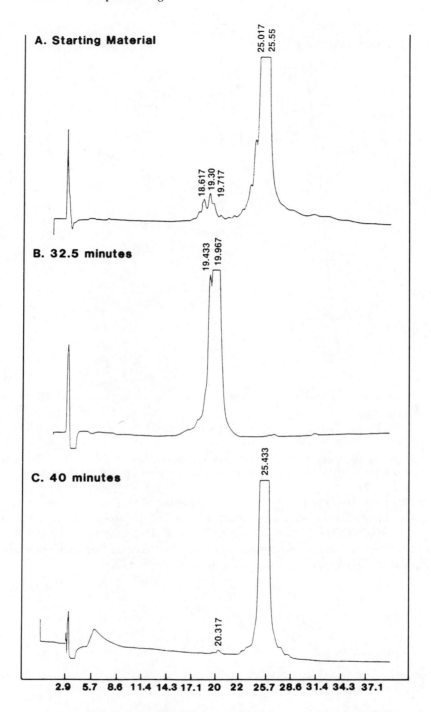

FIGURE 4.   Rechromatography of fractions on an analytical column. Concentrated starting material and two fractions from the chromatogram are shown. The fraction at 32.5
min is 99.8% peak A, with no measurable peak B. The fraction at 40.0 min is 0.3%
peak A and 98.7% peak B.

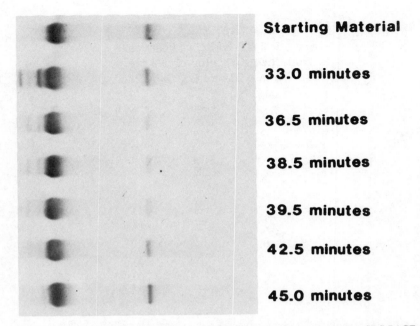

FIGURE 5. Western blot of IFN-β in fractions through the chromatogram. SDS-PAGE and Western blots were done as described in Figure 1. All lanes contain 0.5 μg protein. Low-molecular-weight fragments are concentrated in early fractions.

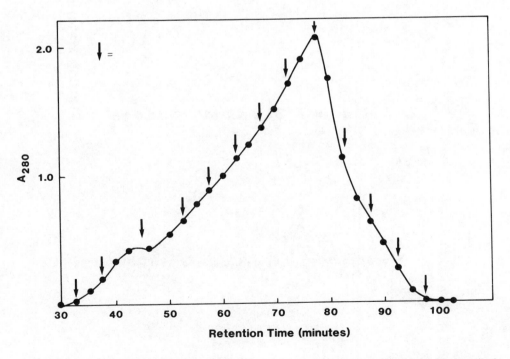

FIGURE 6. Elution of 500 mg IFN-β on a 22-mm C_{18} column. The separation was performed as described in the text. With this amount of starting material, the column is "overloaded".

TABLE 4
Analytical RP-HPLC of Fractions
from the Application of 500 mg
IFN-β to a 22-mm C$_{18}$ Column

Sample (min)	Peak A (%)	Peak B (%)
Starting material	2.3	96.5
32.5	3.3	14.7
37.5	2.9	—
42.5	17.6	—
47.5	3.6	77.0
52.5	0.5	96.7
57.5	—	98.4
62.5	—	98.3
67.5	0.2	98.2
72.5	—	100.0
77.5	—	100.0
82.5	—	97.2
87.5	—	72.1
92.5	3.9	51.6
97.5	—	—

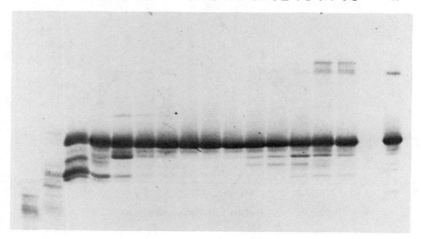

FIGURE 7.   Western blot of IFN-β in fractions through the chromatogram. SDS-PAGE and Western blots were done as described in Figure 1. All lanes contain 0.25 μg protein. Low-molecular-weight fragments are most concentrated in the early fractions, while dimers are more concentrated in late fractions.

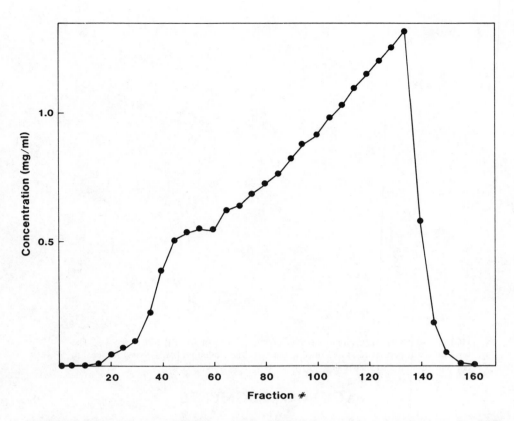

FIGURE 8. Elution of impure starting material by RP-HPLC. The separation was performed as described in the text. The chromatogram is quite similar to that obtained with an equivalent load of purified IFN-β. According to bioactivity measurements, essentially all of the IFN-β is recovered during the gradient elution (data not shown).

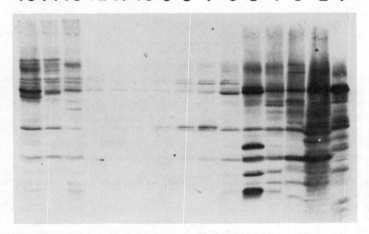

FIGURE 9. Western blot of *E. coli* proteins. SDS-PAGE was performed as described in Figure 1. The Western blot was developed with a polyclonal antiserum to the *E. coli* strain used for expression, containing the plasmid but not the IFN-β gene. The *E. coli* proteins are most concentrated in early fractions, with some increase in late fractions.

FIGURE 10. Chromatogram from pilot production run of the RP-HPLC process. The separation was performed as described in the text. The elution profile is almost identical to previous runs on a smaller scale.

ACKNOWLEDGMENTS

We are grateful for the assistance of Jody Thomson in the experimental work and Candy Jones in the preparation of the manuscript. We also appreciate the advice and support received from James Thomson, Michael Kunitani, and Deborah Johnson. Mark Pemberton and Les Johnson are responsible for the scaleup to the pilot production level.

REFERENCES

1. **Marston, F. A.,** The purification of eukaryotic polypeptides synthesized in *Escherichia coli, Biochem. J.,* 240, 1, 1986.
2. **Prouty, W. F., Karnovsky, M. J., and Goldberg, A. L.,** Degradation of abnormal proteins in *Escherichia coli.* Formation of protein inclusions in cells exposed to amino acid analogs, *J. Biol. Chem.,* 250, 1112, 1975.
3. **Shoemaker, J. M., Brasnett, A. H., and Marston, F. A.,** Examination of calf prochymosin accumulation in *Escherichia coli:* disulphide linkages are a structural component of prochymosin-containing inclusion bodies, *EMBO J.,* 4, 775, 1985.
4. **Mozhaev, V. V. and Martinek, K.,** Inactivation and reactivation of proteins (enzymes), *Enzyme Microb. Technol.,* 4, 299, 1982.
5. **Jaenicke, R. and Rudolph, R.,** Refolding and association of oligomeric proteins, *Methods Enzymol.,* 131, 218, 1986.
6. **Gibson, R., Schlesinger, S., and Kornfeld, S.,** The non-glycosylated glycoprotein of vesicular stomatitis virus is temperature sensitive and undergoes intracellular aggregation at elevated temperatures, *J. Biol. Chem.,* 254, 3600, 1979.
7. **Boss, M. A. et al.,** Assembly of functional antibodies from immunoglobulin heavy and light chains synthesized in *E. coli, Nucleic Acids Res.,* 12, 3791, 1984.
8. **Kunitani, M., Hirtzer, P., Johnson, D., Halenbeck, R., Boosman, A., and Koths, K.,** Reverse-phase chromatography of interleukin-2 muteins, *J. Chromatogr.,* 359, 391, 1986.

9. **Lahn, H. W. and Stein, S. J.**, Characterization of recombinant human interleukin-2 with micromethods, *J. Chromatogr.*, 326, 357, 1985.
10. **Tan, Y. H., Barakat, F., Berthold, W., Smith-Johannsen, H., and Tan, C.**, The isolation and amino/sugar composition of human fibroblastoid interferon, *J. Biol. Chem.*, 254, 8067, 1979.
11. **Derynck, R., Remaut, E., Saman, E., Stanssens, P., DeClerq, E., Content, J., and Friers, W.**, Expression of human fibroblast interferon gene in *Escherichia coli*, *Nature (London)*, 287, 193, 1980.
12. **Goeddel, D. V. et al.**, Synthesis of human fibroblast interferon by *E. coli, Nucleic Acids Res.*, 8, 4057, 1980.
13. **Mark, D., Lin, L., and Yu-Lu, S.**, Human Recombinant Cysteine Depleted Interferon-β Muteins, U.S. Patent 4,588,585, 1986.
14. **Shaked, Z. and Wolfe, S.**, Controlled Oxidation of Microbially Produced Cysteine-Containing Proteins, U.S. Patent 4,530,787, 1986.
15. **Koths, K. E. and Halenbeck, R. F.**, Method for Promoting Disulfide Bond Formation in Recombinant Proteins, U.S. Patent 4,572,798, 1986.
16. **Snyder, L. R. and Kirkland, J. J.**, *Introduction to Modern Liquid Chromatography*, 2nd ed., John Wiley & Sons, New York, 1979.
17. **Tarr, G. E. and Crabb, J. W.**, Reverse-phase high-performance liquid chromatography of hydrophic proteins and fragments thereof, *Anal. Biochem.*, 131, 99, 1983.
18. **Bennett, H. P., Solomon, S., and Goltzman, D.**, Isolation and anlysis of human parathyroid tissue and plasma. Use of reversed-phase liquid chromatography, *Biochem. J.*, 197, 391, 1981.
19. **Burgess, A. W., Knesel, J., Sparrow, L. G., Nicola, N. A., and Nice, E. C.**, Two forms of murine epidermal growth factor: rapid separation by using reverse-phase HPLC, *Proc. Natl. Acad. Sci. U.S.A.*, 79, 5753, 1982.
20. **Heukeshoven, J. and Dernick, R.**, Reverse-phase high-performance liquid chromatography of virus proteins and other hydrophobic proteins, *Chromatographia*, 19, 95, 1984.
21. **Guichon, G. and Colin, H.**, Theoretical concepts and optimization in preparative scale liquid chromatography, *Chromatogr. Forum*, 21, 1986.
22. **Kunitani, M.**, personal communication.

Chapter 3

SENSITIVITY ANALYSIS OF AN INTEGRATED PROCESS OF LIGNOCELLULOSE CONVERSION TO ETHANOL

T. K. Ghose and P. Ghosh

TABLE OF CONTENTS

I. INTRODUCTION

Those who have until recently been substantially involved with experimental and modeling studies of cellulose conversion into sugars, ethanol, and chemicals may be of the view that the subject has now become of historical interest and its future is uncertain. Those of us who live on the other side of North America (in the East, Southeast Asia, and even parts of Eurasia) have additional comments on the subject. It can only be stated that active interest in research and development has slowed down, but it is not yet dead. A reference to the two United Nations Industrial Development Organization (UNIDO) reports[1,2] reveals that eight universities and research centers in the U.S., four in Canada, one in the U.S.S.R., one in Finland, one in Sweden, one in Italy, one in South Africa, one in India, and a few in Japan were until recently actively engaged in studies on cellulose conversion. The gulf oil project report[3] also indicated the economic advantages in the use of cellulosic wastes instead of grain or ethylene as a source of alcohol. Successful crystallization of endoglucanase D (EGD) of *Clostridium thermocellum* from an *Escherichia coli* expression system has recently been reported[4] by a group led by Aubert at the Pasteur Institute in Paris. Overproduction of EGD in the host system made the crystallization of the protein possible. This is a significant advance in the direct conversion of cellulose to ethanol.

Our long-standing interest in the bioconversion of cellulosic substrate which began in 1974 to 1975 has addressed the issues regarding the current status of liquid fuel supply. However, basically our understanding of the biochemical and engineering fundamentals of this complicated system needs much more than what is currently available. A vast area of basic knowledge on enzyme action and inactivation, reaction rates, equilibria, and separation sciences is not yet adequately available. The approach to the bioconversion of lignocellulose to ethanol by the Biochemical Engineering Research Centre (BERC), Indian Institute of Technology, Delhi constitutes a story somewhat different from those in Japan[5] and the U.S.[3] Based on some detailed studies of the process steps such as raw materials survey, cost data for plant equipment, utilities, and products, a sensitivity analysis involving parametric and nonparametric factors has been prepared.

Since rice straw is a dispersed agroresidue having a low bulk density, the base case proposed is a medium-sized plant with an annual capacity of 15,000 m^3 95% (v/v) ethanol. It is envisaged that an adequate supply of rice straw must be ensured from a given region not far away from the plant site. Parametric variables considered are

1.   Impact of plant capacity on product cost
2.   Impact of rice straw cost on product cost
3.   Impact of lignin credit on product cost
4.   Impact of cellulase enzyme source on product cost

Nonparametric effects include

1.   Delignification of rice straw by alkali treatment replacing the solvent process
2.   Ethanol separation by adsorption-desorption replacing distillation

Effects of these variables on the sensitivity of the chosen process have been analyzed in terms of costs of various materials and the sources of enzyme used in the process studies. Replacement of the subsystems, such as solvent delignification of rice straw and distillation separation of ethanol by other methods, has been analyzed.

It is important to mention here that while all cost data are drawn from literature dealing with U.S. market information, the cost of rice straw is taken as $10 (Rs 126)/ton, which is considerably higher than what is currently quoted in India ($2.4 to $4.0/ton). This cost

($10.0) is, however, lower than the available cost of rice straw in the U.S. and is considerably lower than the price which the Ministry for International Trade and Industry (MITI) pilot plant at Hofu (Prefecture-Yamaguchi), Japan is reported to be paying to import from Thailand, Y 10,000 ($72)/ton. We have chosen rice straw as the raw material for the process because of its abundance in many states in India. Nearly 90 million tons of this residue are currently produced, and much of it is burned on the field, while a part is still used as an inferior fodder.

II. THE PROCESS

The process considered in the sensitivity analysis consists of pretreatment of rice straw in two steps, separate production of two enzymes, single cell protein (SCP) production, simultaneous saccharification and fermentation (SSF) of treated straw, and separation of ethanol. The flow diagram is presented in Figure 1.

The solvent pretreatment process developed at the BERC is a two-step system comprising autohydrolysis and solvent delignification.[6,7] The combined operations separate the major components of rice straw for further processing. Rice straw contains roughly 40% cellulose, 30% hemicellulose, 15% lignin, and 15% ash. In the steam autohydrolysis, nearly 70% of the hemicellulose present in straw is removed as soluble mixed sugars, mostly xylose. The pentose sugars (4.5 wt%) present in the water extract from the autohydrolysis reactor are used to produce *Candida utilis,* proposed to be used as animal feed.[8] The autohydrolyzed residue (56.5% cellulose, 13% hemicellulose, 15% lignin, and 15.5% ash) is delignified with 50% (v/v) aqueous ethanol solution in the presence of a catalyst at 170°C for 30 min.[6,10]

Filtered and dried-solvent-treated straw (76% cellulose, 8% hemicellulose, 6% lignin, and 10% ash) is sent to the SSF reactor for its conversion to ethanol. The lignin-containing aqueous ethanol solution is batch-distilled to recover the solvent. During the distillation, the reflux ratio is increased from 0.5 to 10 in order to achieve a constant overhead vapor composition. Lignin present in the bottom of the reboiler is filtered off from the aqueous phase and dried. The adhering ethanol is recovered by drying.

A combined process of cellulose saccharification and fermentation, SSF is carried out under programmed vacuum cycling coupled with intermittent substrate feeding.[9,11] Vacuum cycling is defined as a programmed application of vacuum on a volatile product-forming system operating between the limits of product concentration within which product inhibition is either very low or absent. From the energy economy point of view, application of a continuous vacuum is redundant.

The bioreactor system consists of an SSF reactor, a settler, and a flash vessel. It operates in conjunction with a vapor recompression system, drawing the reactor broth into a flash chamber maintained under vacuum (80 mmHg), while the reactor works under atmospheric pressure. The conversion processes in the SSF reactor are initiated by introducing treated rice straw, nutrients, mixed enzymes, and yeast inoculum. *Trichoderma reesei* E-12,[12] a powerful mutant strain developed at the BERC, is the source of cellulase enzyme. Since a *T. reesei* E-12 culture is deficient in β-glucosidase (filter paper activity, 17.5 IU/ml; β-glucosidase activity, 6.0 IU/ml), this enzyme is supplemented with *Aspergillus wentii* Pt 2820 culture filtrate containing 20 IU/ml of β-glucosidase.[13] The culture filtrates from these two sources are mixed such that the ratio of β-glucosidase to filter paper activity becomes 1.6. A thermotolerant yeast, *C. acidothermophilum,* is used for simultaneous conversion of sugars to ethanol at pH 4.8 and 50°C.

Once the ethanol concentration in the SSF reactor reaches 22 to 23 g/l, the programmed vacuum-cycling process is switched on and the broth is circulated between the reactor and settler through the ethanol separation circuit. The settler is used to remove unconverted, siliceous residue contained in the system. The settler is connected with the vacuum system

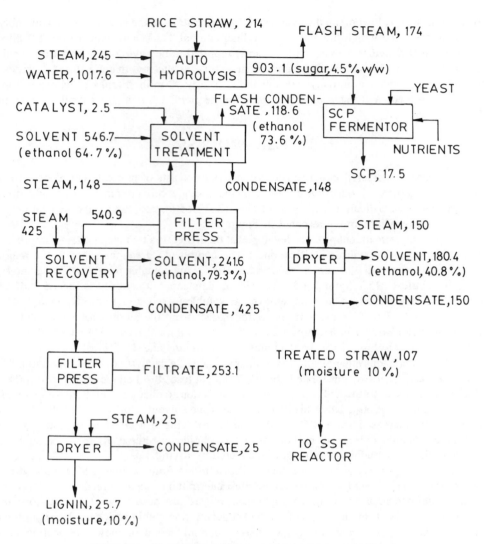

FIGURE 1.   Flow diagram of bioconversion of rice straw to ethanol.

through a flash vessel. During the period of operation of the vacuum cycle, this vessel is maintained at 80 mmHg. The vacuum cycle is automatically terminated as soon as ethanol concentration in the SSF reactor reaches 5 g/l.

Following separation of the ethanol-water vapor mixture from the liquid broth in the flash vessel, the vapors are adiabatically compressed from 80 to 186 mmHg, resulting in a corresponding increase in the temperature of the vapors. These are cooled and condensed by their passage through a heating coil located in the flash vessel to supply the heat of vaporization to ethanol in the solution. The condensate enters a knock-off drum in which the noncondensable vapors are separated from the condensed aqueous ethanol solution. The vapors from the knock-off drum are further compressed to atmospheric pressure by the second stage compressor, followed by the intercooler. The condensed aqueous ethanol solution collected from both knock-off drums, containing 12.4% (w/w) ethanol, is sent to a distillation system to recover the product as 95% (v/v) ethanol.

Following programmed termination of the vacuum cycle, ethanol concentration in the SSF reactor during the second cycle goes up and again reaches a concentration of ca. 22 to

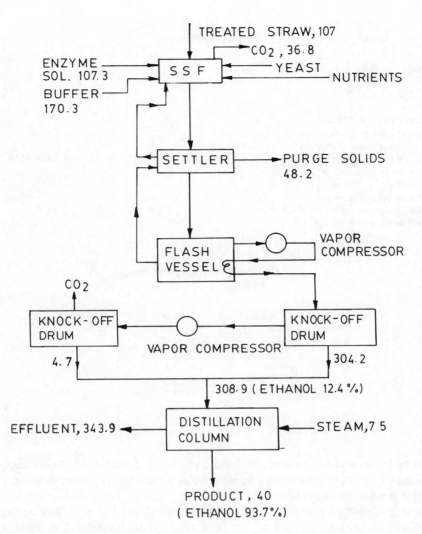

(Values in tonnes / day)

FIGURE 1 (continued).

23 g/l. The vacuum system is again actuated, and the process follows in repeated cycles. Between every two vacuum cycles, a fresh batch of delignified cellulosic substrate is introduced.

In this way the system has been operated with 14 feedings of cellulose and 30 vacuum cycle operations covering a total period of 220 h, and an average ethanol productivity of 4.4 g/l/h has been obtained. During the cyclic process, a portion of the uncoverted residue is removed from the settler every 12 h to limit the buildup of the nonfermentables. The settler placed in series with the SSF reactor increases the number of feeding cycles, besides removing the nonfermentables.

III. ECONOMIC CONSIDERATIONS

Based on the process described above, the product (ethanol) cost is estimated for a plant with an annual capacity of 15,000 m³. Cost estimates are based on the procedure recom-

TABLE 1
Major Equipment

Pretreatment reactors (1 × 120 m³, 1 × 75 m³)
SSF Fermentor (8 × 120 m³)
Settler (8 × 15 m³)
Flash vessel (8 × 10 m³)
SCP fermentor (8 × 120 m³)
Cellulase enzyme production fermentor (2 × 150 m³)
β-Glucosidase enzyme production fermentor (2 × 100 m³)
Storage tanks (2 × 675 m³, 1 × 170 m³, 1 × 310 m³, 1 × 350 m³, 1 × 50 m³, 1 × 20 m³, 1 × 900 m³)
Screw conveyer system (2 × 25 tons/h)
Vapor compression system
Columns for ethanol recovery
Batch-distillation system for solvent recovery
Heat exchanger, filters, pumps, etc.

TABLE 2
Costs of Rice Straw, Utilities,
Credits, and Labor

Rice straw	$10/ton
Utilities	
Steam	$8.50/ton
Electricity	$0.035/kWh
Water	$0.008/1000 m³
Credits	
Lignin	$1000/ton
Animal feed	$228/ton
Steam exhaust	$1.67/ton
Labor	$10/man-hour

mended by Peters and Timmerhaus,[14] Holland et al.,[15] and Guthrie.[16] Cost of land has not been taken into account, and 100% company financing has been considered in the analysis.

Major equipment required for the process is given in Table 1. The costs of major equipment are updated according to the Marshall and Swift Index of September 1986.[17] Costs of rice straw, utilities,[18] labor, and by-product credits considered in the analysis are given in Table 2. Credits from lignin, animal feed, and exhaust steam are considered in the analysis. For profits, a 15% simple after-tax return on total capital investment (RCI) has been considered.

IV. ANALYSIS AND DISCUSSION

Presented below are three elements of the sensitivity analysis. Case 1 is parametric and includes the following steps:

1. Rice straw pretreatment by two-stage autohydrolysis followed by solvent delignification process
2. Mixed enzyme preparation consisting of culture filtrates of *T. reesei* E-12 and *A. wentii* Pt 2820 and β-glucosidase to filter paper activity ratio of 1.6 used in the SSF reactor
3. Simultaneous saccharification and fermentation with vacuum cycling and intermittent substrate feeding for the conversion of cellulose to ethanol
4. Ethanol separation by distillation

TABLE 3
Summary of Capital Investment (Case 1)

Plant capacity 15,000 m³/year ethanol 95% (v/v)

Item	Cost ($ × 10³)
Purchased equipment	4,384
Installation (40% of equipment)	1,752
Instrumentation (10% of equipment)	438
Electrical (10% of equipment)	438
Process piping (20% of equipment)	876
Utilities (20% of installed equipment)	1,226
Building (20% of equipment)	876
Engineering and supervision (20% of total fixed capital)	3,330
Contingency (20% of total fixed capital)	3,330
Total fixed capital	16,650
Working capital (10% of total fixed capital)	1,665
Total capital investment	18,315

Case 2 is nonparametric. Its process is the same as in Case 1, except that ethanol separation by distillation is replaced by adsorption-desorption. Case 3 is also nonparametric. Its process is the same as in Case 1, except that solvent pretreatment is replaced by alkali pretreatment.

A. CASE 1

The summary of fixed capital investment and manufacturing cost for this case is given in Tables 3 and 4. For a plant of 15,000-m³ annual capacity, the total fixed capital is estimated to be $16.6 million. Considering 10% working capital, the total capital investment is $18.3 million. The direct production cost works out to be 52.4¢/l. Utilities (54% of production cost) have a major contribution to the direct production cost. This is mainly due to the high energy requirement of the pretreatment process. Only 20% of the direct production cost is consumed in raw material. In the conventional molasses-to-ethanol process, raw material cost amounts to 75 to 80% of the total production cost. However, in the molasses-to-ethanol process, utilities cost much less, as little or no pretreatment is needed in the process. Taking into consideration the by-product credits, the cost of the product becomes 54.6¢/l.

1. Plant Capacity

The impact of plant capacity on product cost is given in Table 5. Plant capacities are varied from 3,000 to 30,000 m³/year. For the estimation of equipment cost of plants of different capacities, the six-tenth factor rule[14] is applied with 15,000 m³/year as the base. By increasing the plant capacity by tenfold, from 3,000 to 30,000 m³, a 2.7-fold reduction in product cost can be achieved.

In order to keep the product cost below 60¢/l, a plant size exceeding 15,000 m³ of ethanol per year seems to be desirable.

2. Rice Straw (Substrate)

Rice straw cost is taken as $10/ton. In a country like India and in several Southeast Asian countries, the rice straw cost is still lower. In Japan, the U.S., and some European countries the cost is on the higher side. Since rice straw cost has a direct impact on the product cost, we analyzed its sensitivity on ethanol cost. Cost of straw has been varied from $10 to $100/ton. From the computed product cost vs. various rice straw costs, it is evident from the data of Table 6 that with a tenfold increase in straw cost, the product cost increases

TABLE 4
Summary of Product Cost (Case 1)

Item	Cost (¢/l)
Raw material	10.6
Operating labor	4.6
Operating supervision (15% of operating labor)	0.7
Utilities	28.3
Maintenance (6% of total fixed capital)	6.5
Operating supplies (15% of maintenance)	1.0
Laboratory charges (15% of operating labor)	0.7
Direct production cost	52.4
Depreciation (linear 18-year zero salvage)	6.1
Taxes and insurance (4% of total fixed capital)	4.4
Plant overheads (60% of operating labor + super-vision + maintenance)	7.1
Fixed cost	17.6
Manufacturing cost	70.0
Administrative cost (20% of plant overhead)	1.8
Distribution and marketing (5% of total production cost)	3.8
General expenses	5.6
After-tax profit (15% return on investment)	18.3
Income tax (50% of profits)	18.3
Lignin credit	(−)46.2
Animal feed credit	(−) 8.0
Steam exhaust credit	(−) 3.4
Product cost	54.6

TABLE 5
Impact of Plant Capacity on Product Cost

Plant capacity (m³/year)	Ethanol cost (¢/l)
3,000	110.3
6,000	82.0
15,000	54.6
24,000	44.5
30,000	40.2

TABLE 6
Impact of Cost of Rice Straw on Product Cost

Rice straw cost ($/ton)	Product cost (¢/l)
10	54.6
20	59.1
40	67.7
60	77.0
80	86.0
100	95.1

nearly 1.7 times. At $10/ton the raw material accounts for 20% of the total production cost, whereas at $100/ton it amounts to 54%.

3. Lignin (Coproduct)

It is also seen from Table 4 that credit obtained from lignin significantly affects the product cost. In the base case (Case 1), the lignin cost is taken as $1000/ton. However, the value of lignin depends on its quality, determined by chemical makeup as well as the process by which the lignin is recovered. The existing and potential markets of lignin have been classified as polymers, modified polymers, prepolymers, low-molecular-weight chemicals, and fuels.[20] The potential uses of lignin reflect a good number of possiblities of its outlet as an important industrial chemical. However, this is not possible to realize because of its nonavailability in bulk. Depending on its specific application, different values of lignin have been quoted in literature. Thus, Sundstrom and Klei[20] considered lignin value ranging from

TABLE 7
Impact of Lignin Credit on
Product Cost

Lignin cost ($/ton)	Product cost (¢/l)
100	96.2
200	91.6
400	82.3
500	77.7
1000	54.6
1500	31.5

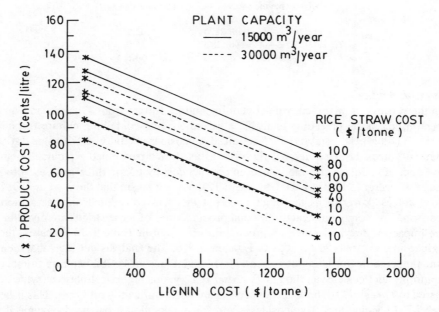

FIGURE 2.   Impact of cost of lignin and rice straw on product cost.

$88 to $1100/ton, depending on its use as fuel or high-molecular-weight polymer. Katzen et al.[21] used $200/ton of lignin in his economic analysis of the alcohol pulping process. In the phenol pulping process Battelle used a lignin value of $330/ton,[22] whereas Myerly et al.[23] used a cost of $176/ton in their economic analysis when its use as modified lignin was envisaged.

The sensitivity of lignin cost (ranging from $100 to $1500/ton) to the product cost is given in Table 7. At $100/ton for lignin the product cost is 96.2¢/l, whereas at $1000/ton it is 1.8 times less. The potential economic benefit of high-molecular-weight lignin is much greater than its value as fuel. The lignin obtained by the solvent process is high molecular weight and almost native in form. The value of solvent-derived lignin at $1000/ton considered in the analysis appears reasonable.

The dependence of product cost on the scale of operation, raw material (substrate) cost, and by-product credit is evident from the analysis. The variations in product cost due to changes in the costs of straw and lignin at two scales of operation, namely, 15,000 and 30,000 m³ ethanol per year, are presented in Figure 2. It illustrates that the product cost could be about the same at straw costs of $10 and $40/ton if the scale of operation is doubled. The data also indicate that even at comparatively higher rice straw costs (say $80/ton), a 30,000-m³ facility could be economically attractive if lignin can fetch a price of $1300/ton.

TABLE 8
Enzyme Activity with Different *T. reesei* Strains

Source	Filter paper activity (IU/ml)	β-Glucosidase activity (IU/ml)	Maximum productivity (IU/l/h)
T. reesei 9414	9.2	3.0	28.6
T. reesei D-1/6	13.5	8.0	62.5
T. reesei E-12	17.5	6.0	97.3

TABLE 9
Impact of Enzyme Source on Product cost

Mixed enzyme source	Product cost (¢/l)
T. reesei 9414—*A. wentii* 2820	62.1
T. reesei D-1/6—*A. wentii* 2820	55.0
T. reesei E-12—*A. wentii* 2820	54.6

4. Enzyme Source

In the process, a mixed-enzyme-containing culture filtrate from *T. reesei* and *A. wentii* has been used. In the preceding analysis, a mixed-enzyme-containing culture filtrate of *T. reesei* E-12 (filter paper activity, 17.5 IU/ml; β-glucosidase activity, 6 IU/ml) and *A. wentii* Pt 2820 (β-glucosidase activity, 20 IU/ml) was used. We performed a parametric analysis of the effect of cellulase enzyme source on product cost by taking three cellulase-producing strains of *T. reesei* (QM 9414, D-1/6, and E-12). We are aware that the *T. reesei* QM 9414 strain is not as powerful as the other two, but this was considered only as a model for comparison. The comparative activities and productivities of these strains are given in Table 8. The impact of enzyme source on product cost is given in Table 9. In all cases, the ratio of β-glucosidase activity to FP activity is taken as 1.6. The analysis indicates that the mixed enzyme preparation consisting of *T. reesei* E-12 and *A. wentii* Pt 2820 culture filtrates gives minimum product cost. Even though *T. reesei* E-12 yields higher cellulase enzyme activity compared to *T. reesei* D-1/6, the product costs are not significantly different. This is because *T. reesei* E-12 yields less β-glucosidase, and for maintaining a balanced enzyme it needs to be supplemented with larger amounts of *A. wentii* culture filtrate. However, the impact of source of cellulase enzyme on product cost is not very significant.

B. CASE 2

Concentration and dehydration of ethanol from fermentation broth are conventionally done by distillation and rectification. In order to reduce the energy requirement for ethanol separation, several nondistillation approaches have recently been reported.[24-29] Ethanol separation by adsorption-desorption, studied at BERC, is one such approach[26-28] used in the analysis. Some resins and activated carbon preferentially adsorb ethanol from aqueous solution. This is documented extensively. Using these methods, it is theoretically possible to develop equilibrium separation stages to achieve a fairly high concentration of ethanol to a near-dehydrated state. This principle has been applied to the use of some resins as sorbents to dehydrate ethanol from the product obtained from the SSF reactor via the vacuum cycling process. The ethanol is then desorbed from the saturated sorbent by elution with CO_2, N_2, or even air, and upon condensation the eluted fraction gives a concentrated ethanol solution. The process has been described in detail elsewhere.[28] We have analyzed the impact of this process of ethanol separation on product cost.

The data of a two-stage adsorption-desorption system for ethanol separation[28] on which

TABLE 10
Adsorption-Desorption Data for Ethanol Separation

	Stage 1	Stage 2
Sorbent	IRC-50 resin	Activated carbon
Sorption capacity	0.12 g ethanol/g dry resin	0.8 g ethanol/g dry activated carbon
Packing density	0.48	0.50
Feed	12.4% (w/w) ethanol at 30°C	50% (w/w) ethanol at 60°C
Elution	N_2 at mass velocity of 6 g/cm²-min; inlet pressure and temperature, 7 kg/cm² and 150°C; outlet pressure and temperature, 3 kg/cm² and 70°C	N_2 at mass velocity of 12.55 g/cm²-min; inlet pressure and temperature, 5 kg/cm² and 150°C; outlet pressure and temperature, 3 kg/cm² and 70°C

TABLE 11
Summary of Product Cost (Case 2)

Item	Cost (¢/l)
Direct production cost	61.7
Fixed charges	25.3
General expenses	7.0
Credits	(−) 60.1
Profits	27.9
Tax	27.9
Product cost	89.7

product cost estimation is based are given in Table 10. A polymeric hydrophilic sorbent with a weak acidic group (IRC-50) has been used in the first stage, activated carbon in the second stage, and nitrogen gas as eluant for both stages. For equipment cost estimation, in place of distillation towers, columns (2 × 60 m³ and 2 × 8 m³) packed with the described sorbents are taken. Additional equipment for the process are compressor and refrigeration systems. Other equipment is the same as in Table 1. Sorbent life of 2 years is considered in the analysis. The product cost for this case is given in Table 11.

It appears that in case 2 the product cost is 64.3% higher than in case 1. This is mainly attributed to the higher capital investment and additional expenditure on sorbents. These additional investments offset savings in the energy for ethanol separation.

C. CASE 3

In this case, solvent pretreatment is replaced by the conventional alkali process. In the alkali process, chopped rice straw is soaked in 1.5% NaOH for 24 h, followed by delignification at 120°C for 1 h.[30] The treated straw, following filtering, washing, and drying, is sent to the SSF reactor. An alkali solution containing lignin, hemicellulose, and the sugars derived from it is concentrated by evaporation to 60% solids. Alkali is recovered by the usual incineration of the solids. The lignin credit available from the process is considered equivalent only to its fuel value. No credit can be taken into account for animal feed, as pentose sugars are not available from the alkali process for SCP production.

In this case, equipment for solvent delignification is replaced by that required for the alkali process. However, the equipment for the batch-distillation system and that associated with solvent recovery and SCP production are not required. One evaporator (42.3 tons/h water removed) is needed. Based on these modifications, the capital cost for this case has been estimated.

The computed product cost for this case (Table 12) works out to be 74.6¢/l. It follows, therefore, that despite lower by-product credit for this case, vis-à-vis case 2, the product

TABLE 12
Summary of Product Cost (Case 3)

Item	Cost (¢/l)
Direct production cost	41.3
Fixed charges	15.0
General expenses	4.6
Credits	(−)15.7
Profits	14.7
Tax	14.7
Product cost	74.6

TABLE 13
Investment and Ethanol Costs Based on Different Substrates

Substrate	Capacity (m³/year)	Total investment ($ × 10⁶)	Product cost (¢/l)	Remarks	Ref.
Molasses	1×10^6	28.74	53.0	Molasses $85/ton	18
Corn	189×10^3	75.80	49.6	Corn $3/bushel, 10-year depreciation	3
Municipal and urban waste	189×10^3	122.20	38.0	Cellulosic waste $15.75/ton, 10-year depreciation	3
Rice straw	15×10^3	18.30	54.6	Case 1	*
Rice straw	30×10^3	27.70	40.2	Case 1	*

Note: Asterick (*) indicates information from this chapter, Section IV.A.

cost is lower. This is due to decreased fixed charges and direct production cost. However, the product cost is higher than in case 1.

V. CONCLUSIONS

The total capital investment and annual manufacturing cost for the three cases discussed above are compared. These costs are found to be higher for case 2. Although capital investment and manufacturing costs for case 3 are the lowest, the minimum product cost is for case 1. The advantages of lower capital investment and manufacturing cost for case 3 are offset by lower by-product credit. This emphasizes the importance of by-product credits in the total process economics, which is the key position of the BERC process. Under the current market situation, ethanol alone may not be sufficient for a profitable operation of such a bioconversion facility, unless ethanol becomes an emerging need. This underlies the necessity of converting all the major components of biomass to valuable products.

Finally, the investment and product costs for ethanol production based on various substrates (molasses sugar, starch, municipal wastes, and rice straw cellulose) calculated by several workers[3,18] and including the work presented here are shown in Table 13. It may be noted that the product costs are estimated using different bases and at different scales of operation. Nevertheless, it gives an indication of the range of variation in the cost of ethanol production from lignocellulosic biomass.

REFERENCES

1. **Linko, M.,** Technoeconomic Study on the Production of Ethanol from Cellulosic Materials in the Philippines, United Nations Industrial Development Organization, Vienna, 1983.
2. **Klyosov, A. A.,** Enzymatic Conversion of Cellulosic Materials to Sugars and Alcohol — The Technology and Its Implications, Rep. No. UNIDO/15/476, United Nations Industrial Development Organization, Vienna, 1984.
3. **Emert, G. H., Katzen, R., and Kaupisch, K. F.,** Economic up-date of the Gulf cellulose to alcohol process, *Chem. Eng. Prog.,* 76(9), 47, 1980.
4. **Joliff, G., Beguin, P., Jay, M., Millet, J., Ryter, A., Poljak, R., and Aubert, J. P.,** Isolation, crystallization and properties of new cellulase of *C. thermocellum* over-produced in *E. coli, Biotechnology,* 4, 896, 1986.
5. **Nagashima, M., Azuma, M., and Noguchi, S.,** Technology development in biomass alcohol production in Japan: continuous alcohol production with immobilized microbial cells, *Ann. N.Y. Acad. Sci.,* 413, 457, 1986.
6. **Ghose, T. K., Pannirselvam, P. V., and Ghosh, P.,** Catalytic solvent delignification of agricultural residues, *Biotechnol. Bioeng.,* 25, 1577, 1983.
7. **Ghose, T. K. and Ghosh, P.,** unpublished data, 1984.
8. **Nigam, M.,** Isolation and Characterization of Lignin Produced by Solvent Delignification from Rice Straw, M.Tech. thesis, Indian Institute of Technology, Delhi, 1984.
9. **Ghose, T. K., Roychoudhury, P. K., and Ghosh, P.,** Simultaneous saccharification and fermentation of lignocellulosics to ethanol under vacuum cycling and step feeding, *Biotechnol. Bioeng.,* 26, 377, 1984.
10. **Ghose, T. K.,** Indian Patent 155,781, 1981.
11. **Roychoudhury, P. K.,** Simultaneous Saccharification and Fermentation of Lignocellulose to Ethanol, Ph.D. thesis, Indian Institute of Technology, Delhi, 1986.
12. **Ghose, T. K., Tyagi, R. D., and Lebeault, J. M.,** Studies on the Production of Highly Active Cellulase Enzyme by the Mutant Strain of *T. reesei,* final rep. submitted to Elf Biorecherches Societe Ann. Comes, Paris, 1982.
13. **Srivastava, S. K.,** Kinetics of β-Glucosidase Production and Its Role in the Enzymatic Hydrolysis of Cellulose, Ph.D. thesis, Indian Institute of Technology, Delhi, 1985.
14. **Peters, M. S. and Timmerhaus, K. D.,** *Plant Design and Economics for Chemical Engineers,* McGraw-Hill, New York, 1980.
15. **Holland, F. A., Watson, F. A., and Wilkinson, J. K.,** *Introduction to Process Economics,* John Wiley & Sons, London, 1974.
16. **Guthrie, K. M.,** Capital cost estimating, *Chem. Eng.,* 76(6), 114, 1969.
17. Marshall and Swift Index. September 1986, *Chem. Eng.,* November 24, 7, 1986.
18. **Maiorella, B. L., Blanch, H. W., and Wilke, C. R.,** Economic evaluation of alternative ethanol fermentation processes, *Biotechnol. Bioeng.,* 26, 1003, 1984.
19. **Bailey, J. E. and Ollis, D. F.,** *Biochemical Engineering Fundamentals,* 2nd ed., McGraw-Hill, New York, 1986.
20. **Sundstrom, D. W. and Klei, H. E.,** Use of by-product lignin from alcohol fuel processes, *Biotechnol. Bioeng. Symp.,* 12, 45, 1982.
21. **Katzen, R. R., Fredrickson, R., and Brush, B. F.,** The alcohol pulping and recovery process, *Chem. Eng. Prog.,* 76(2), 62, 1980.
22. **Bungay, H. R.,** Commercializing biomass conversion, *Environ. Sci. Technol.,* 17, 24A, 1983.
23. **Myerly, R. C., Nicholson, M. D., Katzen, R., and Taylor, J. M.,** The forest refinery, *Chemtech,* 11, 186, 1981.
24. **Eakin, D. F., Donovan, J. M., Cysewski, G. R., Petty, S. E., and Maxham, J. V.,** Preliminary Evaluation of Alternative Ethanol/Water Separation Processes, Rep. No. PNL-3823/UC-98d, Battelle Memorial Institute, Columbus, OH, 1981.
25. **Pitt, W. W., Haag, C. L., and Lee, D. D.,** Recovery of ethanol from fermentation broth using selective sorption-desorption, *Biotechnol. Bioeng.,* 25, 123, 1983.
26. **Malik, R. K., Ghosh, P., and Ghose, T. K.,** Ethanol separation by adsorption-desorption, *Biotechnol. Bioeng.,* 25, 2277, 1983.
27. **Ghosh, S.,** Ethanol Separation by Adsorption, B. Tech. Proj. Rep., Indian Institute of Technology, Delhi, 1985.
28. **Malik, R. K.,** Ethanol Separation by Sorption-Desorption Technique, Ph.D. thesis, Indian Institute of Technology, Delhi, 1985.
29. **Choudhury, J. P., Ghosh, P., and Guha, B. K.,** Separation of ethanol from ethanol water mixture by reverse osmosis, *Biotechnol. Bioeng.,* 27, 1081, 1985.

30. **Pannirselvam, P. V. and Ghose, T. K.,** Pretreatment of agricultural residues for enzymatic saccharification by the culture filtrate of *T. reesei* QM 9414, in Proc. 2nd Int. Symp. Bioconversion, Vol. 1, Indian Institute of Technology, Delhi, 1980, 268.

Chapter 4

THE STORY OF BIOSYNTHETIC HUMAN INSULIN

John A. Kehoe

TABLE OF CONTENTS

I. INTRODUCTION

This chapter presents a look at the operations involved in the manufacture of a human pharmaceutical, namely, human insulin of recombinant DNA (rDNA) origin — the first recombinant pharmaceutical product ever to be marketed. This material was presented at this symposium for the first time outside of Eli Lilly and Company premises. It is, however, derived from a presentation which we have used internally to acquaint visitors with the production of human insulin.

II. ELI LILLY AND COMPANY AND INSULIN

This report will first briefly touch on Lilly's historical involvement with insulin manufacturing. It will then concentrate on the operations being used to produce biosynthetic human insulin.

Insulin was discovered in 1921 by Dr. Frederick B. Banting and his assistant, Charles H. Best. Following the discovery, Lilly was the first company to commercially produce this life-sustaining hormone from an animal pancreas. Marketing began in 1922. Patients who would have died from the progressive effects of diabetes were given new life. Throughout the more than 60 years that Lilly has been manufacturing insulin, the company has continually sought to improve production techniques and to improve the quality and purity of insulin formulations.

Until recently, the only commercially available forms of insulin were extracted from the pancreas glands of swine and cattle. As an average, an 800- to 1000-lb cow will provide enough insulin to make one vial; this will supply the average diabetic for about 25 d. It takes 14 cattle — or 70 pigs — to sustain a diabetic patient for 1 year. However, the diabetic population has been growing much faster than either the general population or the rate at which we consume beef or pork. A future shortage of animal-source insulin was thus a threat.

Another drawback of extracting insulin from the animal pancreas is that the end product is beef or pork insulin, not human insulin. The chemical structures of beef, pork, and human insulins have different amino acid sequences. As early as 1975, Lilly considered alternative ways to duplicate the human hormone and to break the dependence of diabetic care on the meat industry. Among these alternatives were

1. Extraction of insulin from the pancreas glands of human cadavers
2. Total chemical synthesis (obviously lengthy and expensive)
3. Conversion of pork insulin to human insulin using enzymes (again, dependent on animal glands)
4. Production of human insulin using the techniques of rDNA technology

Lilly opted to pursue rDNA technology because it was the only alternative that provided an economically viable and predictable source of human insulin.

III. THE APPROACH THROUGH rDNA TECHNOLOGY

The genetic manipulations involved in obtaining a modified *Escherichia coli* which will express a foreign protein are

1. Isolation of the plasmid DNA from the host
2. Cleavage at the desired site with restriction enzymes
3. Insertion into the plasmid of the desired DNA sequence coding for production of the product

4. Enzymatically rejoining the ends of the plasmid
5. Reintroduction of the modified plasmid into the host through transformation

In the human body, proinsulin is broken down to form insulin. In addition to the normal A- and B-chains of insulin, this molecule contains a connecting peptide. The molecular structures of proinsulin and insulin, therefore, made two recombinant manufacturing methods available. The first option was to ferment *E. coli* containing either the A or B chains in separate batches, isolate the chains, purify them, combine them, and then isolate the resulting human insulin from the combination mixture (the "chain process"). The second option was to begin with the body's own starting material by fermenting *E. coli* that produces proinsulin, isolating the proinsulin, folding it to form the correct bonds, removing the connecting peptide, and isolating the human insulin from the reaction mixture (the "proinsulin route"). Research with the chain process was more advanced, so we initially chose this approach. We now make insulin by the proinsulin route.

IV. PROCESS OVERVIEW

The process begins with fermentation of the *E. coli*. Next, the cells are killed by a heat inactivation process. Purification operations begin with harvesting the cells from the fermentation broth. This is followed by a chemical cleavage, a step that releases the proinsulin from the amino acid leader sequence used to initiate its biochemical synthesis. Next, there is a sulfitolysis reaction to prepare the proinsulin for chromatographic purification. During sulfitolysis, negatively charged ions are chemically bound to the proinsulin molecule at the disulfide bonds. This reaction provides a chemical "handle" that will be used during chromatographic purification. After sulfitolysis, the modified proinsulin is subjected to initial column purification. These steps include hydrophobic interaction chromatography, ion-exchange chromatography, and gel filtration chromatography.

The result of this processing is purified human proinsulin-*S*-sulfonate that is ready for folding. During the folding reaction, conditions are maintained to enable the proinsulin chain to arrange itself in the proper molecular configuration and the disulfide bonds to form in the proper positions. After undergoing an additional hydrophobic interaction chromatography step, the material is ready for transformation. During this step, the connecting peptide is removed enzymatically, and the resulting product is human insulin ready for final purification. Final purification steps include ion-exchange chromatography, crystallization, gel filtration chromatography, and a final crystallization.

V. FERMENTATION

The fermentation process starts with the streaking of a preserved seed culture of the genetically modified *E. coli* onto an agar plate. The organisms grown on the plate are transferred to shake flasks to prepare a vegetative inoculum. The vegetative inoculum is transferred to a seed tank using a specially designed transfer vessel. When sufficient growth has occurred in the seed tank, the culture is transferred to the production fermentor. The production fermentor has been modified to meet the requirements for the containment of recombinant organisms according to the guidelines proposed by the National Institutes of Health (NIH). These modifications include the use of a double mechanical seal on the agitator which is flushed with hot condensate. The condensate flow and pressure are monitored to assure the integrity of the seal during the fermentation. Exhaust air from the fermentor and seed vessel is first demisted in a cyclone separator, then passed through a coalescing filter to remove any last entrained water or organisms, and finally passed through a pair of sterilizing-grade absolute filters in series before being discharged to the atmosphere.

All condensate which could contain organisms, live or dead, is collected and thermally inactivated before being discharged to waste treatment. Samples from the vessels are taken with the use of contained samplers. Environmental sampling is the last line of action taken to assure containment of the organism. From a process point of view, the modifications to a standard production fermentor were directed at achieving the greater oxygen transfer and cooling required to support this bacterial fermentation. Computer control of the fermentation process regulates the process parameters to achieve optimal growth and expression of product.

At harvest the *E. coli* cells, laden with insoluble deposits of proteins, are thermally inactivated under very precise temperature and flow conditions. Computer control assures that appropriate conditions are maintained for inactivation, and magnetically coupled pumps and spiral heat exchangers provide for mechanical containment.

VI. PURIFICATION

The first purification step is harvesting. The *E. coli* cells are separated from the fermentation broth. This is done in disk stack centrifuges generating about 13,000 × *g* at 6700 rpm. Control is maintained by a programmable controller interfaced with a central computer.

Next, the proinsulin is cleaved from the leader amino acid sequence. For safety reasons, these operations are remotely monitored and controlled from a central computer outside the processing area. The cleavage reaction is the classic cyanogen bromide technique that has been in use for years by biochemists for this specific purpose, but in order to generate enough reagent to support our needs we had to devise a unique method of preparation. Since the reaction is carried out in formic acid, our solution was to generate the reagent from the reaction of sodium cyanide with bromine, then to distill the cyanogen bromide from the reaction mixture and absorb directly into the formic acid. At the conclusion of the cleavage reaction, any unreacted cyanogen bromide is distilled from the reaction mixture — again being absorbed directly into formic acid. Obviously, the nature of these reagents dictates the use of glass-lined vessels and Kinar-lined pipe, but beyond that, all air exhausted from the vessels is scrubbed with a solution of hypochlorite which is prepared and maintained *in situ*. Work rules for this totally isolated area require the operators to work in pairs, entering the isolation area only in full, protective, supplied-air suits and limited to a maximum of 2 h in any one work period in the area. Additionally, the rooms are continuously monitored for the presence of halogen and organic vapor.

After cleavage, the acid is removed in evaporators in which it is displaced by the solvent that will be used during the subsequent purification steps. As indicated previously, the first step is to prepare the molecule for purification by subjecting it to a sulfitolysis reaction. This makes the *S*-sulfonate derivative of the disulfide and sulfhydryl groups present, giving those groups very strong negative charges, which are the keys to purification of the proinsulin-*S*-sulfonate. The first chromatographic purification step is by hydrophobic interaction, carried out in columns specifically designed for the packing being used. The ion-exchange and gel filtration chromatography steps that follow are carried out in a cold room using Amicon columns. We are currently using columns to 140 cm in diameter for these operations, but in the initial design stages for this facility the largest columns that were then available were only 100 l in size. Because of our need at that time, Amicon built and we tested the prototype 63-cm-diameter columns which became our mainstay for the next 4 or 5 years. Performance of the columns is monitored by on-line high-performance liquid chromatography, as well as the customary optical density and conductivity measurements. These are interfaced with a process control computer which has completely automated these process steps.

Now the proinsulin-*S*-sulfonate is ready for folding to the proper spatial configuration with the proper disulfide linkages. The folded proinsulin next undergoes another form of hydrophobic interaction chromatography — this time carried out in a reverse-phase mode.

Again, Amicon columns are used for this step. Following evaporation to remove the organic solvent used in the elution of the above step, the enzymatic transformation to biosynthetic human insulin is conducted in a batch mode. The final purification steps of the human insulin are again carried out in a cold room in chromatography columns from Amicon, in the case of the ion-exchange step, and Pharmacia, in the case of the final gel filtration. The final crystallization is carried out in the presence of zinc, and the product is vacuum dried in a shelf dryer.

The final packaging of Humulin® takes place in another facility, one that is designed for the sterile processing of injectable human pharmaceuticals to their final package forms.

VII. TECHNOLOGICAL ADVANCEMENT

From the start-up of production of human insulin in 1981, significant progress had been made toward increasing the yields of the chain process. However, the proinsulin method has provided production yields that are twice that of the chain process. Furthermore, we no longer have to process the A and B chains separately through the plant, which means we now have a significantly shorter and more capital-efficient production method. Thus, the advantages of the proinsulin route can be summarized as follows:

1. It increases the capacity of the plant.
2. The improved yields provide a better opportunity to control costs.
3. Proinsulin can now be eveluted as a therapeutic agent for some forms of diabetes.
4. It allows Lilly to maintain leadership not only in state-of-the-art insulin production, but also in the rDNA field.

VIII. CONCLUSION

While we have emphasized the importance of rDNA methods, the production of a recombinant pharmaceutical product like Humulin® requires extensive capability in other technologies. They are

1. Large-scale fermentation technology with the additional demands for containment of the recombinant organism
2. Complex protein chemistry like that used to produce biologically active chemicals and to modify antibiotics
3. Protein isolation and purification similar to that used in the separation of animal insulin from a pancreas
4. Analytical chemistry essential to monitoring yields and purity as we separate the dilute product from a complex mixture

When Eli Lilly and Company began the human insulin project in the mid-1970s, we had expertise in these four essential technologies. We acquired the fifth essential technology in 1978 from Genentech, Inc.: the successful creation and insertion of the human insulin gene into *E. coli* bacteria. We had the resources and commitment to do 5 years of research and then 4 years of process development. We were able to obtain the first approval from the NIH to scale up production of a recombinant product. We were willing to make the $70-plus million investment for full-scale production facilities to manufacture Humulin® here in the U.S. and at Lilly facilities in the U.K.

The success of the human insulin program is an achievement we are proud of, but we hope there will be others to come. We believe this technology holds promise for the discovery of new pharmaceutical and agricultural products to help mankind.

Chapter 5

PURIFICATION AND SEQUENCING OF INTERFERONS AND OTHER BIOLOGICALLY ACTIVE PROTEINS AND POLYPEPTIDES

T. L. Nagabhushan, R. Kosecki, B. Pramanik, J. Labdon, and P. P. Trotta

TABLE OF CONTENTS

I. INTRODUCTION

The purification and structural characterization of proteins occurring at low levels in natural sources has been a challenging problem to protein chemists. The advent of recombinant DNA technology has provided the basis for the expression of heterologous proteins in a variety of bacterial and eukaryotic systems. Although the levels of expression of recombinant proteins may represent a high percentage of the total cellular protein, techniques must be developed for efficient extraction and purification with good yields. Of special interest is the resolution of electrophoretic variants in the final purified material. We describe here the application of a novel continuous free-flow isoelectric focusing system for the resolution of forms of a recombinant human hybrid α-interferon (IFN α) with distinct isoelectric points. This apparatus was designed as a modification of previously reported recycling isoelectric focusing systems.[1,2] The technique appeared to be comparable in resolving power to chromatofocusing, but was superior with respect to the fact that large quantities of material could be processed continuously.

Structural characterization of the purified protein must also be obtained, especially with respect to its primary sequence and the state of oxidation of cysteine residues, if present. Although conventional mass spectrometry (MS) has been useful for obtaining structural and molecular weight information on nonpolar compounds, its usefulness for proteins and polypeptides is limited since electron ionization requires volatilization before analysis. Fast atom bombardment/mass spectrometry (FAB/MS) is a new technique for obtaining high-quality mass spectra that previously had been difficult or impossible to obtain by ionization techniques.[3,4] Since FAB/MS does not require heating, it avoids potential thermal degradation. FAB/MS may be utilized to determine the sequence of as many as 20 to 30 amino acid residues and requires only 10 to 15 nmol of material. We report here the successful application of this technique to obtain amino acid sequence information and molecular weights on polypeptides of biological significance. The technique has also been employed to identify the state of oxidation of sulfhydryl groups on a polypeptide related to the N-terminus of human recombinant IFN γ. Finally, we have combined gas chromatography (GC) with chemical ionization (CI) MS to determine the amino acid sequence of *N,O*-permethylated, *N*-acetylated leu-enkephalin.

II. EXPERIMENTAL

A. MATERIALS

Ampholytes were obtained from Serva Fine Biochemicals (Westbury, NY). CHAPS (3-[{3-cholamidopropyl} dimethylammonio]-1-propanesulfonate) was purchased from Pierce Chemical Company (Rockford, IL). Urea was purchased from BioRad (Richmond, CA). Synthetic polypeptides were synthesized using solid-phase techniques by Bachem (Torrence, CA).

B. CONTINUOUS FREE-FLOW ISOELECTRIC FOCUSING APPARATUS

The focusing cell consisted of a 0.75-mm-wide channel (38.5 × 5.2 cm) recessed into a Plexiglas® plate (50 × 11.5 × 2.4 cm). Attached to this Plexiglas® plate was an aluminum plate (50 × 11.5 × 1 cm) to which thermoelectric couplings were connected for dissipation of Joule heat. A sensor in the aluminum plate provided monitoring of temperature during the run. Located at the bottom and top of the cell for entry and exit of the sample, respectively, were 40 tubes. The anode and cathode compartments were separated from the adjacent focusing cell by an ion-exchange membrane. Each compartment contained a platinum electrode wire. In the experiment reported here, the cathode and anode solutions consisted of 0.2 *M* histidine and 0.04 *M* glutamic acid, respectively. Circulation was achieved with a peristaltic pump. The sample was applied through a 40-channel Plexiglas® bubble trap to which a sealing bar was attached during focusing. Collection was accomplished with a coupling block, which was unbolted at the completion of focusing and placed over a test-tube rack for simultaneous collection of solution from the 40 channels. Power was provided with a Bio-Rad® 3000 XI high-voltage power supply.

C. FAST ATOM BOMBARDMENT/MASS SPECTROMETRY

Mass spectra were obtained on a VG ZAB-SE mass spectrometer equipped with an FAB source operating at an accelerating voltage of 8 kV. FAB mass spectra were also obtained using a Finnigan MAT® 312 mass spectrometer operating at an accelerating voltage of 2 to 3 keV. Samples were dissolved in 2 to 10 μg/ml dimethyl sulfoxide (DMSO) and deposited on an FAB probe tip. A thin layer of either glycerol or thioglycerol was applied to the probe tip containing the samples and was mixed thoroughly with a Pasteur pipette before insertion into the source. The primary atom (xenon) was produced using a saddle field ion source operating at a tube current of 1 to 1.5 mA at an energy of 8 keV.

D. GAS CHROMATOGRAPHY/CHEMICAL IONIZATION MASS SPECTROMETRY

GC/CI mass spectra were obtained on an Extrel 400-1 GC/MS system. An 8-m SPB-1 column was used; the temperature of the column was programmed from 150 to 305°C at 8°C/min. The temperatures of the MS source and the injector were maintained at 180 and 230°C, respectively. The helium gas pressure was 4 psi, and the MS source pressure with methane was maintained at 1.5×10^{-4} torr.

III. RESULTS

A. CONTINUOUS FREE-FLOW ELECTROPHORESIS

A novel human recombinant hybrid IFN, δ-4 α-2/α-1, has been purified to a high degree of homogeneity by conventional chromatographic procedures.[6] Analytical isoelectric focusing of the purified protein in the pH range of 5 to 7, however, indicated the presence of two major components (Figure 1). The approximate isoelectric point (pI) values of these components were found to be 6.25 and 6.03, respectively, as determined by soaking gel strips in 0.01 M KCl followed by measurement of pH values. Based on these results it was decided to conduct a continuous free-flow isoelectric focusing experiment in 1% Servalyt® carrier ampholytes (pH 5 to 7) for isolation of these components on a preparative scale, utilizing 10 mg of the hybrid IFN as starting material. For solubilization of the hybrid IFN, 1% CHAPS and 6 M urea were added to the solution. Since both of these compounds are electrically neutral, neither interfered with the focusing process. After 4 h of focusing at a constant power of 100 W and a temperature of 15°C, the solutions from the 40 channels were pumped into test tubes (about 5 ml/tube). Each solution was analyzed for IFN antiviral activity, pH, and protein concentration.

As shown in Figure 2, an approximately linear pH gradient was established between pH 5.0 and 6.5, constituting a change of approximately 0.04 pH units per channel. Two main protein peaks were observed at pI values of 6.3 to 6.4 and 5.8 to 6.0, respectively. Analytical isoelectric focusing was performed on fractions corresponding to each of these protein peaks. The first peak eluted (tube numbers 5 to 7) was found to contain mainly the component with a pI value of 6.25, whereas the second peak (tube numbers 16 to 18) was observed to be enriched for the species with a pI value of 6.03 (data not shown). Interestingly, measurements of specific antiviral activity of eluted fractions indicated that the protein peak with the more acidic pI value exhibited a higher specific antiviral activity (Figure 2). It has not yet been established, however, whether this result indicates that two subpopulations of

+

A.   **B.**   **C.**   **D.**

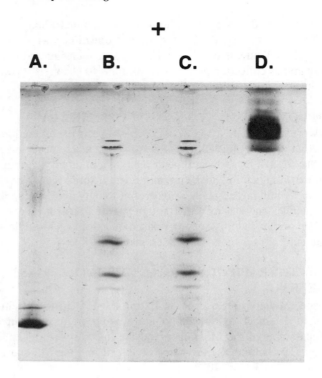

—

FIGURE 1.   Analytical isoelectric focusing of human recombinant IFN δ-4 α-2/α-1 variants. (A) Human carbonic anhydrase (pI = 6.6); (B,C) IFN δ-4 α-2/α-1; (D) β-lactoglobulin (pI = 5.3). Focusing was performed on a Phast electrophoresis system employing Phast Gel IEF media (Pharmacia, Piscataway, NJ). Gels were incubated at room temperature for 30 min in a solution containing 6 M urea, 1% CHAPS, and 4% Servalyt® carrier ampholytes (pH 5 to 7). After the gels were prefocused for 75 V-hours, samples were applied and focusing was allowed to proceed for 525 V-hours. Gels were silver stained. The band appearing at the anodal end was produced artifactually by protein precipitation at the point of sample application.

IFN δ-4 α-2/α-1 molecules with different intrinsic antiviral activity exist or whether the presence of impurities in peak 1 lowers the specific antiviral activity.

We have also attempted to resolve the electrophoretic variants of the hybrid IFN α by chromatography on a monophosphate HR 5/20 chromatofocusing column employing a fast protein liquid chromatography (FPLC) system. The column was equilibrated with 40 mM Tris·HCl (pH 7.4) containing 6 M urea and 1% CHAPS; the sample was eluted with Poly-buffer® 74 (pH 5.0) prepared in 6 M urea, 1% CHAPS. Analysis of the eluted fractions by analytical isoelectric focusing indicated the achievement of a resolution comparable to that observed with continuous free-flow electrophoresis (data not shown).

B. FAST ATOM BOMBARDMENT/MASS SPECTROMETRY

1. Human IFN γ N-Terminal Nonapeptide

The N-terminal sequence of human IFN γ is of special interest, since it contains two cysteine residues at positions 1 and 3 that are separated by a tyrosine residue. The question arises as to whether a disulfide bond can form between two sulfhydryl groups despite potential steric hindrance from the intervening amino acid. In order to investigate this possibility, the following nonapeptide corresponding to the N-terminus of human IFN γ was synthesized by solid-phase techniques:

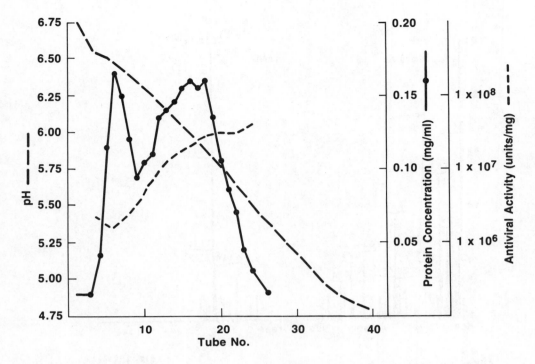

FIGURE 2. Continuous free-flow isoelectric focusing of recombinant human IFN δ-4 α-2/α-1. Antiviral activity was measured by a cytopathic effect inhibition assay.[9] Protein concentration was determined as described by Bradford.[7] Further details are described in Section II.

$$H_2N-C-Y-C-Q-D-P-Y-V-K-COOH$$

The final product was air-oxidized as a dilute sample while still attached to the synthesis gel. Oligomers were subsequently removed by gel filtration. Reverse-phase high-performance liquid chromatography (HPLC) on a C-8 column indicated the presence of a single major peak eluting with a retention time of 22.7 min. After boiling with 2-mercaptoethanol for 10 min, the peptide eluted with a retention time of 18.7 min.

The nonapeptide was dissolved in water or was treated with 5% 2-mercaptoethanol (final concentration >20 mg/ml). The FAB/MS data of the peptide in water are shown in Figure 3A. A strong ion was observed at m/z1116, which represented the molecular weight as NH$^+$ for the disulfide-containing form of the peptide. The FAB mass spectrum of the peptide in a thioglycerol matrix showed an ion at m/z1116 at the earlier time points of the repetitive scanning, rapidly converted to an ion at m/z1118. These data support the conclusion that the disulfide bond was reduced under FAB/MS conditions in a thioglycerol matrix. In other experiments employing a VG ZAB-SE mass spectrometer in Reinhold's laboratory (Harvard University), the presence of an ion at m/z2231 was observed in addition to the observations described above.[13] The ion at m/z2231 may be interpreted as the [2M + H]$^+$ ion.

The FAB/MS data of the 2-mercaptoethanol-treated peptide displayed a very intense ion at m/z1118, corresponding to the completely reduced peptide (Figure 3B). A strong ion at m/z1194, 76 amu higher than the ion at m/z1118, was also observed. These data (m/z1194) indicate that it represents a mixed disulfide between 1 mol of this peptide and 1 mol of 2-mercaptoethanol. It has not yet been established which of the two cysteine residues participates in this mixed disulfide bond.

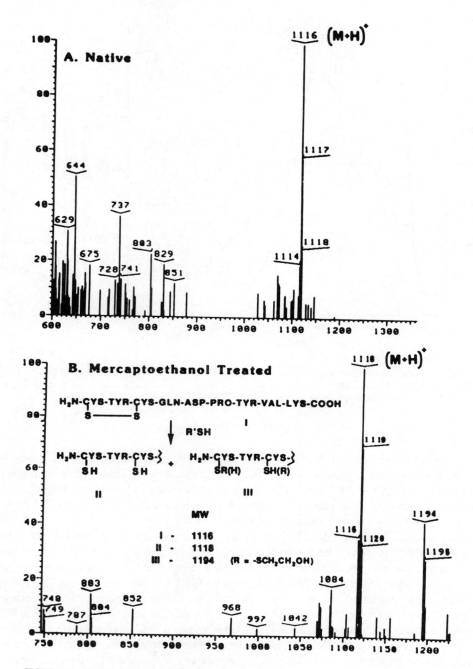

FIGURE 3.   FAB mass spectra of Gif-N9, a synthetic nonapeptide corresponding to the N terminus of human IFN γ. (A) Native peptide in water; (B) peptide treated overnight with 5% 2-mercapto-ethanol at room temperature.

2. Identification of Fragments of Recombinant Human IFN

Treatment of fully reduced recombinant human IFN α-2 with CNBr is predicted from the distribution of methionine residues to result in six polypeptide chains.[8] However, since in the native molecule disulfide linkages exist between Cys_1 and Cys_{98} and between Cys_{29} and Cys_{138}, CNBr cleavage results in four fragments.[9] These fragments are referred to as 1-4, 3-5, 2, and 6. Fragments 1-4 and 3-5 refer to polypeptides in which fragments 1/4 and 3/5, respectively, are covalently linked by disulfide bonds. The fragments were separated by reverse-phase HPLC as described by Lydon and co-workers[9] and subjected to FAB/MS for identification (Figure 4). Fragment 2 displayed a molecular ion at m/z545 [M + H]$^+$, which reflected the conversion of the C-terminal methionine to homoserine during the CNBr cleavage. An intense ion was observed at 527, [M + H−H$_2$0]$^+$, which corresponded to the homoserine lactone. The amino acid sequence of fragment 2 was obtained from a number of fragment ions obtained in its FAB mass spectrum (Figure 4a).

Fragment 6 displayed an intense ion at m/z1983, which corresponded to the predicted molecular weight of the [M + H]$^+$ ion. (Figure 4b). The major peak at 991 represented the doubly charged ion for this polypeptide and indicated the presence of basic amino acid residues (i.e., lysine and arginine). The formation of doubly and triply charged ions in the FAB/MS of polypeptides is consistent with observations on other polypeptides.

The FAB mass spectrum of fragment 1-4 exhibited a very weak ion at m/z7640 (Figure 4c). However, the predicted molecular weight of this fragment is 7735. Thus, there is a difference of 95 amu between the observed and predicted values. However, it must be noted that fragment 1-4 contains two methionine residues at the two C-termini of fragments 1 and 4, respectively. The formation of homoserine lactone from each of the two methionine residues in fragment 1-4 resulted in a predicted molecular ion of 7639, which agreed well with the observed value of 7640. As noted above, a similar result was obtained for fragment 2.

In continuation of our FAB/MS studies on recombinant human IFN α-2, we have analyzed a tryptic digest of this protein without further purification. The mass spectral data of the unpurified digest allowed us to confirm the presence of most of the predicted tryptic peptides in a single scan, with the exception of a few low-molecular-weight fragments which were below the mass range that was actually scanned. In addition, these data confirmed the presence of the $Cys_1 - Cys_{98}$ and $Cys_{29} - Cys_{138}$ disulfide bonds. In a separate experiment, IFN α-2 derivatized with 4-vinylpyridine was digested with trypsin and the digest analyzed by FAB/MS. The FAB/MS data provided the expected tryptic peptides, including some fragments exhibiting an increased molecular weight due to chemical derivatization. The data indicated that the derivatized ions provided strong molecular ions in the FAB mass spectra.

3. Sequencing of the C-Terminal Peptide of Recombinant Human IFN γ

A peptide containing the C-terminal 15 amino acids of recombinant human IFN γ was synthesized with the following predicted structure:

$$H_2N-R-K-R-S-Q-M-L-F-R-G-R-R-A-S-Q-COOH$$

The calculated molecular weight of this peptide is 1875. FAB/MS analysis showed a strong ion at m/z1876, which corresponded to the protonated molecular ion, [M + H]$^+$. As shown in Figure 5, the FAB spectrum contained a large number of fragment ions. These ions permitted confirmation of the sequence noted above.

C. GAS CHROMATOGRAPHIC-MASS SPECTROMETRIC SEQUENCING OF LEU-ENKEPHALIN

Rose et al.[5] have reported a new technique for the permethylation of mixtures of acylated peptides at a level of 2 to 10 nmol. These volatile derivatives can be resolved by GC and

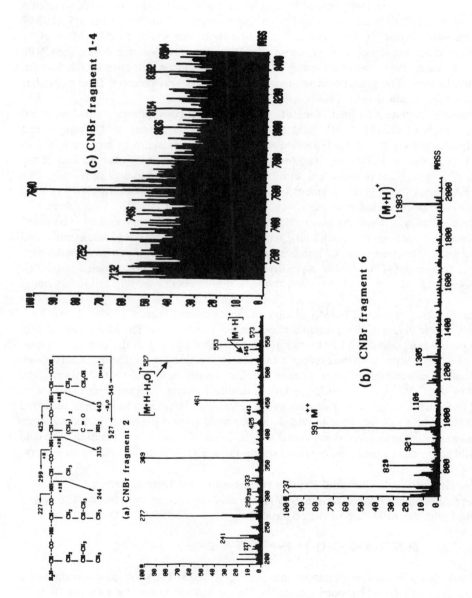

FIGURE 4.   FAB mass spectra of CNBr fragments of recombinant human IFN α-2. (a) Fragment 2; (b) fragment 6; (c) fragment 1-4. The fragments and CNBr reaction conditions have been described previously.[9]

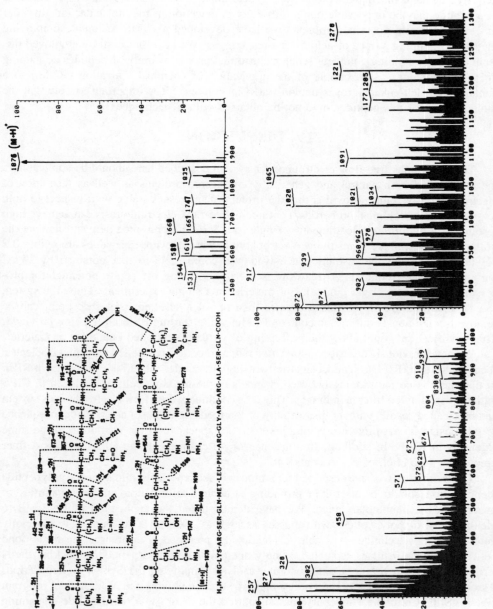

FIGURE 5. FAB mass spectrum of a 15-residue polypeptide corresponding to the C terminus of human IFN γ.

their molecular masses determined by MS. This technique is especially useful for providing sequence surveys of proteins and polypeptides through the generation of proteolytic fragments which can be converted to volatile derivatives as described above and analyzed by GC/MS. Ionization required for MS analysis is frequently achieved by electron impact (EI). However, standard EI techniques result in the formation of an unstable radical cation which spontaneously fragments into daughter ions. Hence, determination of the last residue in the sequence of a peptide may be impossible due to difficulty in detection of the molecular ion. In order to circumvent this potential problem, we have developed a CIMS technique employing methane containing a trace of helium as a reagent gas. We have successfully employed this technique for obtaining complete sequence information on a variety of peptides containing up to five residues. An example of the application of chemical ionization GC/MS to a pentapeptide, leu-enkephalin, is demonstrated in Figure 6. It is important to note that the molecular ion of this peptide could not be obtained with EI techniques.

IV. DISCUSSION

The continuous free-flow electrophoresis system described here should be applicable to the isolation of both natural and genetically engineered proteins as well as resolution of electrophoretic variants. Provided that the protein of interest is stable with respect to both its physical and biological properties, continuous free-flow electrophoresis can achieve high resolution at high protein loadings in a single run. The data presented here substantiate the utility of this technique in resolving electrophoretic variants whose pI values are within 0.2 pH units. Although the resolution achieved was comparable to that obtained by FPLC chromatofocusing, continuous free-flow isoelectric focusing offers the potential for processing significantly larger quantities of material in a single experiment. Thus, the system can be modified readily so that a species focusing at a selected isoelectric point can be removed continuously during the course of a run while starting material is being provided. This modification would allow the processing of virtually unlimited quantities of material.

We have previously described applications for a computer-controlled recycling isoelectric focusing system (RIEF).[2] The isoelectric focusing system described here differs from RIEF in that there is no partitioning between channels in the focusing cell, as achieved in RIEF with the use of nylon filter membranes. Thus, the continuous free-flow electrophoresis system can be cleaned readily without disassembly of the focusing cell. An additional consequence of this design is that protein precipitation does not appear to affect resolution to the same extent as in RIEF. In addition, the presence of 40 channels in the RIEF permits a finer preparative resolution of components.

The techniques we have described for amino acid sequence determination and polypeptide identification should be useful in providing structural information on low quantities of material in the nanomolar range. We have demonstrated the application of FAB/MS to identification of polypeptides with masses as high as 7000 to 8000 amu. It is especially notable that this technique was able to identify the presence of an internal disulfide bond between cysteine residues separated by only one amino acid in a peptide mimicking the N terminus of human IFN γ. Conformational analysis employing the Sybil program (Tripos Associates, St. Louis, MO) and energy minimization employing the Macromodel program (W. C. Still, Columbia University) has supported the theory that the disulfide-containing structure is a stable, low-energy conformation.[12] These calculations have indicated that stabilization is achieved in part by a transannular hydrogen bond between the N-terminal cysteine carbonyl group and the amide NH of the cysteine at position 3. As further substantiation for this conclusion, we observed that treatment of this peptide with 2-mercaptoethanol converted it into a form that eluted earlier on reverse-phase HPLC. Thus, our data strongly support the existence of this disulfide linkage, in distinction to the previous con-

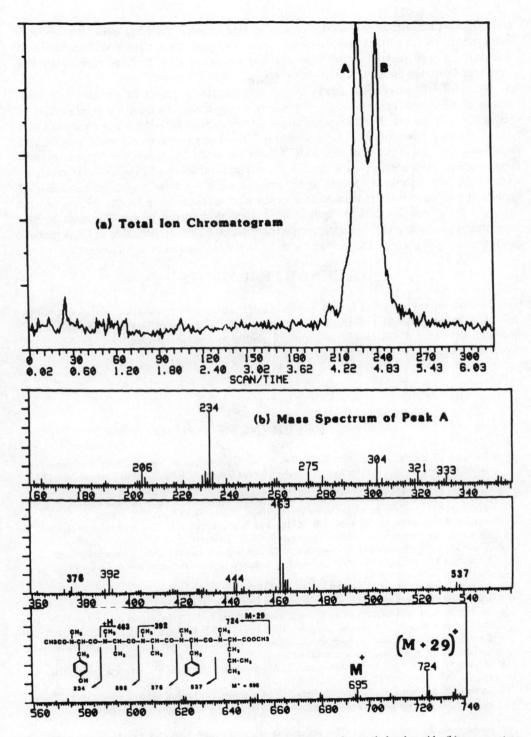

FIGURE 6.   GC/CIMS of leu-enkephalin. (a) Total ion chromatogram of permethylated peptide; (b) mass spectrum of peak A. Permethylation was performed as described by Rose et al.[5]

clusion that such a linkage between closely spaced residues does not exist in proteins.[10] Interestingly, this disulfide linkage may have a regulatory function, since it has been observed that genetically engineered murine IFN γ lacking the first three N-terminal amino acids has significantly higher antiviral activity than the full-length molecule.[11]

The power of mass spectrometry to provide complete sequence information in a moderately sized peptide of 15 residues was clearly demonstrated for the C-terminal peptide of recombinant human IFN γ. Such a technique will be especially useful when combined with a previous step for resolution of peptides, e.g., liquid or gas chromatography. Amino acid sequence analysis of peptides using GC/MS as described by Rose et al.[5] has now been confirmed in our laboratory on a number of test polypeptides containing up to six residues. The GC/CIMS technique appears to be superior since we have noted for a number of polypeptides that it provides abundant molecular ions not observed in GC/EIMS. We foresee a larger future role for MS in rapid identification of unknown proteins through peptide mapping, in providing amino acid sequence information for the construction of DNA probes, and in establishing the correctness of N- and C-terminal regions of recombinant proteins.

ACKNOWLEDGMENTS

The authors acknowledge Schering R & D Engineering for construction of the continuous free-flow electrophoresis and Dr. Milan Bier of the University of Arizona for advice and useful discussions. We thank Jeffrey Lipps for skillful technical assistance and Ms. Margarite Poepoe for typing the manuscript.

REFERENCES

1. **Bier, M., Egen, N. B., Allgyer, T. T., Twitty, G. E., and Mosher, R. A.,** New developments in isoelectric focusing, in *Peptides: Structure and Biological Function,* Pierce Chemical Company, Rockford, IL, 1979, 79.
2. **Nagabhushan, T. L., Sharma, B., and Trotta, P. P.,** Application of recycling isoelectric focusing for purification of recombinant human leukocyte interferons, *Electrophoresis,* 7, 552, 1986.
3. **Barber, M., Bordoli, R. S., Sedgwick, R. D., and Tyler, A. N.,** Fast atom bombardment of solids in (F.A.B.): a new ion source for mass spectrometry, *Chem. Commun.,* p. 325, 1981.
4. **Williams, D. H., Bradley, C. V., Santikarn, S., and Bojesen, G.,** Fast-atom-bombardment mass spectrometry, a new technique for the determination of molecular weights and amino acid sequences of peptides, *Biochem. J.,* 201, 105, 1982.
5. **Rose, K., Simona, M. G., and Offord, R. E.,** Amino acid sequence determination by G.L.C. — mass spectrometry of permethylated peptides, *Biochem. J.,* 215, 261, 1983.
6. **Trotta, P. P., Le, H. V., Sharma, B., and Nagabhushan, T. L.,** Isolation and purification of human alpha, a recombinant DNA protein, *Dev. Ind. Microbiol.,* 27, 53, 1987.
7. **Bradford, M.,** A rapid and sensitive method for the quantitation of microgram quantities of protein utilizing the principle of protein-dye binding, *Anal. Biochem.,* 72, 248, 1976.
8. **Streuli, M., Nagata, S., and Weissmann, C.,** At least three human type α interferons: structure of α 2, *Science,* 209, 1343, 1980.
9. **Lydon, M. B., Favre, C., Bove, S., Neyret, O., Benureau, S., Levine, A. M., Seelig, G. F., Nagabhushan, T. L., and Trotta, P. P.,** Immunochemical mapping of α-2 interferon, *Biochemistry,* 24, 4131, 1985.
10. **Richardson, J.,** The anatomy and taxonomy of protein structure, *Adv. Protein Chem.,* 34, 167, 1981.
11. **Le, H. V., Mays, C. A., Syto, R., Nagabhushan, T. L., and Trotta, P. P.,** in *The Biology of the Interferon System 1985,* Stewart, W. E., II and Schellekens, H., Eds., Elsevier, New York, 1986, 73.
12. **Czarniecki, M.,** unpublished observations.
13. **Reinhold, V.,** personal communication.

Part 2. Advances in Fermentation

Bioreactor research and development has, in recent years, led to increased diversity rather than standardization. This diversity is partly a reflection of the diversity of organisms being developed as "catalysts" and partly a reflection of the rapid commercialization of diverse approaches developed in scientific laboratories to fulfill scientific (and not necessarily commercial) goals. Thus, issues of scaleup and optimization are being attacked through a combination of modeling, cell engineering, and biological experimentation. In this section, fermentation scientists from corporate and academic laboratories discuss their experiences at the frontier of fermentation development. This subject attracted several poster presentations, which introduced a variety of innovations in modeling (including applications of control theory) and process integration.

Chapter 6

KINETICS OF IMMOBILIZED CELLS: A STRUCTURED MODEL

Harold G. Monbouquette and David F. Ollis

TABLE OF CONTENTS

I. INTRODUCTION

The many advantages of immobilized-living-cell processes over traditional suspension-culture fermentations have been canvassed in several reviews.[1-7] A number of research groups have praised the potential of immobilized-living-cell systems as more efficient means for production of virtually the whole spectrum of biologically derived products. However, published descriptions of attempts to scale up immobilized-cell technology are rare.[5,6,8] The absence of an experimentally verified theoretical model of immobilized biocatalyst behavior has probably retarded industrial implementation of these processes.

To address this need, a simple structured model of immobilized cells has been constructed as a qualitative, experimentally verifiable description of the diffusion and reaction phenomena and the physiological state of immobilized cells under steady-state conditions.[9] This two-component description of biomass predicts the RNA concentration per unit dry biomass (roughly proportional to protein synthesizing capability) with depth into a porous carrier. As this structured model has been developed from an experimentally verified model for free-suspension culture,[10] it offers a means of comparing free-suspension and immobilized-culture kinetics on a qualitative basis.

II. MICROSCOPIC OBSERVATIONS

Steady-state biocatalyst performance in continuous-flow bioreactors is normally achieved subsequent to an initial period of *in situ* cell growth. If sufficient nutrients are supplied continuously, the maximum immobilized-biomass concentration that the support can accommodate within the pore space accessible to cells is eventually reached at steady state. Observation of this phenomenon poses some difficulty with friable solid supports, but gels containing immobilized microcolonies in various stages of expansion have commonly been treated with fixing agents, sectioned, and subjected to microscopic examination. *In situ* growth has been observed from the initial stage, where single cells exist within 5- to 10-μm cavities, to the development, expansion, and eventual merging of spheroidal microcolonies near a gel particle surface (see Figure 1).[11-15] In the later stages of biocatalyst activation, cell divisions occur primarily in the surface region of the carrier. Presumably, the uneven distribution of metabolic activity is dictated by limitations on the internal transport of nutrients and inhibitory products.

Over the course of this immobilized-cell biocatalyst activation process, the gel matrix is often severely distorted to accommodate the increase in immobilized biomass, resulting in up to a threefold volumetric expansion of the original carrier particle.[16-19] Subsequent immobilized-cell divisions result in cell leakage from the support matrix through surface fractures[15,18,20-22] at a commensurate rate. At this stage, the biocatalyst is termed fully activated.

Microscopic examination of a fully activated biocatalyst interior has often revealed a dense, clearly defined cell layer extending 50 to 100 μm from the biocatalyst surface, depending on the microorganism and the system-dependent transport properties of the limiting nutrient(s) and inhibitory product(s).[11,23-30] Typically, few or no viable cells are found toward the center of a sufficiently large gel particle.

The expected existence of a gradient in metabolic activity of immobilized cells with depth into a carrier follows directly from the hypothesis of intrabiocatalyst mass transfer limitations. Electron micrographic observations of Inloes et al.[31] and others[20,32,33] strongly suggest that such a gradient is present in both immobilized bacterial and yeast systems. Electron micrographs of *Pseudomonas putida* immobilized in polyacrylamide gel[32] show living cells at high density near the surface, yet apparently dead, metabolically inactive cells and compressed cell walls are found near the center. A photomicrograph published by

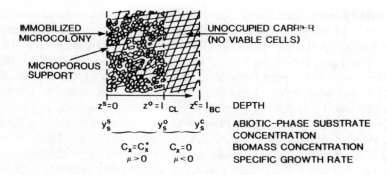

FIGURE 1. Cross section of a slab geometry, immobilized viable cell catalyst. (From Monbouquette, H. G. and Ollis, D. F., *Ann. N.Y. Acad. Sci.*, 469, 230, 1986. With permission.)

Koshcheyenko et al.[32] of immobilized *Saccharomyces cerevisiae* in a polyacrylamide gel granule cross section reveals uniformly shaped, healthy (well-stained) cells at high density to a depth of about 150 μm. Deeper, beyond the 150-μm level there appears to be a rather abrupt change in cell morphology as many irregularly shaped unhealthy cells are evident, but at the same high density. Deeper still within the gel, cell density drops rapidly; yeast cells are relatively scarce and scattered. Similar observations can be made from an electron micrograph of yeasts immobilized in κ-carageenan.[20]

A. PHYSICAL AND CHEMICAL INFLUENCES

Strong evidence exists for the advantageous or disadvantageous partitioning of substances between the carrier and the bulk suspension media. For example, mass transfer of nutrients from dilute media to cells attached to solid surfaces such as clay particles has been shown by several investigators to be enhanced by the tendency of such nutrients as glucose to accumulate at the solid-liquid interface.[1,34] In addition, polyanionic polymer gels may bind free potassium and sodium ions, which have been shown to inhibit yeast growth and ethanol production.[35] It has been reported that often harmful or inhibitory organic solvents such as ethanol and acetone prefer the bulk solution to κ-carageenan gel.[36] Also, phenol-degrading microorganisms adsorbed to the surface of activated carbon particles are protected from harmful, transient increases in phenol concentration due to rapid adsorption by the activated carbon.[37] Fukui and Tanaka[38] have reviewed the usefulness of the more hydrophobic pre-polymer gels for bioconversions of hydrophobic substrates (e.g., steroids) in organic or organic-water solvent systems. pH shift phenomena have been observed for microbes adsorbed to ion-exchange resins[39-41] and two-gel-phase systems.[42,43] Partitioning effects are clearly phenomena that have not yet been studied extensively, but do hold promise for future exploitation in ways analogous to immobilized enzyme systems.

A combination of partitioning and diffusion governs mass transport of nutrients and products in porous immobilized-cell biocatalysts. The illusory quality of diffusional masking effects, particularly with regard to biocatalytic stability, has been firmly established for immobilized enzyme systems and should be included in a proper interpretation of the apparent stability of immobilized-cell systems as well.[44,45] Such effects may be manifested as differing responses of immobilized cells to temperature[43] and increased productive half-lives relative to free cells.[2,46,47]

B. THE "CONTINUOUSLY GROWING IMMOBILIZED-CELL STEADY STATE"

In the continuously growing steady state, cells are constantly growing and dividing throughout a living cell layer, but the growth rate and, therefore, the physiological state of the immobilized microbes change with distance from the support surface. Given a sufficient nutrient supply and adequate product removal rates, cells at the biocatalyst surface may be highly metabolically active and growing rapidly. Proceeding more deeply into the cell layer, a moribund population containing viable, senescent, and dead (or metabolically inactive) cells may be encountered, primarily due to substrate and product diffusional limitations. Here, viability indicates ability to reproduce, whereas the senescent state corresponds to a productive potential of zero which normally results from a cell aging process. At or near the inner edge of a cell layer, where the growth rate is very low, the population may be largely nonviable and cryptogenic growth may be highly important. Thus, the variation in physiological states with depth into a carrier roughly corresponds to changes in the state of a microbial population in batch culture with time; yet, the entire system is a continuous one. Such an arrangement may prove to have some characteristics useful for the scientist studying cellular metabolism, especially at very low growth rates. Indeed, at very low growth rates, prokaryotes exhibit an anomalous, abrupt down step in yield of biomass from substrate coupled with peculiarities in cellular metabolism.[48]

III. INTRABIOCATALYST MASS TRANSPORT

If dependable quantitative mathematical models are to be developed, theoretical and experimental studies of solute transport in immobilized-cell biocatalysts are desperately needed. For the simplest steady-state case, effective diffusivity data are needed to describe the flux of a solute of concentration y, based on free intrabiocatalyst volume not occupied by cells or support matrix, across a total biocatalyst cross-sectional area. Solute concentrations must be on a free intrabiocatalyst volume basis to be consistent with the biokinetic expressions for substrate uptake, product synthesis, and substrate or product inhibition (see Section IV.A). Potential differences in interstitial diffusivities (D) within the free volume between cells in an immobilized microcolony and unoccupied support porespace, which are due to the presence of polymeric cell exudates and/or differences in diffusion path tortuosity, must be included in a theoretical explanation of effective diffusivity data. Here, the local effective diffusivity, D_e, equals the interstitial diffusivity, D, multiplied by the free-volume fraction, since the solute concentration is expressed on a free-volume basis.

IV. THEORY OF STRUCTURED MODELING

Nearly all of the published mathematical models intended to quantitatively or qualitatively simulate the kinetics of immobilized-living-cell biocatalysts have been based on unstructured, distributed, deterministic models of the immobilized biomass.[49] The modeling of immobilized-living-cell systems is an example in which such traditional unstructured models (those that describe biomass only by concentration) may prove inadequate. Unstructured models have been proved unable to predict transients in bioreactor performance due to their inability to describe cell physiological states.[50] Such models are therefore not capable of predicting the distribution of biocatalytic activity (as related to cell physiology) within a porous support in an experimentally verifiable way. This information would prove useful in assessing the true effects of immobilization on microbial physiology. Although simple unstructured models are most common, other avenues for mathematical description of biological systems are available.[51-53]

Unlike typical porous catalysts used in the petroleum refining and chemical synthesis industries, the desired conversion does not take place at a surface in immobilized-living-cell biocatalysts. The biochemical reactions in question typically occur within the *biotic* phase (the immobilized biomass) of variable volume (relative to the carrier microenvironment) and biocatalytic activity. Rather than a two-phase system, a three-phase catalyst system

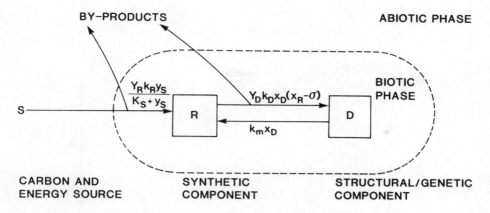

FIGURE 2. Block diagram of the two-compartment model proposed by Esener et al.,[10] with modified rate expressions. (From Monbouquette, H. G. and Ollis, D. F., *Ann. N.Y. Acad. Sci.*, 469, 230, 1986. With permission.)

exists: the carrier material, the pore-filling liquid medium through which most solute diffusion occurs (i.e., the *abiotic* phase), and the biomass itself. Comparing the size of microbial cells and the macropores which they inhabit to the characteristic dimension of a typical support, a simplifying assumption of time-independent catalyst isotropy cannot always be justified. The distinction between a biotic phase and an abiotic phase must always be maintained since, at full activation, the biotic volume is no longer a small fraction of the total as in a typical batch fermentation. Transport of nutrients and excreted products within the carrier occurs within the abiotic phase; thus, the relevant concentrations are intrinsic, abiotic-phase concentrations.

A. BIOTIC-PHASE EQUATION

The structured model is most easily conceptualized in terms of the macromolecules which constitute the bulk of the cell dry weight. The synthetic component (R) consists primarily of ribosomal RNA (rRNA), the machinery for protein synthesis; the structural/genetic component (D) is comprised of protein and DNA.

The block diagram, including rate expressions, presented as Figure 2 illustrates R and D syntheses as sequential, growth-related processes with concurrent formation of by-products and D turnover as a maintenance-related process with perfect yield.[9] The thousands of reactions and transport processes that take place within a typical bacterial cell or across cell walls and/or membranes have been lumped together into three simple conversions. Despite this vast condensation, impressive simulations of continuous *Klebsiella pneumoniae* and *Zymomonas mobilis* fermentations have been obtained.[10,54]

The inability of the Esener model[10] to predict a proper steady-state x_R profile arises from the failure of the model to account for the inefficiency of rRNA at low substrate levels. A significant fraction of the total rRNA remains in an "inactive" state under starvation conditions. That is, μ may approach zero with a significant amount of rRNA still present and mobilized, presumably, for a sudden increase in nutrient availability.[55] Thus, a nonlinear relationship between percent R and μ is observed at low specific growth (see Figure 3), and the minimum steady-state percent RNA content hovers at about 10%.

The specific problem with the original Esener model[10] mathematics rests within the expression given for D-compartment synthesis,

$$r_R(x_R, x_D) = k_D x_R x_D \tag{1}$$

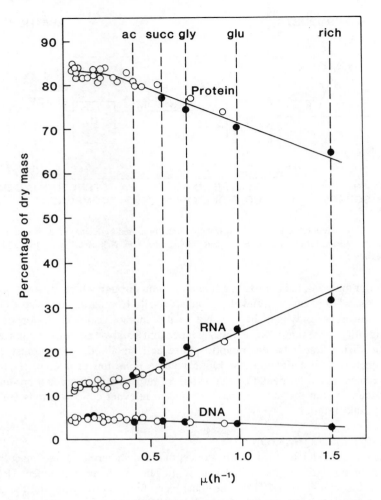

FIGURE 3.   *E. coli* dry weight composition as a function of specific growth rate.[55] Filled circles correspond to maximum specific growth rate in acetate, succinate, glycerol, glucose, and enriched media; open circles refer to glucose-limited growth in a chemostat.

According to this expression, x_R must approach zero for D-compartment synthesis to cease, when, in truth, the minimum R-component concentration is always well above zero. This dilemma may be resolved easily by implementing the suggestion of Roels.[53] The biotic component mass balances may be written assuming an unknown, $f(x_R,x_D)$, for $r_R(x_R,x_D)$, giving

$$\frac{dx_R}{dt} = \frac{Y_R k_R y_s}{K_s + y_s} - f(x_R,x_D) + k_m x_D - \mu x_R \tag{2}$$

$$\frac{dx_D}{dt} = Y_D f(x_R,x_D) - k_m x_D - \mu x_D \tag{3}$$

A simple expression for μ may be obtained by assuming a plot of x_R vs. μ to be linear. Although at very low growth rates this relationship is nonlinear, the approximation is still a good one over most of the range in specific growth rate, such that

$$x_R = m\mu + x_R^0 \tag{4}$$

Solving Equation 4 for μ, substituting this equation into the steady-state D-component balance, and rearranging give the following expression for $f(x_R, x_D)$:

$$f(x_R, x_D) = \frac{x_D}{Y_D m}(x_R + mk_m - x_R^0) \tag{5}$$

If Equation 5 may be assumed to hold over transient conditions as well (a difficult assumption to prove), then $r_R(x_R, x_D) = f(x_R, x_D)$. By setting $k_D = 1/(Y_D m)$ and $\sigma = x_R - mk_m$, Equation 6 is obtained:

$$r_R(x_R, x_D) = k_D x_D(x_R - \sigma) \tag{6}$$

a rate equation predicting zero net D synthesis at a finite RNA level. Substituting this expression into Equation 2, and noting that Equations 2 and 3 are not independent (since $x_D = \rho_b - x_R$), leads to the following balance on the R component, which fully describes the two-component biotic phase:

$$\frac{dx_R}{dt} = \frac{Y_R k_R y_S}{K_S + y_S} - k_D(\rho_b - x_R)(x_R - \sigma) + k_m(\rho_b - x_R) - \mu x_R \tag{7}$$

where

$$\mu = \frac{1}{\rho_b}\left[\frac{Y_R k_R y_S}{K_S + y_S} + (Y_D - 1)k_D(\rho_b - x_R)(x_R - \sigma)\right] \tag{8}$$

and ρ_b = intrinsic biomass density. These equations, coupled with the simple balances on substrate and total biomass for a chemostat, can be solved numerically to adequately simulate all the corresponding steady-state experimental data published by Esener et al.[10]

With the simple modification to account for "inactive" ribosomes at low specific growth rates, this intrinsic structured model appears adequate to describe immobilized growing biomass. The model may be applicable to many species of bacteria growing on a limiting carbon and energy source; it gives some important explicit information concerning cell physiological state through the value of x_R and could be extended easily to include substrate and product inhibition or biological product synthesis. In addition, the substrate concentration at the inner edge of the cell layer (where $\mu = 0$) may be estimated by setting Equation 8 equal to zero and solving for y_S^0:

$$y_S^0 = \frac{K_S(1 - Y_D)k_D(\rho_b - x_R^0)(x_R^0 - \sigma)}{Y_R k_R - (1 - Y_D)k_D(\rho_b - x_R^0)(x_R^0 - \sigma)} \tag{9}$$

B. ABIOTIC-PHASE EQUATION

Consider now the steady-state problem of interest, with C_x (biomass concentration per unit pore volume) assumed constant with respect to position in the cell layer. The reaction-diffusion equation may be put on a total biocatalyst volume basis[9] to give

$$\beta[\epsilon\xi D_m + (1 - \xi)D_c]\frac{d^2 y_S(z)}{dz^2} = \beta\frac{C_x}{\rho_b}\frac{k_R y_S(z)}{K_S + y_S(z)} \tag{10}$$

Here, ξ is defined as the fraction of the pore volume accessible to cells, which is assumed equal to $(C_x/\rho_b)/(1 - \epsilon)$, given that the biomass has filled all accessible pore space at steady state and is constant with respect to position in the cell layer. If an effective diffusivity is then defined as

$$D_e = \beta[\epsilon\xi D_m + (1 - \xi)D_c] \tag{11}$$

the defining equation for steady-state transport and consumption of the limiting carbon and energy source takes on a familiar reaction-diffusion form.

For the purposes of this study, since C_x is assumed constant with respect to position in the cell layer, an effective rate constant, k_e, may be defined as $\beta(C_x/\rho_b)k_R$. Given the expression for D_e (Equation 11) and k_e, the steady-state substrate balance (abiotic-phase equation) takes the following familiar form:

$$D_e \frac{d^2y_S(z)}{dz^2} = \frac{k_e y_S(z)}{K_S + y_S(z)} \tag{12}$$

At the inner edge of the cell layer ($z = z^0$), $\mu = 0$; therefore, $x_R = x_R^0$, and y_S^0 is given by Equation 9. Alternatively, if the steady-state thickness of the cell layer, l_{CL} ($= z^c - z^s$), is greater than half the thickness of the symmetrical slab carrier, l_{BC}, or is unknown, the substrate concentration at the biocatalyst center, y_S^c, may be greater than or equal to y_S^0. In addition, the zero-flux condition holds at the biocatalyst center line (nonzero living biomass at $z = z^c$) or the inner edge of the cell layer (zero viable biomass at $z = z^0$).

Assuming the limiting carbon and energy source concentration to be a constant, y_S^s, at the carrier surface completes the determination of the necessary boundary conditions, which are summarized as follows:

$$\frac{dy_S(z^c)}{dz} = 0 \qquad z = z^c \tag{13a}$$

$$z = z^s \qquad y_S(z^s) = y_S^s \tag{13b}$$

V. SIMULATION AND OPTIMIZATION OF BIOCATALYST PERFORMANCE

The abiotic-phase equation may be integrated once analytically using the no-flux boundary condition to give the following expression for the substrate concentration gradient:

$$\frac{dy_S(z)}{dz} = \left\{ \frac{2k_e}{D_e} \left[y_S(z) - y_S^c + K_S \ln\left(\frac{K_S + y_S^c}{K_S + y_S(z)} \right) \right] \right\}^{1/2} \tag{14}$$

This nonlinear, first-order differential equation is easily integrated from the biocatalyst surface inward using a fourth-order Runge-Kutta computer algorithm. Once the abiotic-phase substrate concentrations are known at each depth, R-compartment concentrations, specific growth rate, and biomass flux outward at each point may be calculated sequentially in that order.

Numerical calculations, given base-case parameter values (Table 1) and assuming a biocatalyst slab thickness greater than twice that of the cell layer, yield three curves rep-

TABLE 1
Base-Case Values for Model Parameters[9]

Biotic-phase equation	Abiotic-phase equation
From Reference 10:	
$k_R = 470.4$ g/l·h	$D_m = 3.5 \times 10^{-6}$ cm²/s
$k_m = 0.06$ h^{-1}	$D_c = 7.0 \times 10^{-6}$ cm²/s
$Y_R = 0.73$	$\epsilon = 0.20$
$Y_D = 0.66$	$\xi = 0.90$
$K_S = 0.07$ g/l	$y_S^s = 10.0$ g/l
$m = 48.7$ h·g/l	$y_S^c = 1.393 \times 10^{-3}$ g/l
$x_R^0 = 23.0$ g/l	
	Calculated:
From Reference 55:	$C_x/\rho_b = 0.72$
$\rho_b = 240.0$ g/l	
Calculated:	
$k_D = 0.0311$ l/g·h	
$\sigma = 20.1$ g/l	
$y_S^0 = 1.393 \times 10^{-3}$ g/l	

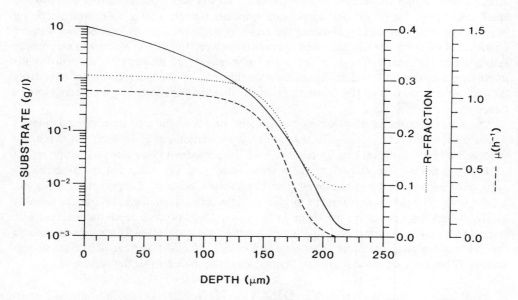

FIGURE 4. Base case substrate, R-compartment, and specific growth rate profiles with depth into the bio-catalyst; $\eta_{CL} = 0.75$. Parameter values: $y_S^s = 10.0$ g/l, $\xi = 0.90$, and $l_{BC} > l_{CL}$. (From Monbouquette, H. G. and Ollis, D. F., *Ann. N.Y. Acad. Sci.*, 469, 230, 1986. With permission.)

resenting substrate, R-fraction, and μ profiles (Figure 4).[9] All of the curves are monotonically decreasing with depth into the biocatalyst. However, the R-fraction and specific growth rate profiles are nearly flat at the higher substrate concentration, where substrate uptake is approximately zero order. The eventual decline in specific growth rate nearly parallels the drop in R-component concentration until the R fraction approaches its minimum value of just under 0.10, where μ is predicted to be zero.

A cell layer effectiveness, η_{CL}, based on immobilized microbial growth may be defined to quantify this effect:

$$\eta_{CL} = \frac{q_b^s/\beta}{\mu^s C_x l_{CL}} \qquad (15)$$

where μ^s and q_b^s are the specific growth rate and biomass flux at the surface, respectively, and l_{CL} is the thickness of the cell layer. The denominator in this equation describes a rectangle on the μ vs. depth plot bounded by $\mu = 0$ and $\mu = \mu_m$, and $z^s = 0$ and $z^0 = l_{CL}$. The numerator, q_b^s, is represented as the area under the μ vs. depth curve.

The ratio of these areas gives η_{CL}, which for the base case ($y_S^s = 10.0$ g/l) is equal to 0.75 (Figure 4). Thus, there is considerable error in usage of the zero-order kinetics assumption under these conditions, even though y_S^s is quite large relative to the Monod constant, K_S ($= 0.07$ g/l). Raising or lowering the surface substrate concentration increases or decreases the thickness of the cell layer and strongly affects η_{CL} (see Figures 5 and 6). However, increasing y_S^s to 100.00 g/l, corresponding to a rather high steady-state surface substrate concentration (except, perhaps, at the entrance to a packed bed), still results in a cell layer effectiveness of 0.90, which is significantly less than one. Therefore, the assumption of zero-order kinetics in a biocatalyst wherein radial substrate concentration actually falls off well into the first-order regime may be a poor one.

Recently, researchers have developed several methods to generate gel beads with diameters from 1 mm down to less than 100 μm. This is accomplished either by stripping small droplets of the pregel cell suspension from needle tips with a concentric sterile air blow-off stream,[56] by rapidly shearing the aqueous suspension in an inert oil to create an emulsion of tiny droplets that may subsequently be induced to gel,[57] or by controlled breakup (vibration or ultrasonic) of a jet of the pregel suspension into droplets.[58,59] Decreasing the width of gel carriers to less than the steady-state thickness of the living cell layer effectively cuts off all or a portion of the first-order regime, thereby increasing η_{CL} to values close to one.

Non-growth-associated product synthesis based simply on the total immobilized biomass would also be expected to be affected directly by variations in ξ. In fact, the maximum immobilized biomass level exactly coincides with the maximum biomass production rate at all D_c/D_m ratios for biocatalyst thickness more than twice l_{CL}. Note that the relative proportions of substrate and growth rate profiles for all values, holding other parameters constant, are the same. The parameter exerts its influence on the substrate gradient through the quantity $2k_e/D_e$, which does not affect the shape of the y_S vs. dimensionless depth (normalized with respect to l_{CL}). For various ξ and all other parameters identical, all plots are superimposable for slab biocatalysts. Therefore, changing η does not affect the average substrate concentration, R fraction, or specific growth rate, μ, over the thickness of the cell layer.

VI. DISCUSSION

In the formulation of this steady-state intrinsic, structured model of an immobilized-cell biocatalyst, several assumptions have been made to avoid excessive detail. Three of these approximations are especially unusual as they pertain only to these porous, living-cell biocatalysts. First, instead of the common assumption of uniform distribution of catalytic sites within a carrier, constant biomass density within the pore volume has been assumed. Biocatalytic activity with depth into the carrier is *not* a constant, however, since cell physiological state or biocatalytic quality often declines with increased nutrient deficiency. Second, the "continuously growing immobilized-cell" condition has been described as the only steady

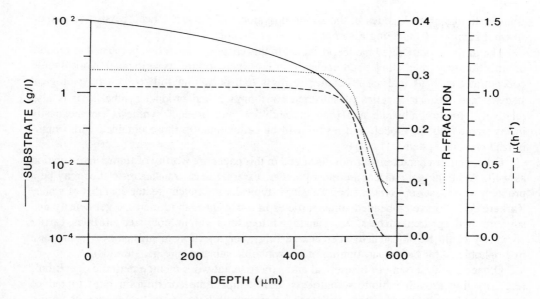

FIGURE 5. Substrate, R-compartment, and specific growth-rate profiles with depth for a high surface substrate concentration; $\eta_{CL} = 0.90$. Parameter values: $y_S^s = 100.0$ g/l, $\xi = 0.90$, $l_{BC} > l_{CL}$.

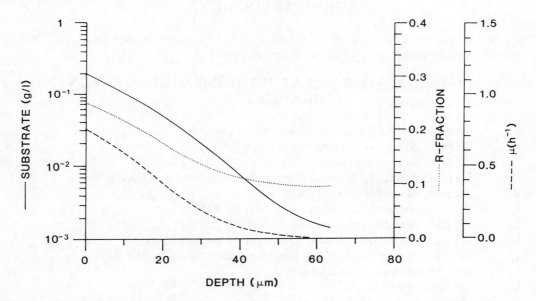

FIGURE 6. Substrate, R-compartment, and specific growth-rate profiles with depth for a low surface substrate concentration; $\eta_{CL} = 0.34$. Parameter values: $y_S^s = 0.2$ g/l, $\xi = 0.90$, $l_{BC} > l_{CL}$.

state possible for nearly all cells in these living immobilized microbial systems. There is clearly no inanimate heterogeneous catalyst analogue to this phenomenon of propagation of active sites. Finally, transport within immobilized-cell carriers involves diffusion in microcolonies and carrier pore space, necessitating a more complex description of an effective diffusion coefficient, somewhat analogous to that derived for macro- and micropore zeolites.[60] The simplest approach was taken here by assuming a unidirectional diffusive flux through both regions, proportional to the same substrate gradient given little or no mass transfer limitation to kinetics in the lateral directions. Clearly, the need persists for experimental support of modeling assumptions and predictions.

The intrinsic, structured model presented in this chapter is attractive because it is capable of quantitatively predicting phenomena in areas where future experimental work should concentrate, such as internal resistance to mass transport, immobilized-cell physiological states,[13] and intrinsic kinetics of substrate consumption and product synthesis. It is also relatively simple. Substrate and product inhibition and product synthesis can be easily incorporated into the model, and results will be comparable to those obtained with immobilized-enzyme systems.

Other carrier characteristics not discussed in this paper are worthy of future investigation as well. Although gel carriers are most popular, external mass transfer resistance may be a problem with these supports, as their density is typically not much greater than that of water. Karkare et al.[58] have embedded silica particles in κ-carrageenan to increase gel density and strength, with apparent success. Also, analogous to efforts with immobilized enzymes, carrier chemistry could be adjusted to alter charge (e.g., for rejection of protons) or to increase hydrophobicity for better partitioning of hydrophobic substrates (e.g., steroids).

Other areas deserving of theoretical and experimental work include periodic operation[61] (e.g., cycling growth inhibitors, nutrients, pH, temperature, or dilution rate) to reduce biomass leakage and/or increase productivity, coimmobilization, and dynamics of immobilized unstable recombinant cultures.

ACKNOWLEDGMENT

We are pleased to thank Prof. S. Peretti for helpful comments and Celgen, Inc. for financial support during the preparation of this manuscript.

APPENDIX: NOMENCLATURE [DIMENSIONALITY IN BRACKETS]*

C_x	Biomass (dry weight) culture concentration $[M/L^3]$
D_c	Substrate diffusivity in carrier micropores $[L^2/T]$
D_e	Effective (or apparent) substrate diffusivity $[L^2/T]$
D_m	Substrate diffusivity in immobilized microcolony abiotic volume $[L^2/T]$
K_S	Substrate saturation constant $[M/L^3]$
k_D	Rate constant for D-component formation $[L^3/MT]$
k_e	Effective rate constant (Equations 10 and 12) $[L^3/MT]$
k_m	Rate constant for maintenance (D-component turnover) $[1/T]$
k_R	Rate constant for R-component formation (substrate uptake) $[M/L^3T]$
l_{BC}	Half the slab-biocatalyst thickness $[L]$
l_{CL}	Cell layer thickness $[L]$
m	Slope of x_R vs. μ $[TM/L^3]$
q_b	Biomass (dry weight) flux $[M/L^2T]$
r	Rate expression of varied definition in text $[M/L^3T]$
r_R	Rate expression for D-component formation $[M/L^3T]$
x_D	D-component biotic-phase concentration (dry weight/wet biotic volume) $[M/L^3]$
x_R	R-component biotic-phase concentration (dry weight/wet biotic volume) $[M/L^3]$
Y_D	Yield of D-component from R-component [dimensionless]
Y_R	Yield of R-component from substrate [dimensionless]
y_S	Abiotic-phase substrate concentration $[M/L^3]$
z	Distance from carrier surface $[L]$

* M = mass units, L = length units, T = time units.

Greek characters

β   Cell carrier porosity [dimensionless]

ε   Abiotic-phase fraction of immobilized microcolony volume

ξ   Fraction of pore volume accessible to cells

η_{CL} Theoretical viable cell layer effectiveness [dimensionless]

μ   Specific growth rate [1/T]

ρ_b Biomass density [M/L^3]

σ   Correction for inactive ribosomes [M/L^3]

Superscripts

c   Value at the slab-biocatalyst center line

0   Value when $\mu = 0$

s   Value at the biocatalyst surface

REFERENCES

1. **Abbot, B. J.,** Immobilized cells, *Annu. Rep. Ferment. Proc.,* 1, 205, 1977.
2. **Cheetham, P. S. J.,** Developments in the immobilization of microbial cells and their applications, *Top. Enzymol. Ferment. Biotechnol.,* 4, 189, 1980.
3. **Venkatasubramanian, K. and Vieth, W. R.,** Immobilized microbial cells, *Prog. Ind. Microbiol.,* 15, 61, 1979.
4. **Chibata, I., Tosa, T., and Fujimura, M.,** Immobilized living microbial cells, *Annu. Rep. Ferment. Proc.,* 6, 1, 1983.
5. **Mattiasson, B.,** Immobilized viable cells, in *Immobilized Cells and Organelles,* Vol. 2, Mattiasson, B., Ed., CRC Press, Boca Raton, FL, 1983, 23.
6. **Kennedy, J. F. and Cabral, J. M. S.,** Immobilized living cells and their applications, *Appl. Biochem. Bioeng.,* 4, 190, 1983.
7. **Blanch, H.,** Immobilized microbial cells, *Annu. Rep. Ferment. Proc.,* 7, 81, 1984.
8. **Nagashima, M., Azuma, M., Noguchi, S., Inuzuka, H., and Samejima, H.,** Continuous ethanol fermentation using immobilized yeast cells, *Biotechnol. Bioeng.,* 26, 992, 1982.
9. **Monbouquette, H. G. and Ollis, D. F.,** A structured model for immobilized cell kinetics, *Ann. N.Y. Acad. Sci.,* 469, 230, 1986.
10. **Esener, A. A., Veerman, T., Roels, J. A., and Kossen, N. W. F.,** Modeling of bacterial growth: formulation and evaluation of a structured model, *Biotechnol. Bioeng.,* 24, 1749, 1982.
11. **Shinmyo, A., Kimura, H., and Okada, H.,** Physiology of α-amylase production by immobilized *Bacillus amyloliquefaciens, Eur. J. Appl. Microbiol. Biotechnol.,* 14, 7, 1982.
12. **Baudet, C., Barbotin, J.-N., and Buespin-Michael, J.,** Growth and sporulation of entrapped *(Bacillus subtilis)* cells, *Appl. Environ. Microbiol.,* 45, 297, 1983.
13. **Monbouquette, H. G. and Ollis, D. F.,** Scanning microfluorimetry of Ca-alginate immobilized *Zymomonas mobilis, Bio/Technology,* 6, 1076, 1988.
14. **Dhulster, P., Barbotin, J.-N., and Thomas, D.,** Culture and bioconversion use of plasmid-harboring strain of immobilized *E. coli, Appl. Microbiol. Biotechnol.,* 20, 87, 1984.
15. **de Taxis du Poet, P., Dhulster, P., Barbotin, J.-N., and Thomas, D.,** Plasmid inheritability and biomass production: comparison between free and immobilized cell cultures of *Escherichia coli* BZ18 (pTG 201) without selection pressure, *J. Bacteriol.,* 165, 3, 871, 1986.
16. **Siess, M. H. and Divies, C.,** Behavior of *Saccharomyces cerevisiae* cells immobilized in polyacrylamide gel, *Eur. J. Appl. Microbiol. Biotechnol.,* 12, 10, 1981.
17. **Lee, T. H., Ahn, J. C., and Ryu, D. D. Y.,** Performance of an immobilized yeast reactor system for ethanol production, *Enzymol. Microbiol. Technol.,* 5, 41, 1983.
18. **Burrill, H. N., Bell, L. E., Greenfield, P. F., and Do, D. D.,** Analysis of distributed growth of *Saccharomyces cerevisiae* cells immobilized in polyacrylamide gel, *Appl. Environ. Microbiol.,* 46, 716, 1983.
19. **Chotani, G. K. and Constantinides, A.,** Immobilized cell cross-flow reactor, *Biotechnol. Bioeng.,* 26, 217, 1984.

20. **Wada, M., Kato, J., and Chibata, I.,** Electron microscopic observation of immobilized growing yeast cells, *Ferment. Technol.,* 58, 327, 1980.
21. **Kuek, C. and Armitage, T. M.,** Scanning electron microscopic examination of calcium alginate beads immobilizing growing mycelia of *Aspergillus phoenicus, Enzymol. Microbiol. Technol.,* 7, 121, 1985.
22. **Bailliez, C., Largeau, C., and Casadevall, E.,** Growth and hydrocarbon production of *Botryococcus braunii* immobilized in calcium alginate gel, *Appl. Microbiol. Biotechnol.,* 23, 99, 1985.
23. **Wada, M., Kato, J., and Chibata, I.,** Continuous production of ethanol using immobilized growing yeast cells, *Appl. Microbiol. Biotechnol.,* 10, 275, 1980.
24. **Eikmeier, H., Westmeier, F., and Rehm, H.,** Morphological development of *Aspergillus niger* immobilized in Ca-alginate and κ-carrageenan, *Appl. Microbiol. Biotechnol.,* 19, 53, 1984.
25. **Kopp, B. and Rehm, H. J.,** Semicontinuous cultivation of immobilized *Calviceps purpurea, Appl. Microbiol. Biotechnol.,* 19, 141, 1984.
26. **Osuga, J., Mori, A., and Kato, J.,** Acetic acid production by immobilized *Acetobacter aceti* cells entrapped in a κ-carrageenan gel, *J. Ferment. Technol.,* 62, 139, 1984.
27. **Bettman, H. and Rehm, H.,** Degradation of phenol by polymer entrapped microorganisms, *Appl. Microbiol. Biotechnol.,* 20(5), 285, 1984.
28. **Robinson, P. K. and Dainty, A. L.,** Physiology of alginate-immobilized *Chlorella, Enzymol. Microbiol. Technol.,* 7(5), 212, 1985.
29. **Gosman, B. and Rehm, H. J.,** Oxygen uptake by microorganisms entrapped in Ca-alginate, *Appl. Microbiol. Biotechnol.,* 23, 163, 1986.
30. **Mahmoud, W. and Rehm, H. J.,** Morphological examination of immobilized *Streptomyces aureofaciens* during chlortetracycline fermentation, *Appl. Microbiol. Biotechnol.,* 23, 305, 1986.
31. **Inloes, D. S., Smith, W. J., Taylor, D. P., Cohen, S. N., Michaels, A. S., and Robertson, C. R.,** Hollow-fiber membrane bioreactors using immobilized *E. coli* for protein synthesis, *Biotechnol. Bioeng.,* 25, 2653, 1983.
32. **Koshcheyenko, K. A., Turkina, M. V., and Skyrabin, G. K.,** Immobilization of living microbial cells and their application for steroid transformation, *Enzymol. Microbiol. Technol.,* 5, 14, 1983.
33. **Somerville, H. J., Mason, J. R., and Ruffell, R. N.,** Benzene degradation by bacterial cells immobilized in polyacrylamide gel, *Eur. J. Appl. Microbiol. Biotechnol.,* 4, 75, 1977.
34. **Martin, J. P., Filip, Z., and Haider, K.,** Effect of montmorillonite and humate on growth and metabolic activity of some actinomycetes, *Soil Biol. Biochem.,* 8, 409, 1976.
35. **King, V. A.-E. and Zall, R. R.,** Ethanol fermentation of whey using calcium alginate entrapped yeasts, *Proc. Biochem.,* 18(6), 17, 1983.
36. **Chibata, I., Tosa, T., and Takata, I.,** Continuous production of L-malic acid by immobilized cells, *Trends Biotechnol.,* 1, 9, 1983.
37. **Ehrhardt, H. M. and Rehm, H. J.,** Phenol degradation by microorganisms adsorbed on activated carbon, *Appl. Microbiol. Biotechnol.,* 21, 32, 1985.
38. **Fukui, S. and Tanaka, A.,** *Adv. Biochem. Eng./Biotechnol.,* 29, 1, 1984.
39. **Hattori, T.,** Growth of *Escherichia coli* on the surface of an anion-exchange resin in continuous flow system, *Appl. Microbiol.,* 18, 319, 1972.
40. **Hattori, T. and Furusuka, C. J.,** Chemical activities of *E. coli* adsorbed on a resin, *J. Biochem.,* 48, 831, 1960.
41. **Hattori, T. and Furusuka, C. J.,** Chemical activities of *Azotobacter agile* adsorbed on a resin, *J. Biochem.,* 50, 312, 1961.
42. **Day, D. F. and Sarkar, D.,** An immobilized yeast cell column for the fermentation of molasses, *Enzyme Eng.,* 6, 343, 1982.
43. **Williams, D. and Munnecke, D. M.,** The production of ethanol by immobilized yeast cells, *Biotechnol. Bioeng.,* 23, 1813, 1981.
44. **Ollis, D. F.,** Diffusion influences in denaturable insolubilized enzyme catalysis, *Biotechnol. Bioeng.,* 14, 871, 1972.
45. **Ooshima, H. and Harano, T.,** Effect of intraparticle diffusion resistance on apparent stability of immobilized enzymes, *Biotechnol. Bioeng.,* 23, 1991, 1981.
46. **Hiemstra, H., Dijkhuizin, L., and Harder, W.,** Diffusion of oxygen in alginate gels related to the kinetics of methanol oxidation by immobilized *Hansenula polymorpha* cells, *Eur. J. Appl. Microbiol. Biotechnol.,* 18, 189, 1983.
47. **Deo, Y. M. and Gaucher, G. M.,** Semicontinuous and continuous production of penicillin-G by *Penicillium chrysogenum* cells immobilized in κ-carrageenan beads, *Biotechnol. Bioeng.,* 26, 285, 1984.
48. **Stouthamer, A. J.,** The relation between biomass production and substrate consumption at very low growth rates, *Prog. Ind. Microbiol.,* 20, 517, 1984.
49. **Karel, S. F., Libicki, S. G., and Robertson, C. R.,** The immobilization of whole cells: engineering principles, *Chem. Eng. Sci.,* 40(8), 1321, 1985.

50. **Fredrickson, A. G., Ramkrishna, D., and Tsuchiya, H. M.,** The necessity of including structure in mathematical models of unbalanced microbial growth, *Adv. Appl. Microbiol.,* 67(108), 53, 1971.
51. **Fredrickson, A. G., Megee, R. D., and Tsuchiya, H. M.,** Mathematical models for fermentation processes, *Adv. Appl. Microbiol.,* 13, 419, 1970.
52. **Harder, A. and Roels, J. A.,** Application of simple structured models in bioengineering, *Adv. Biochem. Eng.,* 21, 55, 1982.
53. **Roels, J. A.,** *Energetics and Kinetics in Biotechnology,* Elsevier, Amsterdam, 1983.
54. **Jobses, I. M. L., Egberts, G. T. C., van Baden, A., and Roels, J. A.,** Mathematical modelling of growth and substrate conversion of *Zymomonas mobilis* at 30 and 35°C, *Biotechnol. Bioeng.,* 27, 984, 1985.
55. **Ingraham, J. L., Maaloe, O., and Neidhardt, F. C.,** *Growth of the Bacterial Cell,* Sinauer Associates, Sunderland, MA, 1983.
56. **Klein, J., Stock, J., and Vorlop, K.-D.,** Pore size and properties of spherical Ca-alginate biocatalysts, *Eur. J. Appl. Microbiol. Biotechnol.,* 18, 86, 1983.
57. **Mosbach, K., Birnbaum, S., Hardy, K., Davies, J., and Bulow, L.,** Formation of proinsulin by immobilized *Bacillus subtilis, Nature (London),* 302, 543, 1983.
58. **Karkare, S. B., Dean, R. C., and Venkatasubramanian, K.,** Continuous fermentation with fluidized slurries of immobilized microorganisms, *Bio/Technology,* 3, 247, 1985.
59. **Tramper, J.,** Immobilizing biocatalysts for use in synthesis, *Trends Biotechnol.,* 3, 45, 1985.
60. **Satterfield, C. N.,** *Mass Transfer in Heterogeneous Catalysis,* MIT Press, Cambridge, MA, 1970.
61. **Forberg, C., Enfors, S.-O., and Haggstrom, L.,** Control of adhesion and activity during continuous production of acetone and butanol with adsorbed cells, *Eur. J. Appl. Microbiol. Biotechnol.,* 17, 143, 1983.

Chapter 7

BIOREACTORS AND MODELS FOR GENETICALLY MODIFIED ORGANISMS

M. L. Shuler

TABLE OF CONTENTS

I. INTRODUCTION

Living cells can be rationally and purposefully designed at the molecular level. This concept, inconceivable in 1970, is a major driving force in bioprocessing technology. Genetic modification of cells is primarily done either to produce a protein as a product or to generate within a cell new metabolic pathways capable of yielding nonprotein products.

The first wave of targets for genetically modified cells have been proteins, particularly therapeutic proteins. The initial promise of production of large quantities of low-cost proteins has been only partially realized. The technical barriers to achieving large-scale production of many proteins have proved to be greater than originally presumed. The extension to metabolic engineering is at an even more primitive state of development than for production of single proteins.

Technical problems that have hindered the large-scale utilization of genetically modified cells are genetic instability, target protein toxicity, target protein instability, sequestering of the target protein in an inactive form, inability to achieve appropriate posttranslational modifications of the target protein, and inability to recover the target protein in a purified form. Many of these problems can be circumvented with the appropriate choice of host organism, vector, bioreactor, and bioreactor operating strategy.

II. SOME CONSIDERATIONS IN CHOOSING HOST-VECTOR SYSTEMS

The selection of host, vectors, and reactors is to a large extent determined by the product. We can broadly consider three classes of products from genetically modified cells: therapeutic proteins, "bulk" proteins, and nonprotein products. Therapeutic proteins are very high in value (about $100 million/kg), but are required in low amounts (a few kilograms per year), while bulk proteins are of moderate price (about $1000/kg), but are required in larger volumes (>1000 kg/year).

Therapeutic proteins are typically used for animal or human health. Proteins and polypeptides made in animal cells typically undergo posttranslational modifications, and in some cases these modifications are necessary for biological activity. Examples of posttranslational modifications include glycosylation, phosphorylation, and acylation. Often complex carbohydrate side chains are involved. For production of therapeutic proteins, the ability to make these posttranslational modifications will be critical to the clinical efficiency of the product in some, but not all, cases.

For bulk proteins, one can imagine producing industrial and food-grade enzymes to be used in processing raw materials into value-added products. Proteins may have value for structural as well as catalytic properties. Proteins can be used as adhesives, filaments for "biomaterials", or as additives in foods to give functionality (e.g., whip-ability), or in personal health care products (e.g., shampoos). The advent of "protein engineering" potentially allows direct design of proteins to improve activity, stability, or specificity properties. Such improved proteins should result in increasing demand. Posttranslational modifications are generally not important for the value of bulk proteins.

With the production of either therapeutic or bulk proteins, it is imperative to obtain high volumetric productivity in the reactor. Such high productivity is typically achieved by constructing the host-reactor system to maximize gene expression. In cells where the objective is to introduce or amplify certain metabolic pathways, such hyperproduction is not nearly as important as the stable maintenance of target protein activity; such stability usually requires a much lower level of gene expression.

With these constraints in mind, let us consider several popular host cells. These hosts are *Escherichia coli* (a Gram-negative bacterium), *Bacillus subtilis* (a Gram-positive bacterium), *Saccharomyces cerevisiae* (a yeast), mammalian cells, and insect cells.

E. coli has been the organism of choice for most initial attempts at protein production. E. coli has a number of very attractive features, the most important being that it is the best understood and characterized organism in the world at the molecular level. A large number of host backgrounds, vectors, and regulated strong promoters are available. Thus, sophisticated genetic manipulations are accomplished much more easily in E. coli systems. The development of strategies to select for those transformants making the desired protein in high yield is greatly aided by the wide variety of readily available mutants.

Recently, it has become common to grow E. coli to high cell densities (50 to 100 g dry weight per liter) in batch culture.[1,2] High cell density can be achieved only if the cells are somewhat restricted in their growth rate. Too rapid growth, particularly on glucose, leads to acetate formation, which is the principle inhibitory metabolic by-product. Even this restricted growth rate ($\mu \approx \frac{1}{2}\mu_{max}$) is still sufficiently high to give attractive volumetric productivities when coupled to high-density cultures.

However, E. coli does have disadvantages. This microorganism was rarely considered for industrial fermentations prior to the development of genetic engineering. As an industrial organism, E. coli is very small and difficult to recover, does not typically excrete products, and is more sensitive to phage attack than many industrial organisms. Also, the cell envelope contains endotoxins (lipopolysaccharides) which can be quite deadly.

For production of plasmid-encoded proteins, the constraint of intracellular protein production magnifies some of these disadvantages. If the target protein is made at a high level within the cellular cytoplasm, it will typically be destroyed by proteases (for example, preproinsulin was reported to have a half-life of less than 2 min) or will form inclusion bodies.[3-5] Inclusion bodies contain biologically inactive protein. Cells must be removed from the broth, broken (which releases endotoxins), the inclusion bodies recovered, solubilized, and the protein refolded to give the biologically active form. If the protein is intended for human or animal health use, it must be purified extensively (e.g., to remove endotoxins, contaminating proteins, or anything else that could trigger an immune response). Although the formation of inclusion bodies greatly concentrates the target protein, the overall effect on the cost of protein purification can be negative and quite significant. Adverse effects on process economics are particularly likely if the protein is large (>60,000 mol wt) and/or has a large number of disulfide bonds, since resolubilization to the biologically active form can then be very difficult.

Many of the problems associated with production of cloned proteins in E. coli are due to the lack of excretion of the protein into the extracellular space. As will be discussed later in this chapter, recent progress has been made in constructing an E. coli-vector system which excretes a plasmid-encoded protein. However, the lack of target protein excretion has led to interest in B. subtilis as a host.

B. subtilis excretes some natural proteases and amylases to give high extracellular concentrations (15 g/l).[6] Since B. subtilis is a Gram-positive organism, it has a single membrane rather than the two membranes present in E. coli. Typically, proteins excreted into the medium by B. subtilis will be secreted through the inner membrane of E. coli.[7,8] However, in E. coli these proteins that pass through the inner membrane are prevented from passing through the outer membrane. B. subtilis is a well-established organism for use in large-scale fermentations; it is robust, grows rapidly, and is more resistant to phage attack than E. coli.

However, B. subtilis has a number of disadvantages which have not been adequately addressed. Plasmid stability is often a much more severe problem in B. subtilis than in E. coli. The presence of proteases and the rapid degradation of foreign proteins remain significant problems. The limited variety of host strains (well-characterized mutants), strong inducible promoters, and cloning vectors makes sophisticated genetic manipulations difficult.

S. cerevisiae has an advantage over both E. coli and B. subtilis in that it can glycosolate proteins, although the type of posttranslational modifications done are not the same as in human cells or those of higher eukaryotes.[9] Furthermore, *S. cerevisiae* is on the GRAS (generally regarded as safe) list, which is significant for food or medical proteins. Yeast can be grown to high concentrations (>100 g dry weight per liter), and their large size simplifies cell recovery. Yeast also can excrete some proteins, although these are usually smaller ones; larger exported polypeptides may become trapped in the complex cell wall.

Even for smaller excreted proteins, the yield has not been very satisfactory. The intrinsic capacity to process such proteins appears to be limited. Expression levels in yeast are typically much lower than in *E. coli* for all types of proteins encoded by foreign genes.[9] This low expression level also implies poor use of resources.

When complete posttranslational modifications to the target protein are thought to be essential to the target protein's efficacious end use, then none of the bacterial or yeast strains are fully satisfactory. Such target proteins are usually mammalian proteins intended for animal or human therapeutic use. In these cases production using animal cell tissue culture becomes important. Animal cell culture is tremendously expensive and difficult. Cells grow very slowly; they are very fragile; it is difficult to obtain high cell densities; and they typically require complex and very expensive media. Contamination can be a severe problem. Thus, animal cell culture is desirable only when the product characteristics absolutely require it.

Production of such proteins is typically accomplished using mammalian cells as hosts, although there is growing interest in insect cell cultures as an alternative. Mammalian cells used for protein production are typically transformed cells. Since cancer cells are transformed cells, there is some concern (but no evidence) that a cancer-causing agent could be transferred from cell cultures to patients as an impurity in the desired protein. Furthermore, the vectors often used in genetically engineered mammalian cells are variants of primate viruses. There exists the remote potential for the back mutation of a vector to a virulent form of the virus. Finally, transfected gene expression in most mammalian cell cultures is very low (approximately 1 to 5% of total cellular protein).

Insect cultures are of interest because of the potentially very high levels of protein expression that can be obtained. Using host cells, such as *Spodoptera frugiperda* (fall armyworm), and the baculovirus *Autographa californica* NPV, Smith et al.[10] have demonstrated the production of human β-interferon at a high level with apparently correct posttranslational modifications. A recent review[11] summarizes proteins that have been made in insect cell systems and the degree of posttranslational modifications obtained. The baculoviruses have a very strong late promoter (for occlusion body protein) which is not essential to infectivity in cell culture. This late promoter presents an attractive site for cloning foreign eukaryotic genes. Other advantages of the insect cell system over mammalian cells are that insect cells are not "transformed" cells, and the final vectors are completely nonpathogenic. However, insect cells are every bit as difficult to grow as mammalian cells. Although serum-free media can support cell growth, good cell growth coupled with good virus replication normally requires supplementation of media with expensive serum, although recent reports suggest that this limitation can be circumvented.[11]

III. BIOREACTOR STRATEGIES

Once a host-vector system has been chosen, an appropriate bioreactor and operating strategy must be selected.

Let us consider a typical strategy for protein production from a plasmid-containing *E. coli* before exploring some current reactor concepts still at the formative stage. Normally, the inoculum is brought through a series of batch "seed" tanks. Typically, a strong, regulated promoter is used, with that promoter maintained at the lowest basal production level possible. In the transfer from seed tank to production vessel, it is imperative that a large portion of the population retains the plasmid. Segregational instability (a cell "losing" the plasmid "outgrows" plasmid-containing cells and displaces plasmid-containing cells in the population) is increased as the average number of plasmids in the population decreases and as the growth rate advantage of plasmid-free cells increases. By keeping target protein synthesis to a minimum, the plasmid-containing cells can grow more rapidly, since no cellular resources are "wasted" on making a nonessential protein; furthermore, some proteins, particularly those which are strongly hydrophobic, can be directly toxic to the host cell. With target protein synthesis repressed, the plasmid-containing cells grow nearly as quickly as plasmid-free cells.

Furthermore, selective pressure is usually placed on the culture, either antibiotic resistance or plasmid-encoded synthesis of an essential nutrient that the host cell is unable to synthesize based on its chromosomal DNA. Often two different selective pressures are applied. Structural changes in the plasmid to disable synthesis of the target protein while maintaining synthesis of a selectivity factor (e.g., an antibiotic-degrading enzyme or production of an essential nutrient) can lead to unproductive cells that will grow in the presence of the selective agent(s). If the host cell contains insertion sequences, then chromosome integration of the plasmid-encoded selectivity factor becomes possible. In the process of chromosome integration the portion of the plasmid encoding for the target protein is often rejected. Even if the plasmid and gene encoding the selective agent are lost, the gene product (e.g., an enzyme to degrade an antibiotic) will be retained for a few generations before being diluted to a level too low to elicit a biological response. Thus, any applied selective pressure is likely to be successful only for a finite number of generations, and even when successful does not guarantee the absence of non-plasmid-containing cells; however, non-plasmid-containing cells will not overtake the culture.

Up to the point of induction of target protein synthesis in the production tank, the whole objective is to place as little stress as possible on the plasmid-containing cells. At a high cell density, usually late exponential growth phase, the culture is induced. At this point, the objective is typically to sustain the production phase for as long as possible. The hyperproduction of the protein often leads to eventual cell death. Induction can be done by starvation for a nutrient (e.g., phosphate and the *Pho*A promoter) with a chemical (e.g., IPTG [isopropyl-β-D-thiogalactoside] and the *lac* or *tac* promoters) or by a temperature shift (e.g., the γP_L promoter). The starvation promoters have the disadvantage of simultaneously affecting growth and, eventually, the protein synthesizing machinery. The large-scale transfer of cells from a growth to an induction medium would be problematic. The use of a chemical induction system offers precise control and can be done easily on a large scale. However, a number of the chemicals used for induction are rather expensive. Temperature induction (involving a shift from a lower temperature to 42°C) also induces nonspecific synthesis of other proteins (e.g., heat shock proteins); some of these proteins are proteolytic.[12] Finally, the rapid shiftup of a large fermentor to a new temperature involves substantial dynamic lags. Since the profile of the temperature shift can alter product formation,[13] this becomes a constraint on scaleup.

The actual timing of the induction can also have a significant effect on total productivity.[14] Induction is often done toward the end of the cycle, when glucose and oxygen levels are likely to be very low in a high-density culture. However, glucose and oxygen starvation in *E. coli* can induce proteolytic enzymes.

Many variations exist on the above "typical" procedure, depending on the nature of the protein product and the host cell. If protein formation is for the development of new pathways to make a nonprotein product, "weaker" promoters are used so that target protein synthesis does not unduly tax the cell. For these nonprotein products, stability is the key issue, and sustained continuous operation would be desirable.

DeTaxis duPoët et al.[15] have shown how cell immobilization can improve reactor stability in a continuous-flow system. If a culture is entrapped in a gel matrix, separate isolated colonies develop within each bead. The possibility of a revertant forming in each of these microcolonies is very low because the number of cells in each microcolony is very small. Even if a nonproductive cell emerges early in a microcolony, it will only dominate that small microcolony. If all of these microcolonies are well mixed, as they would be if they were in suspension, the nonplasmid cell can overtake the whole reactor. DeTaxis duPoët et al.[15] developed a simple mathematical model to predict this increased stability. Actual experiments showed that immobilization improved stability beyond that predicted simply by compartmentalization. The authors then speculated that immobilization might alter cell physiology, perhaps due to stress-induced changes in morphology, and that these altered states led to greater intrinsic plasmid stability.

Cell immobilization and continuous reactor operation are not only attractive for production of nonprotein products, but also are potentially attractive for protein production. However, such a system would require continuous production and excretion of the target protein. Indeed, such operations for *B. subtilis, S. cerevisiae,* and animal cells are quite plausible. Most reactor schemes for protein production from animal cells involve some form of cell immobilization or retention.[16] However, attempts are also being made to extend that approach to protein production from microorganisms.[14,17-19] In our laboratory we have been working on developing a scheme to allow the use of *E. coli* in a continuous-flow immobilized-cell system for the production of plasmid-encoded proteins.

IV. NOVEL HOST-VECTOR AND BIOREACTOR STRATEGIES

Our proposed scheme[14] is outlined in Figure 1. The approach is to couple a strong, regulated promoter, a ribosome binding site, and a signal sequence fused to a target protein. A strong terminator may also be needed to prevent read through. A plasmid containing these control regions and the target protein gene can then be inserted into a robust host cell (a strain of *E. coli*). The culture is grown under noninducing conditions to reduce potential problems of genetic instability. Once a high cell concentration is obtained, the cells can be immobilized by gel entrapment or retention in a membrane reactor. Target protein production is induced in the growing culture just prior to immobilization. We have found host-vector systems where high-level expression of the target protein also induces alterations of the outer membrane. These alterations lead to "leaky" behavior. In many cases the signal sequence will lead to secretion of the overproduced protein across the inner membrane. The target protein, once it passes through the inner membrane, can be released through the "leaky" outer membrane into the extracellular fluid. Such immobilized cultures enable continuous nutrient flow with the target protein being selectively released into the extracellular medium. Because sustained continuous operation of the reactor system is necessary for economic considerations, net cellular growth must be controlled to a low level to prevent disruption of the immobilizing matrix. Control of growth can be done through nutrient starvation, decreased temperature, or controlled bleeding of excess cells from the reactor.

The potential advantages of such a system are several fold. Purification is simplified because relatively few other proteins are in solution and the target protein is present at a far higher level than other proteins. Recovery is made easier since cell separation is not required and the target protein can be available at high concentrations (>1 g/l). Since there is no cell breakage, the release of endotoxins may be circumvented. By having a continuous-flow system with high cell density, high volumetric productivities can be obtained if high cellular metabolic rates are sustained. Ideally, the cell consumes only those nutrients necessary to meet maintenance energy requirements and for synthesis of the target protein, since cell removal from the reactor is nearly zero and other nongrowth cellular functions are suppressed. Consequently, the product yield (or substrate conversion) can be maximized. Since protein is excreted from inside the cell, it may be possible to avoid formation of inclusion bodies.

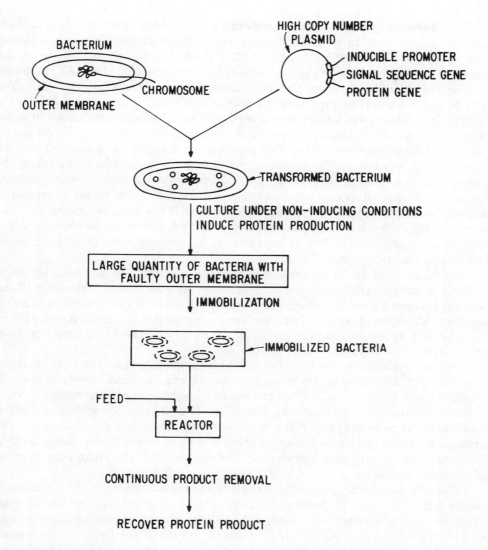

FIGURE 1. A general scheme is shown for production of a plasmid-encoded protein with excretion into the extracellular fluid. The host cell becomes "leaky" only after induction of high level plasmid-encoded protein synthesis. The use of cell immobilization potentially allows the use of a continuous, high-cell-density bioreactor.

Finally, it would be possible to circumvent genetic instability due to segregational losses of plasmid, since cell growth can be greatly reduced when protein synthesis is initiated. It should be emphasized that these are potential benefits, *not* advantages that have been experimentally demonstrated, although we have made some partial progress toward such a demonstration.

Our goal is a general process applicable to a wide range of proteins. Proteins secreted in their native host cell are good candidates. As we learn more about the secretion process itself, we intend to attempt the extension of such an approach to normally cytoplasmic or man-designed proteins. We believe that continuous-flow reactors producing a target protein at extracellular concentrations greater than 1 g/l at high purity (approximately 80%) with a sustainable productivity of about 100 mg/l-h for periods of several weeks would be very attractive for a large variety of protein products.

We are attempting to convert this concept into a reality. Most of our studies so far have utilized a model system for the production of the normally periplasmic enzyme β-lactamase in *E. coli*, mainly strain RB791[20] and a pTAC11 plasmid or derivatives.[21] However, the same general response has been observed in other *E. coli* host strains. The pTAC11 plasmid is essentially pBR322 with the *tac* promoter[21] inserted ahead of the β-lactamase gene. The *tac* promoter is a hybrid *trp-lac* promoter inducible with IPTG.

With this construction we observed a high level of β-lactamase production in shake flasks at 37°C when the culture was induced at mid-exponential growth (optical density = 0.4). Under these conditions, 90 to 95% of the active β-lactamase is excreted into the medium, while very little (<7%) β-galactosidase is found in the medium.[22] The enzyme β-galactosidase is a cytoplasmic enzyme, and its presence in the medium would indicate significant cell lysis rather than protein excretion. If the shake-flask culture is induced in early exponential-phase growth (approximate optical density = 0.2), cell killing and lysis are evident (ca. 35% release of β-galactosidase), while induction in late exponential growth (approximate optical density = 0.8) leads to reduced excretion (about 30%) and reduced total β-lactamase production.

At 37°C and induction at an optical density of 0.4, about 50% of the total β-lactamase produced is inactive and in the form of periplasmic inclusion bodies.[23] However, if the culture is grown at lower temperatures, the formation of inclusion bodies is greatly reduced, with the total amount of active β-lactamase formed being more than twice that at 37°C as measured 24 h after induction. The percent of the active β-lactamase excreted was about 50% at 20°C.

When attempting to use this host-vector system in continuous culture at 37°C, we found it impossible to achieve stable operation while retaining any significant number of plasmid-containing cells. However, at 20°C we can sustain for significant periods cultures actively producing β-lactamase.[37] Productivities averaging 4 U of β-lactamase activity per optical density-hour can be sustained for 400 h with about 50% excretion of the β-lactamase. The β-lactamase in the extracellular solution is about 90% pure. Higher productivities (20 U of β-lactamase activity per optical density-hour) and excretion (ca. 75%) have been observed for significant but shorter periods (ca. 50 h).

These cells can also be grown to high cell densities (optical density ≈ 70) in membrane reactors that retain cells. In membrane reactors we have observed high concentrations of the extracellular β-lactamase (ca. 1.2 g/l) and the ability to maintain high cell densities with no net growth for 1 to 2 weeks.[37] If our more recent experience from low-temperature continuous-culture experiments can be extended to high-density continuous-flow reactors with cell retention, then we will be able to meet most of our goals for system performance.

The ability to extend this approach to other protein products, the ability to sustain the high productivity state for even longer periods, and the need to better understand the mechanisms responsible for induction of "leaky" behavior remain important research objectives for this project. The key to sustaining high expression of protein production, excretion, and cell viability appears to lie in better understanding the balancing of important cellular kinetic processes. These kinetic processes are transcription of the target gene, the rate of mRNA degradation, the rate of translation, the rates of processing the precursor protein and secretion, the rate at which imperfect outer membrane is generated, the rate of release of the mature protein through the outer membrane, and the rate of inclusion body formation.

Since the key to successful system operation is the balancing of these kinetic processes, it would be helpful to be able to predict the quantitative dependence of each of these kinetic processes on important fermentation variables such as temperature, pH, and nutrient levels. Such predictions will require the development of mathematical models. It is our intention to construct such models, although our efforts to do so have just begun.

V. CELL MODELS

A rather large number of mathematical models have been proposed for plasmid-containing bacteria and yeast. These models are usually concerned with the dynamics of populations containing both plasmidless and plasmid-containing cells (e.g., References 24 to 28) or with host-vector interactions. These latter models are usually highly structured. Some of the pioneering models on plasmid replication by Lee and Bailey[29-31] have been called genetically structured. Peretti and Bailey[32] have also adapted models developed by our group[33] to predict aspects of host cell-plasmid interactions by including increased structure concerning transcription and translation. These latter models provide the necessary framework to begin modeling cultures with protein excretion.

The basic Cornell model[33] is a relatively detailed model of an individual cell of *E. coli*. The model is applicable to both fully aerobic and anaerobic growth. The model responds explicitly to changes in external concentrations of glucose and ammonium ion. Population models can be constructed by finite representation techniques. Such models have been used to make accurate predictions of dynamic responses of continuous-flow stirred tank reactors (CFSTRs) to perturbations in flow or substrate concentration.[34,35]

The base model has been constructed so as to allow the insertion of a more detailed description of any cellular subsystem into the model while retaining the base model structure. Thus, the response of such a modified model is a direct test of the plausibility of the mechanism used to construct the description of the cellular subsystem. Using this concept, we have inserted into the model a submodel for the replication of plasmids with the ColE1 origin of replication. Such plasmids, exemplified by the plasmid pBR322, are commonly used as industrial vectors for production of foreign proteins. The inclusion of a mechanistically based model for plasmids with the ColE1 origin of replication results in quite reasonable predictions of plasmid copy number.[35]

Once copy number is known, then the effects of target protein synthesis on host cell physiology must be predicted if we wish to predict the dynamics of reactor performance. The increased detail on transcription and translation inserted in our base model by Peretti and Bailey[32] will be important; however, even without such detail it is possible to make reasonable predictions of plasmid stability in CFSTRs as a function of growth rate (see Figure 2) by simply including the effects of competition for precursors and energy.[36]

Further modifications to these models will be necessary if we wish to model protein excretion. We are currently working on extensions of the model that explicitly recognize amino acids in the medium, an improved model of the *lac* promoter, and population models of plasmid-bearing cells, using a finite representation technique.

VI. SUMMARY

The advent of genetic and molecular-level engineering has greatly expanded our knowledge of cellular physiology as well as given us tools to directly alter cellular mechanisms. Consequently, the engineer can now envision a living cell as a complex catalyst that can be "designed" rationally and whose behavior can be predicted. This chapter has attempted to describe some preliminary but incomplete attempts to fulfill this vision.

ACKNOWLEDGMENTS

We gratefully acknowledge support by the National Science Foundation (grant 8513612) and the Office of Naval Research (contract N00014-85-K-0580). David B. Wilson contributed by both codirecting the experimental work and participating in many helpful discussions.

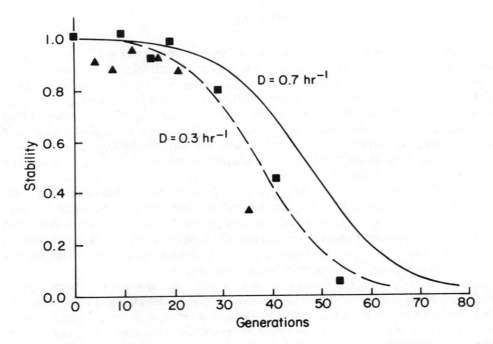

FIGURE 2.   The prediction of plasmid stability of *E. coli* B/r containing a pBR322 derivative is compared to experimental data for a glucose-limited chemostat at 37°C.[36] Stability is the fraction of plasmid-containing cells in the culture. About 15 to 17% of the total protein synthesis is due to the presence of the plasmid. The solid line is the prediction for the system at a dilution rate of 0.7 h^{-1} and the dashed line for 0.3 h^{-1}. Data at D = 0.7 h^{-1} is depicted by ■; at 0.3 h^{-1}, by ▲.

The work reported here is the result of the efforts of many graduate students: George Georgiou, Jeff Chalmers, Paul Togna, Mike Domach, Mohammad Ataai, Jen Shu, Byung-Gee Kim, and Lisa Laffend.

REFERENCES

1. **Zabriskie, D. W. and Arcuri, E. J.,** Factors influencing productivity of fermentations employing recombinant microorganisms, *Enzyme Microb. Technol.,* 8, 706, 1986.
2. **Allen, B. R. and Luli, G. W.,** A gradient feed process for obtaining high cell densities and control of product expression from recombinant *Escherichia coli,* in *World Biotech Report USA, The Proceedings of Biotech '85,* Vol. 2, Online International, New York, 1985, 447.
3. **Talmadge, K. and Gilbert, W.,** Cellular location affects protein stability in *Escherichia coli, Proc. Natl. Acad. Sci. U.S.A.,* 81, 1830, 1982.
4. **Williams, D. C., Van Frank, R. M., Muth, W. L., and Burnett, J. P.,** Cytoplasmic inclusion bodies in *Escherichia coli* producing biosynthetic human insulin proteins, *Science,* 215, 687, 1982.
5. **Langley, K. E., Berg, T. F., Strickland, T. W., Fenton, D. M., Boone, T. C., and Wypych, J.,** Recombinant DNA derived growth hormone from *Escherichia coli.* I. Demonstration that the hormone is expressed in reduced form, and isolation of the hormone in oxidized native form, *Eur. J. Biochem.,* 163, 313, 1987.
6. **Nicaud, J.-M., Macman, N., and Holland, I. B.,** Current status of secretion of foreign proteins by microorganisms, *J. Biotechnol.,* 3, 255, 1986.
7. **Palva, I., Sarvas, M., Lehtovaara, P., Sibakov, M., and Kaarjainen, L.,** Secretion of *Escherichia coli* β-lactamase from *Bacillus subtilis* by the aid of α-amylase signal sequence, *Proc. Natl. Acad. Sci. U.S.A.,* 69, 5582, 1982.

8. **Palva, K., Lehtovaara, P., Kaariainen, L., Sibakov, M., Cantell, K., Schein, C. H., Kashivogi, K., and Weissman, C.**, Secretion of interferon by *Bacillus subtilis, Gene,* 22, 229, 1984.

9. **Kingsman, S. M., Kingsman, A. J., and Miller, J.,** The production of mammalian proteins in *Saccharomyces cerevisiae, Trends Biotechnol.,* 5, 53, 1987.

10. **Smith, G. E., Summers, M. D., and Fraser, M. J.,** Production of human beta interferon in insect cells infected with a baculovirus expression vector, *Mol. Cell. Biol.,* 3, 2156, 1983.

11. **Luckow, V. A. and Summers, M. D.,** Trends in the development of baculovirus expression vectors, *Bio/Technology,* 6, 47, 1988.

12. **Baker, T. A., Grossman, A. D., and Gross, C. A.,** A gene regulating the heat shock response in *Escherichia coli* also affect proteolysis, *Proc. Natl. Acad. Sci. U.S.A.,* 81, 6779, 1984.

13. **Dipasquantonio, V. M., Bettenbaugh, M. J., and Dhurjati, P.,** Improvement of product yields by temperature-shifting of *Escherichia coli* cultures containing plasmid pOU140, *Biotechnol. Bioeng.,* 29, 513, 1987.

14. **Georgiou, G., Chalmers, J. J., Shuler, M. L., and Wilson, D. B.,** Continuous protein production from *E. coli* capable of selective protein excretion: a feasibility study, *Biotechnol. Prog.,* 1, 75, 1985.

15. **DeTaxis duPoët, P., Dhulster, P., Barbotin, J.-N., and Thomas, D.,** Plasmid instability and biomass production: comparison between free and immobilized cell cultures of *Escherichia coli* BZ18 (pTG201) without selection pressure, *J. Bacteriol.,* 165, 871, 1986.

16. **Thilly, W. G.,** *Mammalian Cell Technology,* Butterworths, Boston, 1986.

17. **Mosbach, K., Birmbaum, S., Hardy, K., Davies, J., and Bulow, L.,** Formation of proinsulin by immobilized *Bacillus subtilis, Nature (London),* 302, 543, 1983.

18. **Inloes, D. S., Smith, W. J., Taylor, D. P., Cogen, S. N., Michaels, A. S., and Robertson, C. R.,** Hollow fiber membrane bioreactors using immobilized *E. coli* for protein synthesis, *Biotechnol. Bioeng.,* 25, 2653, 1983.

19. **Karkare, S. B., Burke, D. H., Dean, R. C., Jr., Lemontt, J., Souw, P., and Venkatasubramanian, K.,** Design and operating strategies for immobilized living cell reactor systems, *Ann. N.Y. Acad. Sci.,* 469, 91, 1986.

20. **Brent, R. and Ptashne, M.,** Mechanism of action of the *lex*A gene product, *Proc. Nat. Acad. Sci. U.S.A.,* 78, 4204, 1981.

21. **De Boer, H., Comstock, H. L., and Vasser, M.,** The *tac* promoter: a functional hybrid derived from the *trp* and *lac* promoters, *Proc. Natl. Acad. Sci. U.S.A.,* 80, 21, 1983.

22. **Georgiou, G.,** Inducible Overproduction and Excretion of a Periplasmic Protein (β-Lactamase) in *Escherichia coli,* Ph.D. thesis, Cornell University, Ithaca, NY, 1987.

23. **Georgiou, G., Telford, T. N., Shuler, M. L., and Wilson, D. B.,** Localization of inclusion bodies in *Escherichia coli* overproducing β-lactamase or alkaline phosphatase, *Appl. Environ. Microbiol.,* 52, 1157, 1986.

24. **Imanaka, T. and Aiba, S.,** A perspective on the application of genetic engineering: stability of recombinant plasmid, *Ann. N.Y. Acad. Sci.,* 369, 1, 1981.

25. **Ollis, D. F. and Cheng, H. T.,** Batch fermentation kinetics with (unstable) recombinant cultures, *Biotechnol. Bioeng.,* 24, 2583, 1982.

26. **Lauffenburger, D. A.,** Stability of colicin plasmids in continuous culture: mathematical model and analysis, *Biotechnol. Prog.,* 1, 53, 1985.

27. **Parker, C. and DiBiasio, D.,** Effect of growth rate and expression level on plasmid stability in *Saccharomyces cerevisiae, Biotechnol. Bioeng.,* 29, 215, 1987.

28. **Koizame, J. and Aiba, S.,** Some consideration on plasmid number in a proliferating cell, *Biotechnol. Bioeng.,* 28, 311, 1986.

29. **Lee, S. B. and Bailey, J. E.,** A mathematical model for λdv plasmid replication: analysis of a wild-type plasmid, *Plasmid,* 11, 151, 1984.

30. **Lee, S. B. and Bailey, J. E.,** A mathematical model for λdv plasmid replication: analysis of copy-number mutants, *Plasmid,* 11, 166, 1984.

31. **Lee, S. B. and Bailey, J. E.,** Analysis of growth rate effects on productivity of recombinant *Escherichia coli* populations using molecular mechanism models, *Biotechnol. Bioeng.,* 26, 66, 1984.

32. **Peretti, S. W. and Bailey, J. E.,** Simulations of host-plasmid interactions in *Escherichia coli:* copy number, promoter strength, and ribosome binding site strength effects on metabolic activity and plasmid gene expression, *Biotechnol. Bioeng.,* 29, 316, 1987.

33. **Shuler, M. L.,** On the use of chemically structured models for bioreactors, *Chem. Eng. Commun.,* 36, 161, 1985.

34. **Domach, M. M. and Shuler, M. L.,** A finite representation model for an asynchronous culture of *E. coli, Biotechnol. Bioeng.,* 26, 877, 1984.

35. **Ataai, M. and Shuler, M. L.,** Mathematical model for the control of ColE1 plasmid, *Plasmid,* 16, 204, 1986.

36. **Ataai, M. and Shuler, M. L.**, A mathematical model for prediction of plasmid copy number and genetic stability in *Escherichia coli, Biotechnol. Bioeng.*, 30, 389, 1987.
37. **Chalmers, J. J.**, Ph.D. thesis, Cornell University, Ithaca, NY, 1988.

Chapter 8

MAMMALIAN CELL CULTURE: AN OVERVIEW OF THE PROBLEMS AND VARIOUS APPROACHES FOR THE DEVELOPMENT OF LARGE-SCALE SYSTEMS

Joseph Feder

TABLE OF CONTENTS

I. INTRODUCTION

Many biologically important proteins of therapeutic interest can only be produced by animal cells because of necessary posttranslational processing, such as glycosylation, or other complex events, such as proper folding and secretion. This has motivated the development of large-scale animal culture systems for products such as those listed in Table 1.

Animal cells have distinct characteristics of growth and productivity in contrast to bacteria and yeast which present problems for scaleup. Animal cells in culture require a complex nutrient medium supplemented with growth factors, hormones, and often serum. For example, Dulbecco's Modified Eagle's Medium (DMEM)[1] contains 15 different essential and nonessential amino acids, 6 water-soluble vitamins and coenzymes, glucose, pyruvate, choline, inositol, bicarbonate, and a variety of inorganic ions.[1] This is supplemented with 5 to 10% fetal bovine serum or defined components such as insulin, transferrin, and a specific growth factor. The addition of 10% serum in the culture medium introduces approximately 3.5 g of protein per liter of medium. This represents a significant protein contaminant which must be separated by the subsequent purification process from the desired product, which is often present in milligram per liter concentrations or less. Animal cells exhibit both anchorage-dependent and anchorage-independent growth. Most normal tissue-derived cells require attachment to a solid substrate for growth. Transformed cells, however, often will grow in suspension. The scaleup of culture systems for anchorage-dependent cells must address the issue of adequate surface area to volume for efficient large-scale cell production. The material properties of such substrates are critical, since not all types of materials support cell attachment and growth. Animal cells lack a strong cell wall and therefore are fragile and sensitive to shear forces. This imposes restrictions on the types of agitation that can be employed in such reactors and, hence, affects the mass transfer problems normally encountered in scaleup. Shear forces can affect not only the integrity of a cell, but also the attachment of a cell to a substrate material. For anchorage-dependent cells, detachment results in eventual cell death. This can occur even at shear forces that are not sufficient to disturb cell integrity.

Animal cells are derived from organisms which have developed complex mechanisms for providing a homeostatic growth environment. Hence, they are sensitive to small changes in the physical and biochemical medium environment. Closed batch culture systems result in a continuous departure from a controlled homeostatic state. Consequently, steady-state continuous growth systems have been developed which optimize cell growth and viability. These will be discussed below. Furthermore, a safe, sterile environment is required. Since animal cells grow slowly compared to microorganisms, such cultures are subject to contamination.

The conventional large-scale culture of anchorage-dependent cells in vaccine production, for example, has employed great numbers of roller bottles (Figure 1). The cells are grown on the inner surface of the roller bottles, and slow rotation exposes the layer of cells attached to the bottle surface alternately to medium and air. The surface area of such culture bottles is limited to a few square feet, and although automatic medium-feed devices have been employed successfully, the multiplicity of culture vessels presents a very labor-intensive batch process.

The conventional scaleup approach to suspension-grown cells has been to construct larger, more carefully controlled, agitated vessels operated in batch or semibatch mode.

In addition to these cell growth-related issues, serious consideration must also be given to the effects of such large-scale systems on the downstream purification technology, often the limiting step of a production process. Conventional culture systems tend to provide low cell densities, low product concentrations, and significant purification problems due to high serum supplement contamination. It is valuable, therefore, to examine what constitutes an ideal product stream and how this might be achieved in addressing the design of an animal cell culture system.

TABLE 1
Biologicals Produced by
Mammalian Cell Culture

Tissue plasminogen activator
Pro-urokinase
Urokinase
Protein C
Factor VIII
Erythropoietin
Human growth hormone
Monoclonal antibodies
   Therapeutics
   Diagnostics
   Purification ligands
Anti-inflammatory peptide (Lipocortin)
Tumor necrosis factor
Epidermal growth factor
Insulin-like growth factors
Cartilage induction factor
Colony stimulating factor-1
Interleukin-1
Interleukin-2
Interferons

The ideal cell culture medium stream contains a maximum product concentration with a minimum amount of contaminating protein. This limits the volume of material for processing and provides an initial crude product with a high specific activity. Anything that reduces the serum concentration in the medium directly affects the purification problems. However, serum is not the only source of protein contaminants. Cell disruption also contributes to the level of contaminating proteins in the medium.

To achieve an optimum product stream, it is necessary to maximize both the cell density and the cell productivity. Although genetic engineering and careful cell selection provide a major approach for effecting high expression levels, understanding the specific medium requirements and identifying feedback metabolite inhibition are critical.

Minimization of serum requirements, particularly by use of a defined medium, is critical. Ideally, a defined medium can be developed which employs growth factors and hormones, the identification of which provides an opportunity for devising novel procedures for their removal in the downstream process. A culture system which maintains high cell viability reduces that component of the contaminating protein derived from intracellular materials and cell debris.

A major approach for achieving these goals has been the development of continuous culture systems that attempt to maintain a homeostatic culture environment. Under such steady-state conditions, high cell density cultures are achieved with increased cell and product yields per unit volume of medium used and with significant reduction of the serum concentration in the medium.[2,3]

A number of approaches have been employed to maintain a homeostatic culture environment for suspended cells, including fed batch, Monod-type chemostat, cytostat, and perfusion reactors.

II. STEADY-STATE CULTURE SYSTEMS

A. FED BATCH CULTURE

Fed batch culture is characterized by continuous feeding of medium to a culture of cells in a closed system at a rate which both maintains a nearly constant nutrient level and dilutes

FIGURE 1.   Photograph of roller bottle deck assembly.

metabolic product levels. Efficient operation requires methods for monitoring and controlling the concentration of the limiting component. Throughout the growth process, the culture volume increases until the reactor is filled. The last phase of cultivation then resembles a closed batch system. Nevertheless, over the greater portion of the cell growth phase, a near steady-state culture environment is maintained, and higher cell densities and cell yields per unit volume of medium used are obtained than with conventional closed batch reactors. Successful production of monoclonal antibodies in a fed batch reactor has been reported.[4] Although fed batch cultures can be operated in a semicontinuous mode, they are essentially noncontinuous cultures.

B. CHEMOSTAT CULTURE

In contrast, chemostat culture provides an opportunity for maintaining steady-state culture conditions and cell density in a continuous mode of operation. In chemostat culture, the nutrient medium contains saturating concentrations of all of the growth components with the exception of one, the concentration of which determines both the cell density and the growth rate. Systems have been developed for chemostat culture of animal cells.[5]

The steady state in chemostat operation is maintained by a continuous dilution of the culture. The specific growth rate, μ, is equal to the dilution rate, D, defined as the number of volume changes per unit time. Monod[6] showed that the specific growth rate was a function of the concentration of the limiting growth component:

$$\mu = \mu_{max}\left(\frac{S}{S + K_s}\right) \tag{1}$$

where μ_{max} is the maximum specific growth rate, S is the steady-state concentration of the limiting component, and K_s describes a dissociation constant for the cell-substrate association. Thus, for a Monod-type chemostat culture, the specific growth rate at steady state is less than the maximum possible. It has been shown by Herbert et al.[7] that for a given substrate concentration and dilution rate there is both a unique steady-state substrate concentration, S_{out}, and a unique cell density, given by the following equations:

$$S_{out} = K_s\left(\frac{D}{\mu_{max} - D}\right) \tag{2}$$

and

$$\text{cell concentration} = Y(S_{in} - S_{out}) = Y\left[S_{in} - K_s\left(\frac{D}{\mu_{max} - D}\right)\right] \tag{3}$$

where Y is a yield constant, defined as the amount of cells obtained per unit of substrate consumed, and S_{in} is the substrate concentration of the feed stream. They also showed that the maximum output of a continuous culture is given by the following expression:

$$\text{Output}_{maximum} = \mu_{max} \cdot Y \cdot S_{in}\left(\sqrt{\frac{K_s + S_{in}}{S_{in}}} - \sqrt{\frac{K_s}{S_{in}}}\right)^2 \tag{4}$$

When $S_{in} \gg K_s$, the maximum output obtained is a function of the substrate concentration and μ_{max}. This type of culture system provides an efficient means for continuous production of cells and derived products in a homeostatic environment. Stable steady-state chemostat cultures which exhibited significant cell yield and reactor productivity were reported by Tovey.[5]

C. CYTOSTAT CELL CULTURE

In a cytostat culture, a constant cell density is maintained by continuous culture dilution at a rate equal to the cell growth rate. In contrast to a chemostat, however, all of the nutrient components are maintained at saturating concentrations. Consequently, the maximum specific growth rate is maintained.[8] Under these conditions, a range of steady-state cell densities can be established at the same dilution rate. However, as the cell density increases, the substrate concentration will be decreased to a limiting concentration. At this point, the condition $S \gg K_s$ is not satisfied and the specific growth rate will drop. As will be shown below, the limiting cell density can be estimated from nutrient consumption rates.

D. PERFUSION CULTURE

Perfusion introduces another degree of freedom for increasing the steady-state cell concentrations and, hence, the maximum cell or product output of a continuous culture system. In perfusion systems, the cells are physically retained in the reactor such that continuous nutrient medium feed and spent medium removal is achieved without dilution of the cell density. This process can be coupled with a cytostat mode of operation to provide a perfused continuous culture. Like the cytostat, all substrates are provided in the feed medium at saturating concentrations. The steady-state substrate concentrations, cell density, and uptake rate functions can be calculated from the mass balance equations shown below, assuming saturating substrate concentration and maximum uptake rates.

$$\text{Substrate}_{in} = \text{Substrate}_{consumed} + \text{Substrate}_{out} \tag{5a}$$

$$S_{in} \cdot (D + P) \cdot V = k_s \cdot \text{cell number} + S_{out} \cdot (D + P) \cdot V \tag{5b}$$

$$k_s = \frac{(S_{in} - S_{out})(D + P)}{(\text{cell}/V)} \tag{6}$$

Both the perfusion rate, P, and the harvest dilution rate, D, are expressed as the number of volume changes per unit time. The specific consumption constant, k_s, is defined in terms of the amount of substrate consumed per unit time per cell. From the values of D, P, S_{in}, k_s, and the estimated minimum steady-state substrate level, S_{out}, which still satisfies the condition $S_{out} \gg k_s$ ($S_{out} \sim 10k_s$), one can estimate the maximum cell density from the expression

$$(\text{cells/l}) = \frac{(S_{in} - S_{out})(D + P)}{k_s} \tag{7}$$

Thus, perfusion increases the limiting cell density that can be maintained at maximum growth in a cytostat by the factor

$$\frac{(S_{in} - S_{out})P}{k_s} \tag{8}$$

The maximum output for the perfusion cytostat would be equal to the product of the maximum cell density and (P + D). Suspension-grown cells present a particular challenge for perfusion. Mechanical filtration devices must be devised that allow cell-free medium removal from the reactor without becoming clogged by the cells and cell debris. The large-scale perfused suspension culture systems developed in our laboratory[2,3] employ a spin filter device first described by Himmelfarb et al.[9] A diagram of the early cytostat reactor is shown in Figure 2.

III. MAMMALIAN CELL GROWTH IN A CONTROLLED PERFUSION REACTOR

A. PERFUSION REACTOR SYSTEM

The perfusion reactor was operated in the following manner. Suspended medium, withdrawn through the filter in the satellite vessel, and a capacitive level monitor controlled the fresh medium pump. The pH was monitored by an autoclavable INGOLD® electrode and controlled by varying the CO_2 concentration of the overlay gas and by adding sodium bicarbonate when necessary. Reservoirs for fresh and expended media were connected to

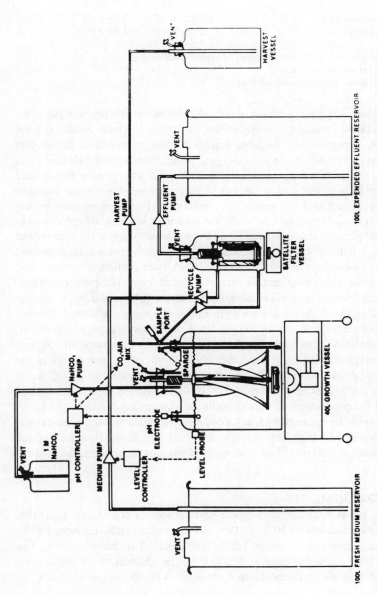

FIGURE 2. A diagram of the perfusion cytostat system is shown. Cell-free expended medium was removed from the interior of the porcelain filter at a rate set on the effluent pump to provide the required perfusion of fresh medium. Cell suspension was recycled to and from the satellite vessel by a double-headed pump at a rate sufficient to maintain similar component concentrations in the two vessels. While operated as a chemostat, a harvest pump removed cell suspension directly to a harvest vessel contained in an adjacent 4°C cold room. The resulting reduction of the liquid level in the growth vessel was compensated by an exterior capacitive level probe which actuated the medium pump to supply fresh nutrient medium. An autoclavable pH electrode penetrating the growth vessel allowed control of the pH by varying the CO_2 in the mixture of overlay gas and by the addition of sodium bicarbonate. A sterile air-shielded sample port and a means for low-rate oxygen sparge through a sintered glass dispersion tube were also provided. Except for the harvest vessel, the entire system was contained within a 37°C warm room. (From Tolbert, W. R. and Feder, J., in *Annual Reports on Fermentation Processes*, Vol. 7, Tsao, G. T., Ed., Academic Press, Orlando, FL, 1983, 35. With permission.)

TABLE 2
Comparison of Production of Rat Walker Carcinosarcoma Cells
in Conventional and Perfusion Suspension Culture

	Total medium[a] (l)	Total cells[a]	Yield
Conventional suspension culture[b]	14,150	$13,422 \times 10^{12}$	0.95×10^{12}
Perfusion culture system[c]	612	$1,705 \times 10^2$	2.80×10^{12}

[a]   The sum of medium and cells from a number of experimental runs.
[b]   100-l Vibromixer suspension reactor.
[c]   4- and 40-l perfusion suspension culture reactors.

the reactor through peristaltic pumps. Agitation of the cell culture suspension was provided by the use of large-area, slowly rotating flexible sheets, or "sails". These minimized the shear effects on the cells, a particularly valuable feature for the microcarrier suspension cultures to be discussed later. Providing adequate oxygen, particularly at high cell densities, is a major problem. The early perfusion reactors provided a gas sweep over the culture surface of a few liters per minute and a slow sparge of pure oxygen. To prevent foaming problems, the sparge was limited to a few milliliters per minute. For oxygen consumption rates of about 0.045 to 0.060 cm^3 (stp)/min/10^9 cells for a number of cells, oxygen quickly becomes limiting when supplied in this manner. To address this problem, a silicone rubber tubing (O.D., 2 mm; I.D., 1.0 mm) "lung" permeator was developed. This was incorporated into the design of a new 14-l computer-controlled research model perfusion reactor. The initial studies were carried out with a lung surface area of about 100 cm^2/l of culture volume. During cell growth the dissolved oxygen was maintained at about 0.11 atm using computer control. The inlet and outlet gases from the reactor were analyzed using gas chromatography. Other parameters monitored were the dissolved oxygen level and headspace gas composition. This yielded oxygen consumption rates and provided control data for regulating the oxygen content of the feed gas mixture to maintain the established dissolved oxygen levels in the medium. Temperature control was maintained by locating the reactor system in a 37°C warm room. Reactor sizes from a few-liter research model to 100-l production models have been developed and employed for production of various cells and expressed products. The perfusion suspension system could be operated both in batch and continuous cytostat-type modes. The cultures were characterized by high cell density, achieved significant increases in cell yield per unit volume of medium, and often had reduced requirements for serum concentration in the medium.

B. CELL GROWTH EXPERIMENTS

In early experiments, rat Walker carcinosarcoma cells grown in continuous perfusion culture reached cell densities as high as 30.8×10^6 cells/ml, with viabilities near 100%, compared to maximum cell densities of about 1.5×10^6 cells/ml in batch culture. The exponential growth rates were about the same (~ 0.042 h^{-1}). In addition, the yield of cells per liter of medium used was about threefold that obtained without perfusion. Data are shown in Table 2.

No attempt was made to optimize the perfusion rate in these studies. However, at rates below 5 ml/h/10^9 cells, the growth rate decreased significantly. The glucose consumption rate for the rat Walker cells was about 50 to 80 μg/h/10^6 cells. Since the DMEM used in these studies contained 4.5 mg/ml glucose, a feed rate of 18.5 ml/h/10^9 cells would be required to maintain a steady-state glucose level.

Another observation regarding perfusion growth was the reduction in required serum

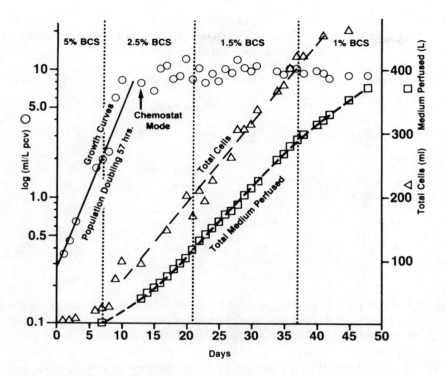

FIGURE 3. Growth curve for human hepatoma cells grown initially in 5% bovine calf serum-supplemented DMEM without antibiotics. Serum concentration was reduced as indicated with chemostat operation initiated on day 10 to maintain level of cell density. (From Feder, J. and Tolbert, W. R., Eds., *Large-Scale Mammalian Cell Culture*, Academic Press, Orlando, FL, 1985. With permission.)

concentration in the medium. The human hepatoma cell line, SK-Hep-1, was grown in continuous perfused cytostat culture with the cell density maintained at 10 ml/l (packed cell volume). The cells were initially grown in DMEM with 5% bovine calf serum, but during the cytostat operation the serum concentration was reduced to 1% with minimal effect on cell growth and viability (see Figure 3).

Perfused cytostat culture has been used to produce both cells and products in a long-term continuous operation. These include monoclonal antibodies, growth factors, tissue plasminogen activator (tPA), and cell-derived molecules such as membrane proteins. Experiments on the growth of Bowes melanoma cells in the 14-l computer-controlled perfusion cytostat reactor yielded the data shown in Table 3. Various metabolic parameters including rates of tPA production were studied.

In runs 1 and 2, the effects of decreasing the perfusion rates were compared. The steady-state cell density was maintained at about 2.7×10^6 cells/ml. However, since these cells have a large cell volume, $3.75 \text{ cm}^3/10^9$ cells, the actual cell volume was about 10 ml/l, or nearly 1.0% of the reactor volume. At the lower perfusion rate, the steady-state glucose concentration dropped to 1.65 mg/ml from a level of 2.68 mg/ml at the higher perfusion rate and the lactate concentration almost doubled from 1.07 to 1.92 mg/ml. However, this did not affect the specific rates of glucose uptake, oxygen consumption, or tPA production. The uptake rate functions were calculated from the mass balance relationships discussed earlier. Assuming that at a steady-state glucose concentration of 0.15 mg/ml the conditions of $S \gg K_s$ still hold, maximum cell densities of 7.08×10^9 and 4.3×10^9 cells/l could be maintained at the higher and lower perfusion rates, respectively. This, of course, assumes that glucose depletion is the limiting process.

TABLE 3

**Perfusion Cytostat Production of Bowes Melanoma Cells and Tissue
Plasminogen Activator (tPA)**

Variable	Run 1	Run 2	Variable	Run 1	Run 2
pH	7.14	7.00	Glucose uptake (mg/h/ 10^9 cells)	25.3	24.8
Dissolved O_2 (atm)	0.11	0.11	Lactate production (mg/ h/10^9 cells)	18.4	19.1
Cell density					
ml/l (PCV)	9.7	10.2	[Glucose]$_{in}$ (mg/ml)	4.15	4.15
cells/l $\times$ 10^{-4}	2.6	2.7	[Glucose]$_{out}$ (mg/ml)	2.68	1.65
Total cells $\times$ 10^{-9}	39.0	40.8			
Doubling time (h)	51.2	51.2	[Lactate]$_{out}$ (mg/ml)	1.07	1.92
D = μ (h^{-1})	0.0135	0.0135	Oxygen uptake (cm^3 [stp]/h/10^9 cell)	2.7	2.9
D + P (h^{-1})	0.0448	0.0268	tPA production (mg/d/ 10^9 cells)	0.28	0.26
P (h^{-1})	0.0313	0.0133			

Optimization of perfusion culture systems requires detailed rate information for a number
of metabolic functions, including both specific nutrient depletion and accumulation of metabolic products that might inhibit growth and expression. Such studies are in progress in
our laboratory.

Perfusion suspension culture is not limited to the described reactor systems. First, the
filtration can be provided by other systems such as membrane filters. Agitation and aeration
can be provided by other reactor configurations such as the air-lift reactor. However, the
principle of a perfusion cytostat system provides significant advantages for efficient large-
scale suspension cell culture.

C. ANCHORAGE-DEPENDENT CELLS

Anchorage-dependent cells present additional challenges. As indicated earlier, a substrate
configuration that maximizes the surface area to volume ratio of the reactor is critical for
efficient scaleup. A number of approaches have been investigated, including the use of
hollow fibers[10-11] and microcarriers.[12]

Anchorage-dependent cells that grow attached to a substrate provide a ready configuration
for perfusion. A number of such systems have been reported, including perfused tissue
flasks, roller bottles, hollow fibers, and microcarriers. Generally, perfusion of such cultures
resulted in increased rates of proliferation, significant increases in cell yields, and multi-
layered cultures. Studies carried out in our laboratory with perfused hollow-fiber reactors
with different configurations also yielded high cell densities.[11] The basic design of these
reactors incorporated delivery of oxygen and carbon dioxide through the lumen of the fiber
and medium perfusion on the shell side of the fibers. A flat bed, hollow-fiber reactor system
was designed which addressed some of the limitations of the cartridge-type configuration
and provided an opportunity for scaleup.

A 30 $\times$ 30 cm fiber bed consisting of about 3000 fibers (O.D., 340 μm) would provide
about 1 m^2 of surface area for cell attachment and growth. A number of cells have been
grown in these reactors, including SV3T3, WI-38, HeLa, rhesus monkey kidney (MK-2),
baby hamster kidney (BHK), Vero, and primary fetal kidney. An important feature of the
perfused hollow-fiber reactor is the capacity to maintain a population of cells after growth
in a nonproliferating state for extended periods of time expressing a specific protein.

Unlike the shear stress to which cells on microcarriers are continuously exposed, the
fiber-attached cells are in a relatively undisturbed homeostatic environment under gentle

perfusion. The development of other such maintenance reactors will be described briefly below.

The results from these studies show that even anchorage-dependent cells, the proliferation of which is substrate area dependent, achieve significantly higher cell densities with higher viabilities under perfusion. Microcarrier systems are one of the most effective methods for large-scale culture of anchorage-dependent cells. There have been various reports on perfused microcarrier culture. The microcarrier culture system developed in our laboratory employed a modification of the perfusion culture reactor described earlier. This involved the addition of a settling chamber in the recycle line between the cell reactor and the satellite filter vessel (see Figure 4).

As cell-microcarrier suspension is pumped to the satellite filter vessel, it flows through the settling chamber, where the relatively dense slurry of cells and microcarriers is allowed to settle outside the agitated volume of the reactor vessel. Essentially cell-microcarrier-free medium continues in the recycle circuit to the satellite filter vessel. This operation could probably be maintained by directly removing a fraction of this spent medium without need for the filter. During the settling procedure, beads and cells at high density are brought into close contact, resulting in a bridging of cells from one carrier to another and the formation of aggregates of microcarriers. Under perfusion conditions the cell density increased significantly beyond that projected from the available bead surface area. Figure 5 shows a photomicrograph of the human foreskin fibroblast on the Bio-Carriers® microcarriers. Cell densities of 10^7 cells/ml were achieved, and the cell-bead aggregates were readily dissociated to yield free cells for growth reinitiation by a brief trypsin/EDTA treatment.

Using this procedure, a total of 3.4×10^{11} cells were produced in a 44-l reactor containing 400 g of Bio-Carriers® with a total surface area of 188 cm². As shown in Table 4, a fourfold increase in cell yield was achieved over that obtained in unperfused roller bottle production.

A number of different nontransformed human fibroblast cells, as well as transformed and genetically engineered animal cells and their secreted products, have been produced efficiently in these perfused microcarrier reactors. Growth vessels of up to 100 l have been developed, equipped with lung permeator oxygen delivery and computer control.

D. PRODUCTS FROM NONPROLIFERATING CELLS

Many biologically important molecules are produced by nonproliferating cells. This has been observed with a variety of cells, particularly those types of molecules that are not involved in regulating growth (e.g., autocrines). In practice, a population of cells is grown in a serum-supplemented nutrient medium, the medium removed, and a serum-free medium added for the production phase of the process. This conditioned medium is free of contaminating serum proteins and consequently simplifies the purification processes that must follow. The production of biomolecules by nonproliferating cells and the maintenance of such productive cells in a biochemically healthy and active state have been the focus of research efforts in our laboratory in recent years. Since the cell growth phase is partitioned from the maintenance and expression phase of the process, the medium component requirement for the latter is often quite simplified. A computer-controlled perfusion maintenance reactor system has been developed which can maintain both anchorage-dependent and anchorage-independent cells at high densities for extended periods of time.[13]

Cells are maintained in a semirigid matrix, within a reactor vessel chamber (Figure 6). Fresh medium is supplied through relatively low-porosity tubes dispersed throughout the reactor chamber, and expended medium is withdrawn through relatively high-porosity tubes which are positioned to minimize the flow path and maximize the medium perfusion. A silicone rubber tube lung permeator is provided to supply oxygen. Anchorage-dependent cells are grown on microcarriers to high density and inoculated as such into the reactor. This provides a ready matrix for the cell packing. Suspension-grown cells are mixed with

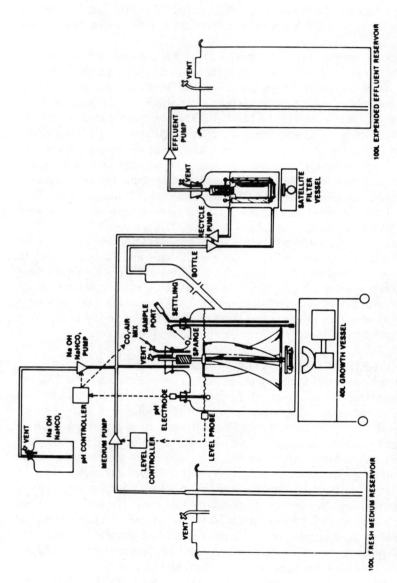

FIGURE 4.   Diagram of the perfusion microcarrier reactor system. It differs from that shown in Figure 2 by the addition of a settling bottle in the recycle circuit between growth and satellite filter vessels. This sequestering of the cells and microcarriers outside of the agitated volume of the growth vessel enhances formation of cell-bead aggregates and allows essentially cell-carrier-free medium to be pumped to the filter. (From Tolbert, W. R. and Feder, J., in *Annual Reports on Fermentation Processes*, Vol. 7, Tsao, G. T., Ed., Academic Press, Orlando, FL, 1983, 35. With permission.)

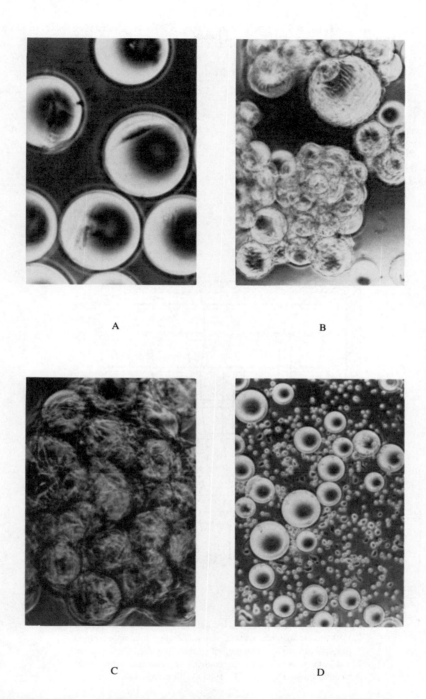

A

B

C

D

FIGURE 5.   Photomicrographs of human diploid foreskin fibroblasts on Bio-Carriers®. (A) 1 d after inoculation of the 4-l reactor. (Magnification × 240.) (B) After 15 d growth. (Magnification × 80.) (C) Higher magnification of aggregate. (Magnification × 160.) (D) After short trypsin treatment. (Magnification × 80.) (From Tolbert, W. R. and Feder, J., in *Annual Reports on Fermentation Processes*, Vol. 7, Tsao, G. T., Ed., Academic Press, Orlando, FL, 1983, 35. With permission.)

TABLE 4

Production of Human Foreskin Fibroblast Cells in Perfusion Microcarrier Reactors

	Surface area (m^2)	Cell density		Total cells	Yield (cells/l medium)[a]
		Cells/ml	Cells/cm^2		
Roller bottle	0.069	3.0×10^5	4.3×10^4	3.0×10^7	3.0×10^8
4-l reactor	21.6[b]	10.0×10^7	1.8×10^5	4.0×10^{10}	1.4×10^9
44-l reactor	188.0[b]	8.0×10^6	1.8×10^5	3.4×10^{11}	1.3×10^9

[a]   DMEM, 10% fetal bovine serum.
[b]   Bio-Carriers®.

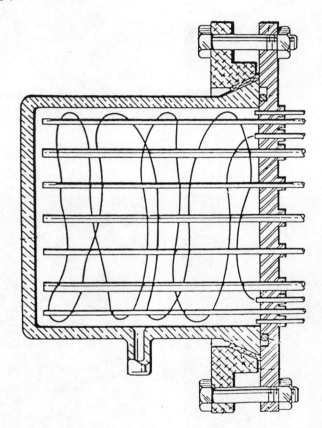

FIGURE 6.   Diagram of cross section of static maintenance reactor (SMR) used to maintain cells in nonproliferating state approaching 1/10 tissue density for harvesting of secreted products. (From Feder, J. and Tolbert, W. R., Eds., *Large-Scale Mammalian Cell Culture*, Academic Press, Orlando, FL, 1985. With permission.)

microcarriers or other material to provide a perfusion matrix. Cell densities of up to 10^8 cells/ml have been achieved in such a reactor. A 16.5-l perfused maintenance reactor has been developed with a total capacity in excess of 10^{12} cells. At these high cell densities, depletion of medium nutrients and accumulation of metabolic products occur rapidly. Nevertheless, successful maintenance of growth-arrested cells in a homeostatic environment has been achieved for extended periods of time.

Among the various cells studied were anchorage-independent hybridomas and anchorage-dependent fibroblasts producing monoclonal antibodies and tPA, respectively.

Although impressive results have been obtained in early studies using serum-free or low-serum media, little is known about the maintenance requirements of cells *in vitro*. Generally, metabolic rates are significantly lower for growth-arrested cells than for growing cells. However, little is known about whether there are specific maintenance factors that are required for continued cell viability and expression that are distinct from growth factors. This is an important area for research that can impact on the utility of such an approach for the production of important molecules.

IV. CONCLUSION AND SUMMARY

To summarize, the introduction of perfusion provides significant advantages for the large-scale culture of both anchorage-dependent and suspension-grown cells. High cell densities with good viability are achieved, cell yields per unit volume of medium are maximized, and generally lower concentrations of medium components (including serum) are required under steady-state conditions. The increase in cell density, particularly when coupled with continuous cytostat operation, and the reduction in serum provide a concentrated crude product stream having a significantly greater specific activity. This can simplify the downstream purification process and, thus, have an impact on the overall production efficiency and yield.

REFERENCES

1. **Morton, H. J.**, A survey of commercially available tissue culture media, *In Vitro*, 6, 89, 1970.
2. **Feder, J. and Tolbert, W. R.**, The large-scale cultivation of mammalian cells, *Sci. Am.*, 248, 36, 1983.
3. **Tolbert, W. R. and Feder, J.**, Large-scale cell culture technology, in *Annual Reports on Fermentation Processes*, Vol. 7, Tsao, G. T., Ed., Academic Press, Orlando, FL, 1983, 35.
4. **Reuveny, S., Velez, D., Miller, L., and Macmillan, J. D.**, Comparison of cell propagation methods for their effect on monoclonal antibody yield in fermentors, *J. Immunol. Methods*, 86, 61, 1986.
5. **Tovey, M. G.**, The cultivation of animal cells in continuous-flow culture, in *Animal Cell Biotechnology*, Vol. 1, Spier, R. E. and Griffiths, J. B., Eds., Academic Press, London, 1985, 195.
6. **Monod, J.**, La technique de culture continue theorie et applications, *Ann. Inst. Pasteur Paris*, 79, 390, 1950.
7. **Herbert, D., Elsworth, R., and Telling, R. C.**, The continuous culture of bacteria; a theoretical and experimental study, *J. Gen. Microbiol.*, 14, 601, 1956.
8. **de St. Groth, S. F.**, Automated production of monoclonal antibodies in a cytostat, *J. Immunol. Methods*, 57, 121, 1983.
9. **Himmelfarb, P., Thayer, P. S., and Martin, H. E.**, Spin filter culture: the propagation of mammalian cells in suspension, *Science*, 164, 555, 1969.
10. **Knazek, R. A., Gullino, P. M., Kohler, P. O., and Dedrick, R. L.**, Cell culture on artificial capillaries: an approach to tissue growth *in vitro*, *Science*, 178, 65, 1972.
11. **Ku, K., Kuo, J., Delente, J., Wildi, B. S., and Feder, J.**, Development of a hollow-fiber system for large-scale culture of mammalian cells, *Biotechnol. Bioeng.*, 23, 79, 1981.
12. **van Wezel, A. L.**, Growth of cell-strains and primary cells on microcarriers in homogeneous culture, *Nature (London)*, 216, 64, 1967.
13. **Tolbert, W. R., Feder, J., and Lewis, C., Jr.**, Static cell culture maintenance system, U.S. Patent 4,537,860, 1985.

Chapter 9

USE OF FUNGAL ELICITORS TO INCREASE THE YIELD OF SECONDARY PRODUCTS IN PLANT CELL CULTURES

Peter E. Brodelius

TABLE OF CONTENTS

I. INTRODUCTION

Higher plants are rich sources of natural products. In fact, over 80% of the ca. 30,000 known natural products are of plant origin. Many substances are isolated from plant tissue on a commercial basis (e.g., pharmaceuticals, flavors, and dyes). The gradual incursion into the natural environment by modern civilization has made the discovery and exploitation of new substances of such economically important commodities increasingly more difficult. The supply of certain raw plant material is today (or in the near future may be) limited. It has become increasingly important to find alternative resources.

The biotechnological application of plant cell cultures for the production of natural products during the last decade has received increasing attention. Plant cells may be grown in submerged cultures analogous to microbial cells. However, plant and microbial cells differ in many respects. A comparison of the two cell types is made in Table 1. Major differences are the size of the cells and the fermentation time. Furthermore, plant cell suspension cultures are normally heterogeneous and contain cell aggregates of various sizes, as illustrated in Figure 1. It is in principle possible to produce any compound found in a plant by cultivating cells originating from this species. This can be done in a fermentor, with the following major advantages:

1.   Production under controlled conditions
2.   Possible continuous production
3.   Constant supply (no seasonal variation)

However, the progress of utilizing plant cell cultures for the production of phytochemicals has been relatively slow. Some of the reasons for this slow development are

1.   Low productivity of the target substance
2.   Slow growth of plant cells
3.   Instability of plant cells in culture
4.   Difficulties in large-scale cultivation

Despite these problems some processes based on plant cell cultures are today in operation on a commercial or near-commercial scale. Some examples are listed in Table 2. However, before plant cell cultures may be generally employed for the production of phytochemicals, a better understanding of cell metabolism and metabolic regulation is required.

II. PLANT CELL CULTURE TECHNIQUES

Various applications of plant tissue cultures are shown schematically in Figure 2. Plant cells in culture have been demonstrated to express "totipotency", which means that any living nucleated parenchyma cell is capable of complete genetic expression, independent of its origin. Entire plants may, therefore, be regenerated from cultured plant cells. Root and/or shoot formation is induced by altering the growth medium, in particular the hormone concentrations.

The standard procedure to establish a plant cell suspension culture is as follows. A piece of plant tissue (explant) is surface sterilized and placed on a medium solidified with agar. After the explant has been exposed to the medium for some time, a callus, which consists of meristematic cells, starts to form if the medium composition is right. Newly formed callus tissue is transferred to fresh medium, and this procedure is repeated until a relatively uniform tissue is obtained. In the next step, a friable callus tissue is transferred to a liquid medium, and the suspended callus pieces are incubated on a gyratory shaker. When required, the suspension is transferred to fresh medium to sustain growth.

TABLE 1
A Comparison of Some Characteristics of Plant and Microbial Cells

	Microbial cell	Plant cell
Size	~2 μm³	>10⁵ μm³
Shearing	Insensitive	Relatively sensitive
Water content	~75%	~90%
Doubling time	<1 h	>20 h
Fermentation time	Days	Weeks
Oxygen consumption	1—2 vvm	0.2—0.4 vvm
Product	Extracellular	Intracellular
Media cost	~$6/m³	~$50/m³
Mutation	Possible	Requires haploid cells

Note: Dollar figures are 1984 U.S. dollars; vvm = volume/volume/min.

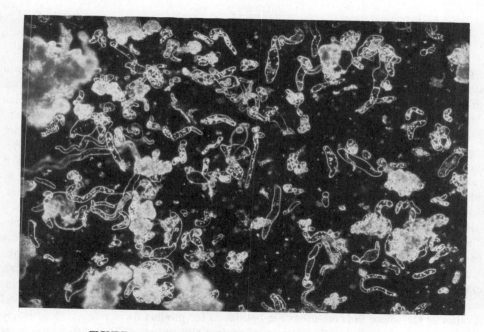

FIGURE 1. Suspension cultured cells of *Daucus carota* (carrot).

TABLE 2
Large-Scale Production of Phytochemicals with Plant Cell Suspension Cultures

Substance	Plant species	Company	Fermentor size (l)
Shikonin	*Lithospermum erythrorhizon*	Mitsui Petrochemical Ind.	750
Berberine	*Coptis japonica*	Mitsui Petrochemical Ind.	2 × 3,000
Yellow pigment	*Rubia akane*	Mitsui Petrochemical Ind.	?
Ginseng	*Panax ginseng*	Nitto Electric Ind.	20,000
Rosmarinic acid	*Coleus blumei*	A. Nattermann & Cie	450

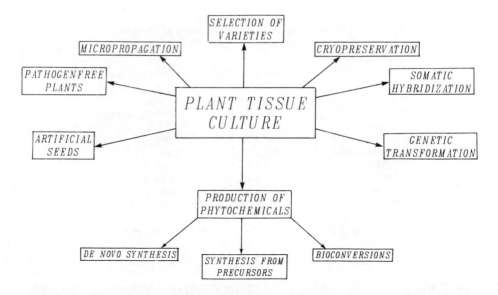

FIGURE 2.   Schematic diagram of some selected applications of plant tissue cultures.

The composition of a plant cell medium is generally somewhat more complex than a medium used for the cultivation of microbial cells. The ingredients of such a medium may be grouped into three classes, i.e., macro-, micro-, and organic nutrients. The macronutrients include nitrogen (in the form of ammonium and/or nitrate), phosphate, sulfate, calcium, and magnesium. The micronutrients include various trace elements (e.g., iron, copper, zinc, manganese, and cobalt). The organic nutrients include vitamins and growth regulators (phytohormones) in addition to a carbon source (most commonly sucrose). The major classes of phytohormones are the auxins (e.g., indole-3-acetic acid, naphthalene acetic acid, and [2,4-dichlorophenoxy]acetic acid) and the cytokinins (e.g., zeatin, kinetin, and benzyladenine), which are often used in combination with one another.

Figure 3 demonstrates various alternative routes from plant to product. The first steps preferably involve the establishment of a high-producing cell culture. Subsequently, biomass is generated by cultivating the cells on a large scale in a medium optimized for growth. When the product formation is linked to growth (growth-associated products), the biomass is extracted and the product isolated. For non-growth-associated products, a different approach may be employed with certain advantages. The biomass is transferred to a second cultivation stage where the conditions are optimized for product formation. This step may be carried out either in free suspension or in the immobilized state. Two different process designs are possible for immobilized cells, depending on whether the product is extra- or intracellular. In the former case a continuous process may be used and in the latter a semicontinuous process with intermittent release of product. For certain bioconversions of plant products, it may be appropriate to use more or less purified enzyme preparations. These can be extracted from plant tissue or plant cell cultures as well as from other sources (e.g., microbial cells).

III. BIOSYNTHETIC CAPACITY OF CULTIVATED PLANT CELLS

The number of natural products isolated from plant cell cultures is large and steadily increasing. From the concept of totipotency it should be possible to produce any compound present in the intact plant in the corresponding culture. Most substances are only produced

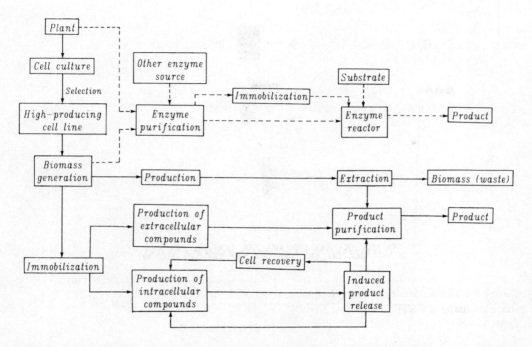

FIGURE 3. Flow-sheet diagram of various alternative routes to produce phytochemicals with plant cell culture systems.

TABLE 3
Some Selected Natural Products Formed in Plant Suspension Cultures with a Yield Close to or Higher than the Parent Plant

Plant species	Compound	% of DW[a] Culture	% of DW[a] Plant[b]	Culture (g/l)	Ref.
Morinda citrifolia	Anthraquinone	18	2.2	2.5	17
Vitis sp.	Anthocyanins	16	n.s.	0.83	18
Coleus blumei	Rosmarinic acid	15	3.0	3.6	19
Lithospermum erythrorhizon	Shikonin	12	1.5	1.4	20
Coptis japonica	Benzylisoquinoline alkaloids	11	5—10	1.7	21
	Berberine	10	2—4	1.2	22
Galium mollugo	Shikimic acid	10	n.s.	1.2	23
Nicotiana tabacum	Cinnamoyl putrescine	10	n.s.	1.0	24
Berberis stolonifera	Jatrorrhizine	7	n.s.	2.7	25
Panax ginseng	Saponins	2.2	0.5	n.s.	26
Peganum harmala	Harmane	2	n.s.	0.12	27
Eschscholtzia californica	Benzophenanthridines	1.7	n.s.	0.15	28
Catharanthus roseus	Ajmalicine	1.0	0.3	0.26	29
	Serpentine	0.8	0.5	0.16	29
Nicotiana tabacum	Ubiquinone-10	0.5	0.003	n.s.	30
Tripterygium wilfordii	Tripdiolide	0.05	0.001	0.00	31

[a] Dry weight.
[b] n.s. = not stated.

in minute quantities, if at all, in cultures. However, an increasing number of cell lines producing higher amounts of the target substance than the parent plant have been established through clonal selection during recent years. Some examples of high-producing cell lines are given in Table 3. Furthermore, a number of plant cell cultures producing compounds

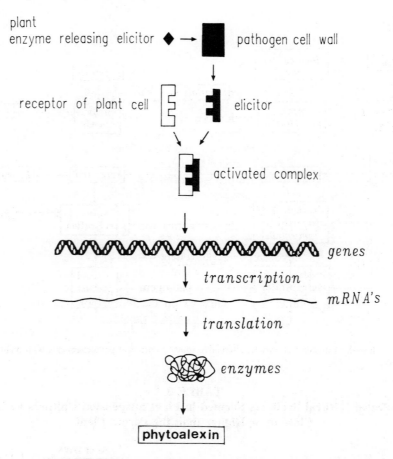

FIGURE 4.   Schematic representation of the principal steps involved in phytoalexin biosynthesis in plants after elicitation.

not found in the parent plant have been described. Consequently, cultivated plant cells may be used as a source of new natural products. The yield of the target substance may be improved by feeding appropriate precursors to the culture (see Figure 2). Finally, plant cell cultures may be employed for bioconversions of various types (e.g., hydroxylation or glycosylation; see Figure 2).

IV. INDUCED PRODUCT FORMATION IN PLANT CELL CULTURES

Methods to increase the productivity of plant cell cultures are most valuable for the further development of plant cell biotechnology. Cell metabolism may be influenced by external factors. Product formation can be induced by certain cultivation conditions; examples are nutritional, light, or hormonal stress. Under such stress conditions, certain enzyme systems are induced, leading to the biosynthesis of secondary products. In practice, this can be achieved in a two-stage process (see Figure 3). In the first stage the cultivation conditions are optimized for growth, and in the second stage product formation is induced.

A. ELICITOR-INDUCED PRODUCT FORMATION

During recent years microbial elicitors, which induce specific enzyme synthesis in whole plants, have been widely investigated. The principal events of microbial elicitation in plants are depicted in Figure 4. Upon infection of a plant by a pathogenic microorganism (e.g., a fungus), a plant enzyme starts to digest the microbial cell wall, releasing various components (elicitors). Next, the elicitor binds to a receptor on the plant cell and, through a mechanism yet to be fully evaluated, the transcription of certain genes coding for enzymes involved in the biosynthesis of phytoalexins is induced. Phytoalexins are defined as secondary products with antimicrobial activity which are present in the plant after infection.

It is generally believed that elicitors can also induce enzymes of secondary metabolism in plant cell cultures. *In vivo* protein-labeling experiments involving ^{35}S-methionine have, in fact, shown *de novo* synthesis or increased synthesis of several proteins within elicitor-treated cells.[1] Figure 5 shows a two-dimensional electrophoresis of extracts of noninduced and elicitor-induced suspension cells after ^{35}S-labeling. Table 4 lists some compounds which have been used as elicitors to induce product formation in plant cell cultures. From Table 4 it is clear that various compounds can act as elicitors. These are classified as biotic (natural) or abiotic elicitors.

In one study, a biotic elicitor (oligosaccharin) was characterized extensively.[2] It was demonstrated that the smallest active glucan from *Phytophthora megasperma* which induced phytoalexin production in soybean plants was a heptaglucoside. The structure of this elicitor is shown in Figure 6. Any change in the structure resulted in loss of elicitor activity, indicating a very specific interaction between plant and pathogen.

In most studies on the effects of elicitors on the biosynthetic capacity of plant cell cultures, the elicitor has been obtained from a pathogenic microorganism (most often a fungus). In practice, this may be achieved by autoclaving a culture of the pathogen and using the supernatant obtained after centrifugation as an elicitor preparation. The elicitor (glucan) preparation may also be partially purified before it is used.

Some examples of the induction of secondary metabolites in plant cell suspension cultures with microbial elicitors are listed in Table 5. In certain cases, the treatment with elicitor results in an increased synthesis of a product also present in nontreated culture. In these cases, the product should not be classified as a phytoalexin, even though it shows antimicrobial activity. Furthermore, it is clear from Table 5 that elicitors may also be prepared from nonpathogenic microorganisms (e.g., yeast). A wide variety of compounds are inducible in plant cell cultures by elicitation; some of their structures are shown in Figure 7. Several of the examples listed in Table 5 will be discussed in some detail.

1. Induction of Glyceollin Production in Cultures of *Glycine max* (Soybean)

In most of our studies we have used a glucan preparation isolated from yeast as the elicitor. This elicitor preparation was first tested on soybean cultures.[3] It is well known that the phytoalexin glyceollin is induced in whole plants and cell cultures after treatment with elicitors isolated from the natural pathogen *P. megasperma* var. *glycinea*.[4,5] This phytoalexin is synthesized via the phenylpropionic pathway, and a key enzyme between primary and secondary metabolisms is phenylalanine ammonia lyase (PAL). An increased PAL activity with a subsequent occurrence of glyceollin is observed after treatment of a soybean culture with the yeast elicitor. The response is highly dependent on the amount of elicitor added to the culture, as illustrated in Figure 8. No product whatsoever can be found in nontreated cells, as expected for a phytoalexin. A transient induction of PAL is observed, as illustrated in Figure 9, and the product is released into the cultivation medium.

From a biotechnological point of view, a repeated elicitation would be of great importance, since the same biomass could be employed more than once for production of an inducible product. If this product is released into the medium, as in the case of glyceollin, it would be beneficial to use immobilized cells. However, attempts to induce glyceollin biosynthesis in alginate-entrapped soybean cells have not been successful.[6] Alginate itself can induce PAL, but no glyceollin is formed. Furthermore, the action of the elicitor is inhibited by alginate, as summarized in Table 6. However, the number of methods available

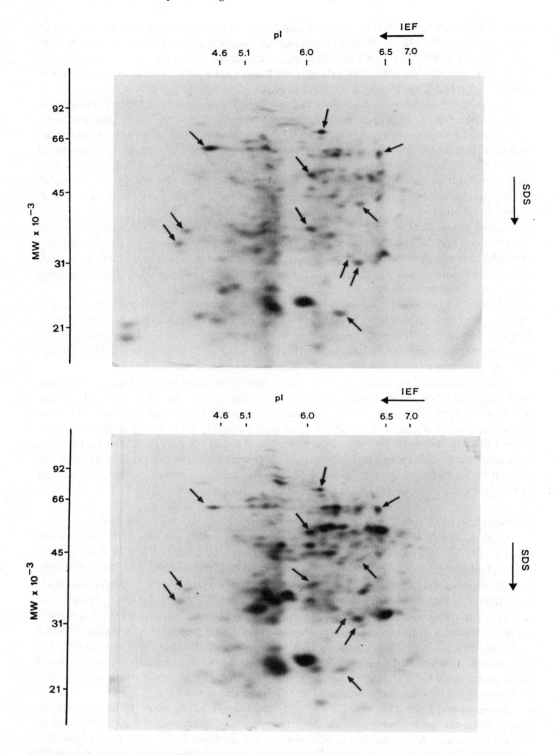

FIGURE 5.   Two-dimensional electrophoresis of ^{35}S-labeled proteins extracted from nontreated (A) and elicitor-treated (B) cells of *Thalictrum rugosum*. The first dimension was isoelectric focusing (separation according to charge) and the second, SDS-PAGE (separation according to size). The arrows indicate proteins present in both extracts at approximately the same levels. (From Gügler, K., Funk, C., and Brodelius, P., *Eur. J. Biochem.*, 170, 661, 1988. With permission.)

TABLE 4
Some Examples of Elicitors
Used to Induce Secondary
Metabolism in Plant Cell
Cultures

Biotic elicitors	Abiotic elicitors
Chitosan	Heavy-metal ions
Glucans	Organic solvents
Glycoproteins	Detergents
Conidia	Pesticides
Enzymes	

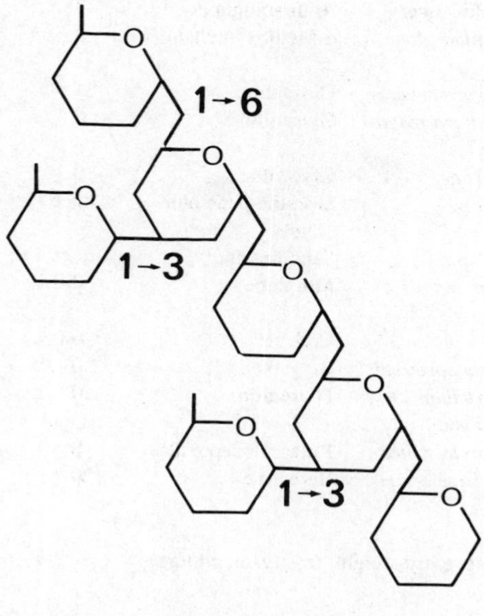

FIGURE 6.   Structure of the smallest active glucan (isolated from *Phytophthora megasperma*) inducing phytoalexin formation in soybean plants.

for the immobilization of plant cells is large,[7] and it is likely that some of these can be employed in combination with elicitation.

Apparently, the yeast elicitor and alginate induce different pathways involving PAL in soybean cells. In this respect, it may be mentioned that different products are induced in parsley cultures by different treatments. After treatment with a fungal elicitor preparation, the culture produces furanocoumarins, while after exposure to UV light, flavonoids are formed.[8,9] The initial three enzymes are the same for the two biosynthetic pathways.

2. Induction of Berberine Production in Cultures of *Thalictrum rugosum*

The yeast elicitor can also be used to induce an increased synthesis of the benzylisoquinoline alkaloid berberine in suspension cultures of *Thalictrum rugosum* (Table 5).[3] The cells produce berberine during growth, but when the carbon source is consumed, both growth and alkaloid synthesis cease. However, if the cells at this stage are treated with the elicitor, more berberine is synthesized. The response of the culture is dependent on elicitor concentration during treatment, as illustrated in Figure 10. At relatively high concentrations, a lower production of berberine is observed due to cell death. Consequently, it is important to determine the lowest elicitor concentration resulting in maximum yield.

TABLE 5
Some Examples of Secondary Products That Have Been Induced in Plant Cell Cultures by Treatment with Microbial Elicitors

Plant species	Microbial species[a]	Product induced	Concentration[b] Before	Concentration[b] After	Incubation time (h)	Ref.
Bidens pilosa	*Pythium aphanider-matum*	Phenylheptatriene	0	3.2 (a)	48	32
Canavalia ensiformis	*Pithomyces chartarm*	Medicarpin	0	0.43 (a)	36	33
Cephalotaxus bar-ringtinia	*Verticillium dahliae*	Harringtonine alka-loids	0.01	0.51 (b)	120	13
Cicer arietinum	*Saccharomyces cere-visiae*	Medicarpin	0.3	18 (a)	24	34
Cinchona ledgeriana	*Aspergillus niger*	Anthraquinones	3	15 (b)	600	15
Daucus carota	*Chaetomium glo-bosum*	6-Methoxymellein	0	1 (c)	48	35
Dioscorea deltoides	*Rhizopus arrhizus*	Diosgenin	25	72 (b)	72	16
Glycine max	*Phytophthora mega-sperma*	Glyceollin	0	0.05 (b)	100	5
	S. cerevisiae	Glyceollin	0	0.2 (b)	10	3
Gossypium arboreum	*V. dahliae*	Sesquiterpene alde-hyde	trace	96 (b)	120	13
Papaver somniferum	*Botrytis* sp.	Sanguinarine	trace	6.6 (b)	24	14
	Fusarium monili-forme	Morphine	0.07	1.40 (b)	n.s.[c]	13
		Codeine	0.08	1.44 (b)	n.s.	13
Petroselium hortense	*Alternaris carthami*	Bergapten	0	1.6 (c)	48	36
Phaeolus vulgaris	*Colletotrichum linde-muthianum*	Phaseollin	0	170 (a)	48	37
Ruta graveolens	*Rhodotorula rubra*	Rutacridonepoxides	0	0.23 (b)	72	38
Thalictrum rugosum	*S. cerevisiae*	Berberine	20	50 (b)	96	3

[a]   Source of elicitor.
[b]   (a) μg/g fresh weight; (b) mg/g dry weight; (c) μg/ml culture.
[c]   n.s. = not stated.

The response of the culture to elicitor treatment is also dependent on the age of the cells. When the elicitor is added to a rapidly growing suspension culture (early exponential phase = day 2 to 4 of a batch culture) of *T. rugosum*, alkaloid synthesis appears to be unaffected, as shown in Figure 11. However, if the elicitor is added to the culture during late exponential or stationary phase (day 6 and thereafter), a pronounced increase in berberine synthesis is obtained.

For production purposes, it is of great importance to establish the earliest possible addition of elicitor resulting in maximum product yield. In the case of berberine production by a batch culture of *T. rugosum*, this result is achieved by exposing the cells to elicitor on day 6 (see Figure 11). Under these conditions the fermentation time is not prolonged, as shown in Figure 12. The induced cells can be harvested 5 d after addition of elicitor. Furthermore, cell growth is somewhat restricted by the elicitor treatment (Figure 12).

All of the enzymes involved in the biosynthesis of berberine from the precursor nor-laudanosoline have been isolated and partly characterized.[10] Norlaudanosoline is a common precursor of isoquinoline alkaloids of various types (e.g., morphine, sanguinarine, and papaverine). This common precursor is synthesized from two molecules of tyrosine,[10] and the immediate precursors to norlaudanosoline are dopamine and 3,4-dihydroxyphenylacet-aldehyde. However, the intermediates have not been fully characterized. It is likely that elicitor treatment of isoquinoline-alkaloid-producing cell cultures will result in the induction

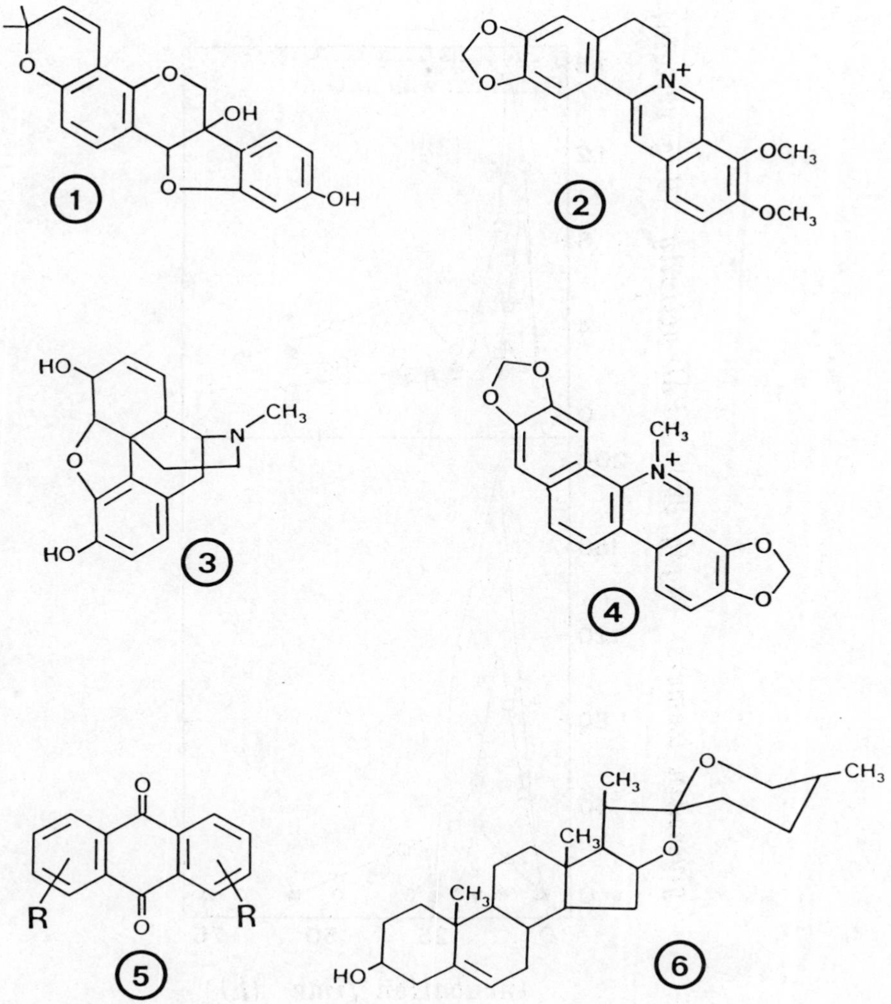

FIGURE 7. Structures of some compounds produced in plant cell cultures after elicitation — (1) glyceollin I; (2) berberine; (3) morphine; (4) sanguinarine; (5) anthraquinone; (6) diosgenin.

of one or more enzymes involved in norlaudanosoline biosynthesis. Consequently, elicitation may be utilized to determine the enzymatic steps involved in norlaudanosoline biosynthesis.

Treatment of isoquinoline-alkaloid-synthesizing plant cell cultures (e.g., *T. rugosum* and *Eschscholtzia californica*) with yeast elicitor results in the induction of tyrosine decarboxylase (TDC).[1,11] A good correlation between induced TDC activity and enhanced berberine biosynthesis has been established for *Thalictrum* cell cultures.[1] The responses of the two cell cultures differ somewhat, as illustrated in Figure 13. Cells of *E. californica* appear to require considerably less elicitor for maximum TDC activity (Figure 13A). Furthermore, this culture responds to elicitor treatment more rapidly (Figure 13B). These observations indicate that TDC may play an important role in norlaudanosoline biosynthesis.

TDC has been purified from the two cell cultures and partially characterized.[11,12] The enzyme has a molecular weight of 112,000 Da and contains two identical subunits. In addition to tyrosine, it decarboxylates DOPA. The K_m values are 0.25 and 0.27 mM for tyrosine and DOPA, respectively, at optimum pH (pH = 8.4). Phenylalanine and tryptophan are not substrates of the enzyme. The coenzyme pyridoxal-5-phosphate is important for enzyme activity.

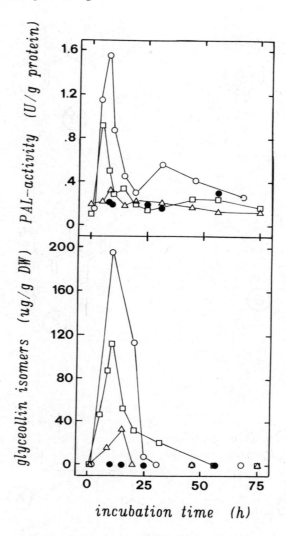

FIGURE 8.   PAL activity and glyceollin isomers formation
in cell suspension cultures of *Glycine max* after addition of
various concentrations of yeast elicitor. (●) 0, (△) 2, (□)
10, (○) 20 μg elicitor per mg dry weight.[3]

3. Induction of Isoquinoline Alkaloid Production in Cultures of *Papaver somniferum*

Morphine, an important pharmaceutical compound, is obtained from *Papaver somniferum* plants (poppy). Cell cultures of this plant only produce trace amounts of morphinan
alkaloids. However, attempts to increase the productivity of suspension-cultured cells by
elicitation have been successful.

The yields of codeine and morphine in suspension cultures of *P. somniferum* could be
increased 8- to 9-fold and 15- to 18-fold, respectively, by treating the cells with elicitor
preparations, as summarized in Table 7.[13] Two different elicitor preparations were used,
i.e., autoclaved *Verticillium dahliae* conidia or autoclaved *Fusarium moniliforme* var. *sub-
glutinans*. The yields of codeine and morphine were essentially the same for both elicitor
preparations.

In another study, the isoquinoline alkaloid sanguinarine could be induced by fungal
elicitors in cell suspension cultures of *P. somniferum*.[14] Homogenized and autoclaved mycelia

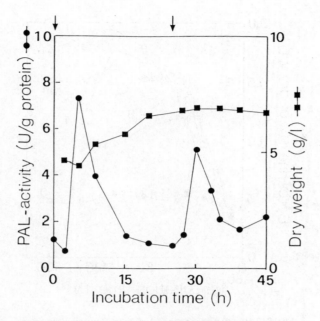

FIGURE 9.   PAL activity in a cell suspension culture of *Glycine max* after repeated treatment with yeast elicitor. The arrows indicate the addition of 20 μg elicitor per mg dry weight. (From Funk, C., Guegler, K., and Brodelius, P., *Phytochemistry,* 26, 401, 1987. With permission.)

TABLE 6
Phenylalanine Ammonia Lyase Activity and Glyceollin Production in Suspension Cells of *Glycine max* after Various Additions[6]

Addition[a]	PAL activity (mU/mg protein)	Glyceollin content (μg/g dry weight)
None	0.25	0
Elicitor	1.25	170
Alginate	1.35	0
Elicitor + alginate	1.20	0

[a]   Amounts added: 20 μg elicitor/mg dry weight and 0.5% (w/v) alginate.

from different microorganisms were tested as elicitors, as summarized in Table 8. It is interesting to note that mycelia from *Phytium* could not elicit in the plant cells, while mycelia from the other two fungi induced a considerable increase in sanguinarine production (up to 29-fold).

Sanguinarine is normally stored within the cells, but after elicitation a part of the alkaloid produced is found in the medium. The fraction found in the medium (up to 63% of total alkaloid) is highly dependent on which elicitor is being used and on the amount of alkaloid produced.[14] The alkaloid is most likely actively transported out into the medium, since no cell damage could be established.

The morphinan alkaloids and sanguinarine are synthesized via norlaudanosoline in cells of *P. somniferum*. It would be of great interest to determine if TDC is induced by the various elicitors (see Tables 7 and 8). A correlation between induced TDC activity and alkaloid

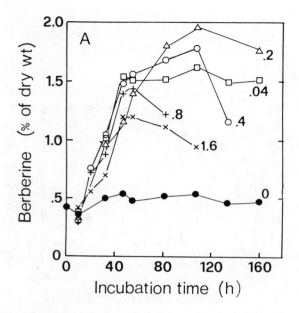

FIGURE 10.   Berberine concentration in *Thalictrum rugosum* suspension cells as a function of incubation time after addition of various concentrations of yeast elicitor as indicated (mg glucan per gram wet weight of cells). (From Funk, C., Guegler, K., and Brodelius, P., *Phytochemistry, 26,* 401, 1987. With permission.)

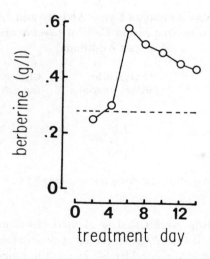

FIGURE 11.   Final berberine concentration in cultures of *Thalictrum rugosum* after elicitor treatment at various stages throughout a batch fermentation. The broken line represents the concentration in nontreated cells.

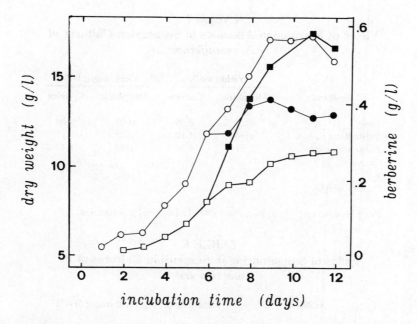

FIGURE 12. Dry weight (○ ●) and berberine (□ ■) content of *Thalictrum rugosum* suspension cultures. Open symbols, nontreated cells; solid symbols, addition of elicitor on day 6.

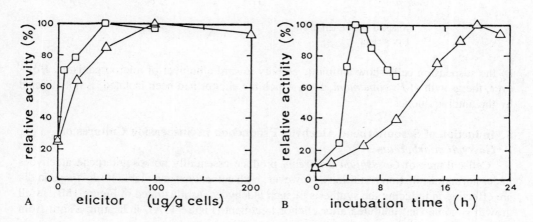

FIGURE 13. Relative tyrosine decarboxylase activity in cells of *Thalictrum rugosum* (△) and *Eschscholtzia californica* (□). (A) Enzyme activity as a function of elicitor concentration; (B) enzyme activity as a function of incubation time.

productivity would support the assumption that TDC is involved in the regulation of isoquinoline alkaloid biosynthesis.

4. Induction of Anthraquinone Production in Suspension Cultures of *Cinchona ledgeriana*

Cell suspension cultures of *Cinchona ledgeriana* produce low amounts of anthraquinones (3 to 4 mg/g dry weight of cells). After treatment of the cells with autoclaved mycelia from *Phytophthora cinnamoni*, anthraquinone production is increased (7 to 8 mg/g dry weight of cells), as illustrated in Figure 14.[15] Cell growth is not affected to any great extent.

Anthraquinones are not recognized as phytoalexins, but the anthraquinones produced

TABLE 7
Yields of Morphinan Alkaloids in Suspension Cultures of
P. somniferum

Addition	Yield (mg/l)		Yield (mg/g DW[a])	
	Morphine	Codeine	Morphine	Codeine
None	0.25	0.46	0.07	0.08
Verticillium dahliae	4.64	4.10	1.25	1.11
Fusarium monili-forme	3.91	4.04	1.40	1.44

[a] Dry weight.

From Heinstein, P. F., *J. Nat. Prod.*, 48, 1, 1985. With permission.

TABLE 8
Yields of Sanguinarine in Suspension Cultures of *P.*
***somniferum*[a]**

Addition	Yield (mg/l)	Yield (mg/g DW[b])
None	0.6	0.10
Pythium aphanidermatum	0.6	0.08
Rhodotorula rubra	5.1	0.65
Botrytis sp.	17.2	2.64

[a] Adapted from Eilert et al.[14]
[b] Dry weight.

by the suspension cells show antibiotic activity toward a number of microorganisms. However, the growth of *P. cinnamoni*, from which the elicitor had been isolated, is not affected by the anthraquinones.

5. Induction of Sesquiterpene Aldehyde Production in Suspension Cultures of *Gossypium arboreum*

Cell cultures of *Gossypium arboreum* produce essentially no sesquiterpene aldehydes under normal incubation conditions. However, both autoclaved and viable *V. dahliae* conidia are effective in inducing the synthesis of such aldehydes, as illustrated in Figure 15B.[13] Cell growth is somewhat inhibited after elicitor treatment (Figure 15A), indicating a shift from primary to secondary metabolism.

6. Induction of Diosgenin Production in Cell Suspension Cultures of *Dioscorea deltoidea*

Diosgenin is a very important starting substance for the synthesis of various steroidal drugs. Attempts to produce the compound in cell cultures have been partially successful. However, for a commercial process the yield must be improved. This may be achieved by the addition of fungal elicitors to cell suspension cultures of *Dioscorea deltoidea* in stationary growth phase.[16] Various autoclaved mycelia have been used, and the responses of the cultures are summarized in Table 9. From Table 9 it is clear that plant cell growth is generally inhibited by addition of mycelia to the culture. Furthermore, some of the fungal preparations do not induce diosgenin synthesis, and the yield is lower than for nontreated cells. The

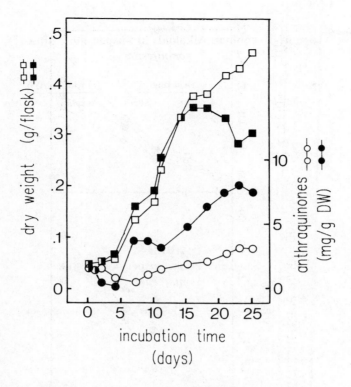

FIGURE 14. Growth and anthraquinone production of *Cinchona led-geriana* cell suspension cultures. Open symbols, nontreated cells; solid symbols, cells treated with autoclaved mycelia from *Phytophthora cin-namoni*. (From Wijnsma, R., Go, J. T. K., van Weerden, I. N., Harkes, P. A. A., Werpoorte, R., and Baerheim-Svendsen, A., *Plant Cell Rep.*, 4, 241, 1985. With permission.)

highest product formation (172% of control) was obtained with mycelia from *Rhizopus arrhizus*. On a dry weight basis, an almost threefold increase in diosgenin production is achieved within 72 h (see Table 5).

V. CONCLUDING REMARKS

Secondary metabolism of plant cells is highly regulated, but the mechanism of this metabolic regulation is not fully understood. Treatment of plant cell cultures with elicitors leading to deregulation of a secondary pathway offers a powerful tool to obtain more information on the regulation of secondary metabolism in plants. Various elicitation studies have shown that metabolic regulation is maintained on a transcriptional level.

As exemplified above, the yield of secondary products in plant cell suspension cultures may be increased considerably by the addition of fungal elicitors. Optimization of product yield is highly empirical. Parameters of importance are the age of the cells, elicitor concentration, and incubation time after elicitation. Plant cells appear to be most susceptible to elicitor treatment in late exponential or early stationary growth phase. Furthermore, the origin of the elicitor appears to be very important. Pathogenic fungi seem to be the best sources for biotic elicitors. However, in our own studies we have used a carbohydrate preparation from a nonpathogenic fungus (i.e., yeast) as the elicitor. An advantage of using

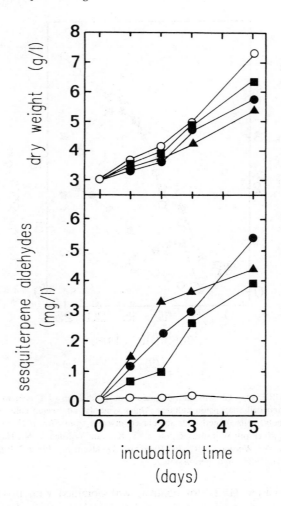

FIGURE 15.   Effect of viable and autoclaved *Verticillium dahliae* conidia on growth (A) and sesquiterpene aldehyde formation (B) of *Gossypium arboreum* cell suspension cultures. (○) Nontreated cells; (■) cells treated with viable conidia (10^4/20 ml); (●) autoclaved conidia (10^4/20 ml); (▲) autoclaved conidia (2×10^5/20 ml). (From Heinstein, P. F., *J. Nat. Prod.*, 48, 1, 1985. With permission.)

this type of elicitor is the fact that yeast can be found on the GRAS (generally regarded as safe) list. Under optimum conditions, an increased product yield may be obtained without increasing the fermentation time.

Further studies on the use of elicitors to enhance the productivity of plant cell cultures are of great importance. The broader biotechnological utilization of plant cell cultures for the production of phytochemicals may be dependent on marginal increases in productivity. Elicitation is one approach to the achievement of the productivity increase required.

TABLE 9
Dry Weight and Levels of Diosgenin in Cell
Suspension Cultures of *Dioscorea deltoidea* after
Addition of Various Autoclaved Fungal Mycelia

Fungal species	Dry weight (g/l)	Diosgenin	
		(mg/l)	(%)
Control	5.4	134	100
Sclerotium rolfsii	4.4	82	61
Fusarium oxysporum	3.0	85	63
Macrophomina phaseolina	4.7	118	88
Rhizoctonia solani	5.4	131	98
Aspergillus niger	3.3	147	110
Giberella fujikuroi	3.4	151	113
Fusarium oxysporum melonis	3.6	155	116
Selerotinia selerotiorum	2.7	177	132
Rhizopus arrhizus	3.2	230	172

From Rokem, J. S., Schwarzberg, J., and Goldberg, I., *Plant Cell Rep.*, 3, 159, 1984. With permission.

REFERENCES

1. **Gügler, K., Funk, C., and Brodelius, P.**, Elicitor-induced tyrosine decarboxylase in berberine synthesizing suspension cultures of *Thalictrum rugosum, Eur. J. Biochem.*, 170, 661, 1988.
2. **Sharp, J. E., Valent, B., and Albersheim, P.**, Purification and partial characterization of a β-glucan fragment that elicits phytoalexin accumulation in soybean, *J. Biol. Chem.*, 259, 11312, 1984.
3. **Funk, C., Guegler, K., and Brodelius, P.**, Increased secondary product formation in plant cell suspension cultures after treatment with a yeast carbohydrate (elicitor), *Phytochemistry*, 26, 401, 1987.
4. **Grisebach, H. and Ebel, J.**, Phytoalexins, chemical defense substances of higher plants?, *Angew. Chem.*, 17, 635, 1978.
5. **Ebel, J., Schmidt, W. E., and Loyal, R.**, Photoalexin synthesis in soybean cells: elicitor induction of phenylalanine ammonia lyase and chalcone synthase mRNAs and correlation with phytoalexin accumulation, *Arch. Biochem. Biophys.*, 232, 240, 1984.
6. **Funk, C., Mosbach, K., and Brodelius, P.**, unpublished data, 1986.
7. **Brodelius, P.**, Immobilized plant cells, in *Enzymes and Immobilized Cells in Biotechnology*, Laskin, A. I., Ed., Benjamin/Cummings, Menlo Park, CA, 1985, 109.
8. **Hahlbrock, K., Knobloch, K.-H., Kruetzaler, F., Potts, J. R. M., and Wellmann, E.**, Coordinated induction and subsequent activity changes of two groups of metabolically interrelated enzymes, *Eur. J. Biochem.*, 61, 199, 1976.
9. **Hahlbrock, K., Lamb, C. J., Purwin, C., Ebel, J., Fautz, E., and Schafer, E.**, Rapid response of suspension-cultured parsley cells to the elicitors from *Phytophthora megasperma* var. *sojae, Plant Physiol.*, 67, 768, 1981.
10. **Zenk, M. H., Rueffer, M., Amann, M., and Deus-Neumann, B.**, Benzylisoquinoline biosynthesis by cultivated plant cells and isolated enzymes, *J. Nat. Prod.*, 48, 725, 1985.
11. **Marques, I. A. and Brodelius, P. E.**, Elicitor-induced L-tyrosine decarboxylase from plant cell suspension cultures. I. Induction and purification, *Plant Physiol.*, 88, 46, 1988.
12. **Marques, I. A. and Brodelius, P. E.**, Elicitor-induced L-tyrosine decarboxylase from plant cell suspension cultures. II. Partial characterization, *Plant Physiol.*, 88, 52, 1988.
13. **Heinstein, P. F.**, Future approaches to the formation of secondary natural products in plant cell suspension cultures, *J. Nat. Prod.*, 48, 1, 1985.
14. **Eilert, U., Kurz, W. G. W., and Constabel, F.**, Stimulation of sanguinarine accumulation in *Papaver somniferum* cell cultures by fungal elicitors, *J. Plant Physiol.*, 119, 65, 1985.
15. **Wijnsma, R., Go, J. T. K. A., van Weerden, I. N., Harkes, P. A. A., Werpoorte, R., and Baerheim-Svendsen, A.**, Anthraquinones as phytoalexins in cell and tissue cultures of *Cinchona* spec., *Plant Cell Rep.*, 4, 241, 1985.

16. **Rokem, J. S., Schwarzberg, J., and Goldberg, I.,** Autoclaved fungal mycelia increase diosgenin production in cell suspension cultures of *Dioscorea deltoidea, Plant Cell Rep.,* 3, 159, 1984.

17. **Zenk, M. H., El-Shagi, H., and Schulte, U.,** Anthraquinone production by cell suspension cultures of *Morinda citrifolia, Planta Med. Suppl.,* p. 78, 1975.

18. **Yamakawa, T., Kato, S., Ishida, K., Kodama, T., and Minoda, Y.,** Production of anthocyanins by *Vitis* cells in suspension cultures, *Agric. Biol. Chem.,* 47, 2185, 1983.

19. **Zenk, M. H., El-Shagi, H., and Ulbrich, B.,** Production of rosmarinic acid by cell suspension cultures of *Coleus blumei, Naturwissenschaften,* 64, 585, 1977.

20. **Fujita, Y., Hara, Y., Suga, C., and Morimoto, T.,** Production of shikonin derivatives by cell suspension cultures of *Lithospermum erythrorhizon.* II. A new medium for the production of shikonin derivatives, *Plant Cell Rep.,* 1, 61, 1981.

21. **Fukui, H., Nakagawa, K., Tsuda, S., and Tabata, M.,** Production of isoquinoline alkaloids by cell suspension cultures of *Coptis japonica,* in *Proc. 5th Int. Congr. Plant Tissue and Cell Culture,* Fujita, A., Ed., Japanese Association for Plant Tissue Culture, Tokyo, 1982, 313.

22. **Sato, F., Endo, T., Hashimoto, T., and Yamada, Y.,** Production of berberine in cultured *Coptis japonica* cells, in *Proc. 5th Int. Congr. Plant Tissue and Cell Culture,* Fujita, A., Ed., Japanese Association for Plant Tissue Culture, Tokyo, 1982, 319.

23. **Amrhein, N., Deus, B., Gehrke, P., and Steinrocken, H.-C.,** The site of the inhibition of the shikimate pathway by glyphosate. II. Interference of glyphosate with chorismic acid formation *in vivo* and *in vitro, Plant Physiol.,* 66, 830, 1980.

24. **Berlin, J., Knobloch, K.-H., Holfle, G., and Witte, L.,** Biochemical characterization of two tobacco cell lines with different levels of cinnamoyl putrescines, *J. Nat. Prod.,* 45, 83, 1982.

25. **Hinz, H. and Zenk, M. H.,** Production of protoberberine alkaloids by cell suspension cultures of *Berberis* species, *Naturwissenschaften,* 67, 620, 1981.

26. **Furuya, T., Yoshikawa, T., Orihara, Y., and Oda, H.,** Saponin production in cell suspension cultures of *Panax ginseng, Planta Med.,* 48, 83, 1983.

27. **Sasse, F., Heckenberg, U., and Berlin, J.,** Accumulation of β-carboline alkaloids and serotonin by cell cultures of *Peganum harmala.* II. Interrelationship between accumulation of serotonin and activities of related enzymes, *Z. Pflanzenphysiol.,* 105, 315, 1982.

28. **Berlin, J., Forsche, E., Wray, V., Hammer, J., and Hosel, W.,** Formation of benzophenanthridine alkaloids by suspension cultures of *Eschscholtzia californica, Z. Naturforsch.,* 38C, 346, 1983.

29. **Zenk, M. H., El-Shagi, H., Arens, H., Stockigt, J., Weiler, E. W., and Deus, B.,** Formation of the indole alkaloids serpentine and ajmalicine in cell suspension cultures of *Cathatanthus roseus,* in *Plant Tissue Culture and Its Biotechnological Applications,* Barz, W., Reinhard, E., and Zenk, M. H., Eds., Springer-Verlag, Berlin, 1977, 27.

30. **Matsumato, I., Kanno, N., Ikeda, T., Obi, Y., Kisaki, T., and Noguchi, M.,** Selection of cultured tobacco cell strains producing high levels of ubiquinone-10 by a cell cloning technique, *Agric. Biol. Chem.,* 45, 281, 1981.

31. **Kutney, J. P., Choi, L. S. L., Duffin, R., Hewitt, G., Kawamura, N., Kurihara, T., Salisbury, P., Sindelar, R., Stuart, K. L., Townsley, P. M., Chalmers, W. T., Webster, F., and Jacoli, G. G.,** Cultivation of *Tripterygium wilfordii* tissue cultures for the production of the cytotoxic diterpene tripdiolide, *Plant Med.,* 48, 158, 1983.

32. **Dicosmo, F., Norton, R. A., and Towers, G. H. N.,** Fungal culture-filtrate elicits aromatic polyacetylenes in plant tissue culture, *Naturwissenschaften,* 69, 550, 1982.

33. **Gustine, D. L., Sherwood, R. T., and Vance, C. P.,** Regulation of phytoalexin synthesis in jackbean callus cultures. Stimulation of phenylalanine ammonia-lyase and *O*-methyltransferase, *Plant Physiol.,* 61, 226, 1978.

34. **Kessman, H. and Barz, W.,** Accumulation of isoflavones and pterocarpan phytoalexins in cell suspension cultures of different cultivars of chickpea *(Cicer ariethinum), Plant Cell Rep.,* 6, 55, 1987.

35. **Kurosaki, F. and Nishi, A.,** Isolation and antimicrobial activity of the phytoalexin 6-methoxymellein from cultured carrot cells, *Phytochemistry,* 22, 669, 1983.

36. **Tietjen, K. G., Hunkler, D., and Matern, U.,** Differential response of cultured parsley cells to elicitors from two non-pathogenic strains of fungi. I. Identification of induced products as coumarin derivatives, *Eur. J. Biochem.,* 131, 401, 1983.

37. **Robbins, M. P., Bolwell, G. P., and Dixon, R. A.,** Metabolic changes in elicitor-treated bean cells. Selectivity of enzyme induction in relation to phytoalexin accumulation, *Eur. J. Biochem.,* 148, 563, 1985.

38. **Eilert, U., Engel, B., Reinhard, E., and Wolters, B.,** Acridone epoxides in cell cultures of *Ruta* species, *Phytochemistry,* 22, 14, 1982.

Chapter 10

USING A STRUCTURED KINETIC MODEL FOR ANALYZING INSTABILITY IN RECOMBINANT BACTERIAL CULTURES

William E. Bentley and Dhinakar S. Kompala

TABLE OF CONTENTS

I. INTRODUCTION

Structural and segregational instabilities have been identified which can result in the dramatic loss of a recombinant strain from continuous cultures.[1,2] In addition, the recombinant strain can be eliminated from continuous cultures due to a growth rate differential between the recombinant (plasmid-bearing) strain and the plasmid-free host. Several experimental studies have established that as plasmid copy number and foreign protein overexpression increase, the growth rate of the recombinant cell decreases.[3-7] It is generally recognized that the replication of high copy number plasmids and the overproduction of plasmid-encoded proteins represent an additional metabolic burden on the normal chromosome-directed metabolism of the bacterial cell, thereby reducing the cell's growth rate.

Segregational instability is described as the generation of a small fraction of plasmid-free cells upon cell division of the recombinant strain due to unequal plasmid partitioning between the daughter cells. Coupling these two effects (plasmid segregation and growth rate differential) results in the loss of plasmid-containing cells and their replacement by plasmid-free cells in continuous fermentations.

In the present work, we have extended the simple structured kinetic model, capable of predicting the growth rate of the bacterial cell mass as a dynamic function of the additional metabolic burdens,[8] to include the kinetics of segregational instability. We associate segregational and structural instabilities with plasmid instability, since they refer to actions directly involving the plasmid. Incorporating the growth rate differential in our mathematical characterization necessitates the use of the term "culture instability" when referring to the presence or absence of the desired recombinant strain in continuous or extended batch cultures. We demonstrate the model by predicting experimentally observed culture instability phenomena.

II. MODEL OVERVIEW

The structured kinetic model which directly calculates the instantaneous specific growth rate has previously been discussed in detail[8] and is briefly reviewed here. A key feature of the modeling framework is the uncommon representation of the state of a microorganism in terms of the fractional mass levels of its intracellular constituents.

Our lumped metabolic model of recombinant cells includes eight major intracellular constituent pools. These pools provide our model with sufficient metabolic detail without becoming numerically burdensome. Included are the following: protein, P; foreign protein, P_f; chromosomal DNA, G; plasmid DNA, G_f; ribosomes, R; lipids, L; nucleotides, N; and amino acids, A. The level of each of the internal constituent pools is expressed in our model equations as mass fraction or gram constituent per gram dry cell mass. Table 1 lists the specific net synthesis rates for each constituent pool, whereas Table 2 contains differential mass balance equations which stoichiometrically link appropriate pools and consequently describe the overall dynamic response of each. The biochemical bases for the synthesis rate expressions and the differential mass balances are elucidated in the previous work.

The summation of all eight equations in Table 2 yields

$$\sum_{i=1}^{8} \frac{dC_i}{dt} = \sum_{i=1}^{8} \sum_{j=1}^{8} r_{ij} - \mu \left[\sum_{i=1}^{8} C_i \right] \tag{1}$$

where C_i denotes each constituent pool: A, N, P, P_f, G, G_f, L, and R; r_{ij} corresponds to the synthesis and depletion terms shown in Table 2. This mass balance equation, which is subsequently used for calculating the instantaneous specific growth rate, was first derived in general form by Fredrickson.[9] Since the entire cell mass is divided into lumped constituent

TABLE 1
Synthesis Rate Expressions for Constituent Pools

$$\left[\frac{dA}{dt}\right]_s = k_1 \left[\frac{K_A}{K_A + A}\right]\left[\frac{S}{K_{AS} + S}\right]\left[\frac{A}{K_{2A} + A}\right] \tag{1}$$

$$\left[\frac{dN}{dt}\right]_s = k_2 \left[\frac{K_N}{K_N + N}\right]\left[\frac{A}{K_{NA} + A}\right]\left[\frac{S}{K_{NS} + S}\right] \tag{2}$$

$$\left[\frac{dP}{dt}\right]_s = \mu_1 \left[\frac{A}{K_{PA} + A}\right]RG - K_{TP}P \tag{3}$$

$$\left[\frac{dP_f}{dt}\right]_s = \mu_4 \left[\frac{A}{K_{P_fA} + A}\right]RG_f - K_{TP}P_f \tag{4}$$

$$\left[\frac{dG}{dt}\right]_s = \mu_2 \left[\frac{N}{K_{GN} + N}\right] \tag{5}$$

$$\left[\frac{dG_f}{dt}\right]_s = \mu_5 \left[\frac{N}{K_{G_fN} + N}\right] \tag{6}$$

$$\left[\frac{dL}{dt}\right]_s = \mu_3 \left[\frac{S}{K_{LS} + S}\right]\left[\frac{A}{K_{LA} + A}\right] \tag{7}$$

$$\left[\frac{dR}{dt}\right]_s = \mu_6 \left[\frac{N}{K_{RN} + N}\right]\left[\frac{A}{K_{RA} + A}\right] - K_{TR}R - K'_{TR}R\left[\frac{K_{TR_s}}{K_{TR_s} + S}\right] \tag{8}$$

Note: Constituent synthesis is equal to the specific rate of synthesis, $\hat{r}_M$, where $M = A, N, P, P_f, G, G_f, L, R$ minus the rate of turnover. Turnover is included explicitly for both protein pools and the rRNA pool, and implicitly for the amino acid pool, since all amino acids and their precursors are included in the same pool.

pools, the sum of all their fraction mass levels is unity at all times. Furthermore, the time derivative of this sum is zero at all times. By noting these results, it can be readily shown that

$$\mu = \sum_{i=1}^{8}\sum_{j=1}^{8} r_{ij} \tag{2}$$

The instantaneous specific growth rate for a recombinant bacterial culture, as represented by the equations listed in Tables 1 and 2, is then

$$\mu = \left[\frac{dA}{dt}\right]_s + (1 - \epsilon_1)\left[\frac{dN}{dt}\right]_s + (1 - \epsilon_2)\left[\frac{dL}{dt}\right]_s + (1 - \gamma_1)$$

$$\left\{\left[\frac{dP}{dt}\right]_s + \left[\frac{dP_f}{dt}\right]_s\right\} + (1 - \gamma_2)\left\{\left[\frac{dG}{dt}\right]_s + \left[\frac{dG_f}{dt}\right]_s + \left[\frac{dR}{dt}\right]_s\right\} \tag{3}$$

The concept of fractional mass units was employed for a corrected form of Williams' two-compartment model[10] by Bailey and Ollis.[11] To our knowledge, this was the only attempt at using this powerful approach for the structured modeling of biological systems.

TABLE 2
Dynamic Equations for Constituent Pools

$$\frac{dA}{dt} = \left[\frac{dA}{dt}\right]_s - \epsilon_1\left[\frac{dN}{dt}\right]_s - \epsilon_2\left[\frac{dL}{dt}\right]_s - \gamma_1\left[\frac{dP}{dt}\right]_s - \gamma_1\left[\frac{dP_f}{dt}\right]_s - \mu A \quad (1)$$

$$\frac{dN}{dt} = \left[\frac{dN}{dt}\right]_s - \gamma_2\left[\frac{dG}{dt}\right]_s - \gamma_2\left[\frac{dG_f}{dt}\right]_s - \gamma_2\left[\frac{dR}{dt}\right]_s - \mu N \quad (2)$$

$$\frac{dP}{dt} = \left[\frac{dP}{dt}\right]_s - \mu P \quad (3)$$

$$\frac{dP_f}{dt} = \left[\frac{dP_f}{dt}\right]_s - \mu P_f \quad (4)$$

$$\frac{dG}{dt} = \left[\frac{dG}{dt}\right]_s - \mu G \quad (5)$$

$$\frac{dG_f}{dt} = \left[\frac{dG_f}{dt}\right]_s - \mu G_f \quad (6)$$

$$\frac{dL}{dt} = \left[\frac{dL}{dt}\right]_s - \mu L \quad (7)$$

$$\frac{dR}{dt} = \left[\frac{dR}{dt}\right]_s - \mu R \quad (8)$$

III. PLASMID SEGREGATION

One major contributing factor to culture instability in continuous systems is the growth rate differential between the competing populations. The existing model successfully calculates this dynamic growth rate differential in both batch and continuous cultures.

A second contributing factor is the unequal partitioning of plasmids from mother to daughter cells. A comprehensive description of culture instability must include these plasmid segregation kinetics. The magnitude of this instability depends upon the plasmid copy number at the time of cell division and presence or absence of a partitioning function. Most plasmids control their own replication genetically, through negative control. Inhibitors include proteins, RNA, or series of direct repeats.[12] When the inhibitor (e.g., RNA I in ColE1)[13] reaches an appropriate concentration, replication terminates, leaving the cell with a defined copy number.

If no mechanism exists for active segregation of plasmid molecules to the daughter cells during division, then the probability, δ, that either daughter cell will fail to inherit a plasmid is given by the binomial distribution

$$\delta = 2_n C_0 \left(\frac{1}{2}\right)^n \left(\frac{1}{2}\right)^0 \quad (4)$$

where n is the copy number per cell at division. If the mechanism which controls plasmid replication is able to correct anomalies in plasmid copy number produced by this random segregation, then all cells will contain equivalent numbers of plasmids at the next division, irrespective of the precise number of plasmids that each daughter cell has inherited. Therefore, the probability of producing a plasmid-free cell at the next division will remain unchanged, and plasmid-free cells will be produced at a constant frequency.[14]

Since the intracellular constituents are expressed in units of mass fraction in our model, the plasmid copy number is expressed as copies per chromosome equivalent, which provides the relative significance of plasmid activity to the cell. Furthermore, this quantity is readily determined experimentally. In order to incorporate plasmid segregation using the above probabilistic approach, plasmid content on a per cell basis must be evaluated. This can be accomplished using the Cooper-Helmstetter model, which provides an estimate for the number of genome equivalents per cell in steadily dividing cell cultures:[15]

$$\overline{G} = \frac{\tau}{C \ln 2} [2^{(C+D)/\tau} - 2^{D/\tau}] \tag{5}$$

This equation is based on the cell-number-derived growth rate, $\omega = \ln 2/\tau$, and the C and D periods (min) of the *Escherichia coli* growth cycle.

By combining the Cooper-Helmstetter result with our modeling framework, the average cell size, $\overline{m}$, is calculated as

$$\overline{m} = \frac{\overline{G}MW_g}{GN_{av}} \tag{6}$$

where MW_g is the molecular weight of the genome, G is the genome content per cell mass, and N_{av} is Avogadro's number. The plasmid copy number per average cell then becomes

$$\overline{N}_p = \frac{\overline{m}G_f N_{av}}{MW_p} \tag{7}$$

where G_f is the plasmid content as grams plasmid per gram dry cell mass and MW_p is the plasmid molecular weight.

In order to calculate the number of plasmids in a single cell that is about to divide, we make the approximation that the plasmid copy number increases linearly over the age of a single cell. In this way, the copy number at cell division, $N_{p(a=1)}$, is

$$N_{p(a=1)} = [1 - \overline{a}] \frac{dN_p}{da} + \overline{N}_p \tag{8}$$

where $\overline{a}$ is the average cell age and $\frac{dN_p}{da}$ is the rate of plasmid replication as calculated in the model after the transformation $\frac{dN_p}{da} = \frac{dN_p}{dt} \frac{dt}{da}$ (where $\frac{dt}{da} = 1$) is made. The average age in a steadily growing culture can be found by integrating

$$\overline{a} = \int_0^1 aW(a)da \tag{9}$$

where W(a) is the cell age distribution as shown by Seo and Bailey[16] as

$$W(a) = \left(\frac{2-\delta}{1-\delta}\right)\omega e^{-\omega a} \tag{10}$$

IV. REACTOR DYNAMICS

The equations listed in Table 2 describe the kinetics of all intracellular constituents of the recombinant bacteria and consequently determine the instantaneous specific growth rate. This set of kinetic equations was incorporated into continuous-flow stirred tank reactor (CFSTR) material balances. CFSTR dynamics are described by adding the following equations to those in Table 2:

$$\frac{dS}{dt} = -\frac{1}{Y_s}\mu^+X^+ - \frac{1}{Y_s}\mu^-X^- + D(S_f - S) \tag{11}$$

$$\frac{dX^+}{dt} = \mu^+X^+ - \gamma\mu^+X^+ - DX^+ \tag{12}$$

$$\frac{dX^-}{dt} = \mu^-X^- + \gamma\mu^+X^+ - DX^- \tag{13}$$

The cell mass, X, is in units of grams dry weight per liter, and the yield coefficient Y_s is in units of grams cell mass per gram substrate. The calculated growth rate, μ, is of dimension h^{-1}. The plasmid-bearing and plasmid-free populations are represented by a superscript plus and minus, respectively. In these equations, the dilution rate, D, and substrate feed concentration, S_f, are in units h^{-1} and grams per liter, respectively. Koizumi et al.[17] have shown that the segregation coefficient, γ, can be calculated as a function of δ, the probability of producing a plasmid-free cell upon division:

$$\gamma = 1 - \frac{\ln(2 - \delta)}{\ln 2} \tag{14}$$

In order to maintain simplicity, we assumed that the cell growth rates based on mass and cell number were identical ($\mu = \omega$). Alternatively, the mean cell size changes as slowly as the growth rate so that in continuous cultures, the process can be described as a series of steady-state increments. This is reasonable for the description of plasmid instability in slowly evolving chemostat cultures.

V. RESULTS AND DISCUSSION

A. STEADY-STATE DETERMINATION OF REPLICATION CONSTANTS

Before predicting behavior in chemostat cultures, the plasmid-associated replication constants were determined. The model simulations of the variation in steady-state plasmid copy number as a function of growth rate are shown in Figure 1. We have taken the data of Seo and Bailey[18] and Siegel and Ryu[19] and have expressed their results as number of plasmids per genome equivalent. Like Seo and Bailey, we assumed that chromosomal content varied with growth rate according to the Cooper-Helmstetter model.[15] The values of μ_5 and K_{GfN} which best represent the data are $1.8 \times 10^{-3} \frac{g \text{ plasmid}}{g \text{ cell mass} \cdot h}$ and $10^{-6} \frac{g \text{ } G_f}{g \text{ cell mass}}$ for Seo and Bailey's pDM247 (5.85 MD) and $1.07 \times 10^{-3} \frac{g}{g \cdot h}$ and $7.5 \times 10^{-4} \frac{g}{g}$ for Siegel and Ryu's pPLc23trpAl (6.5 kb). Since these model parameters were found by fitting experimental data, simulations using these values are limited to the systems from which they were obtained. However, the functional form of the plasmid synthesis expression

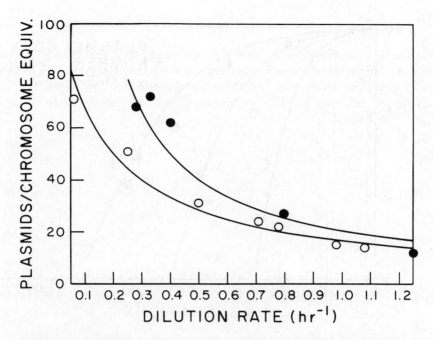

FIGURE 1. Plasmid dependence on growth rate. Experimental data included as open circles are from Siegel and Ryu[19] and closed circles from Seo and Bailey;[18] solid lines are model simulations with $\mu_5 = 1.07 \times 10^{-3}\,h^{-1}/K_{GjN} = 7.5 \times 10^{-4}$ g/g cell mass and $\mu_5 = 1.8 \times 10^{-3}\,h^{-1}/K_{GjN} = 1 \times 10^{-6}$ g/g cell mass, respectively. (From Bentley, W. and Kompala, D. S., *Biotechnol. Bioeng.*, 33, 49, ©1989. With permission of John Wiley & Sons.)

(Equation 6, Table 1) is general and can be modified to describe plasmid content in a wide variety of host/vector systems. Consequently, the effects of plasmid copy number on culture stability can also be determined for a variety of host/vector systems.

B. MIXED POPULATIONS: STABILITY OF CHEMOSTAT CULTURES

We have predicted the culture stability of chemostats containing mixed cultures of two competing populations: the plasmid-free host and the plasmid-bearing recombinant strain (Figures 2 and 3). In these simulations, we assumed that each population could be represented by an "average" cell. Simulations were performed by coupling two chemostat models, one simulating the plasmid-free host population ($\mu_4 = \mu_5 = 0$) and the other simulating the plasmid-bearing recombinant population. The initial population mixture was 99% recombinant cells and 1% plasmid-free cells.

When describing the results of Siegel and Ryu[19] (Figure 2), the plasmid replication constants were selected as described above, and the additional constants which describe foreign protein dynamics were selected so that the growth rate differential gave the appropriate response for a single dilution rate. Comprehensive foreign protein data were not given, precluding independent constant determination. The dilution rates were then varied while plasmid and foreign protein constants remained fixed. Model simulations show remarkably close agreement with the experimentally determined results.

In both cases (Siegel and Ryu[19] [Figure 2] and Bron and Luxen[20] [Figure 3]), the plasmid copy number at cell division was sufficiently high to preclude plasmid segregation as a mechanism for the washout of the recombinant strain ($\gamma = 0$). Hence, the population composition was strictly dependent upon the growth rate differential between the plasmid-free and recombinant strains. Also, in Figure 2 it is apparent that our simulations predict higher chemostat stability at higher dilution rates, which is in good quantitative agreement

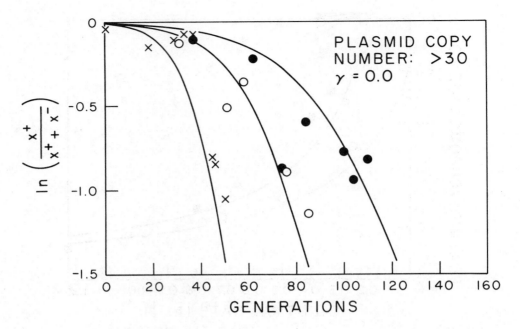

FIGURE 2.   Chemostat instability. Using the plasmid replication constants from Figure 1, we simulated a chemostat failure due to the dynamic growth rate differential between the plasmid-bearing and plasmid-free cells. The segregation extension to the model was used, but the plasmid copy number was sufficiently high to preclude influence from segregation ($\gamma = 0$). With the replication and translation constants fixed, the dilution rate was varied as in the experimental results of Siegel and Ryu.[19] (Dilution rates: ● = 1.08 h^{-1}; ○ = 0.79 h^{-1}; x = 0.48 h^{-1}). The predictions were remarkably close.

with these experimental results and in good qualitative agreement with other experimental studies finding similar phenomena.[4,21,22]

Segregational instability is significant, as described here, when the average number of plasmids per cell is less than approximately ten (Figure 4). The calculated plasmid copy number at cell division can be nearly twice the average, which tends to stabilize the recombinant population. In Figure 4, we have illustrated the expected increasing instability as the plasmid copy number approaches four. In general, low copy number plasmids (<4) contain a *par* locus sufficient for stabilizing partitioning dysfunction.[12]

A third mechanism for culture instability, which has not been considered here, results from a continual selective pressure for subpopulations with decreasing plasmid copy numbers. In plasmids with negative, autonomously controlled replication, however, there will always be a defined plasmid copy number at cell division, thus obviating this mechanism from consideration.

ACKNOWLEDGMENT

The authors wish to acknowledge the National Science Foundation (grant no. ECE-8611305) for support of this work.

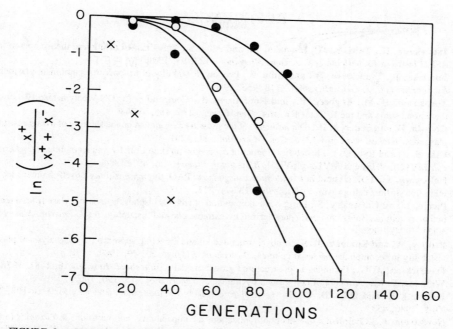

FIGURE 3. Chemostat instability. The plasmids from Bron and Luxen[20] were constructed by inserting segments of DNA into the parent plasmids, pLB5 and pLB2. Replication constants were fixed to yield the reported plasmid copy numbers for the pLB5 plasmids. The translational constant, μ_4, was selected to yield the appropriate response for each plasmid. In this way one can determine the significance of translation from the DNA insert. As in Figure 2, the segregation coefficient remained insignificant; thus, the reported instability was due to the growth rate differential. We cannot account for the dramatic instability observed for pLB5-3C. (pLB5 = upper ●, pLB5-1C = ○, pLB5-1S = lower ●, and pLB5-3C = x.)

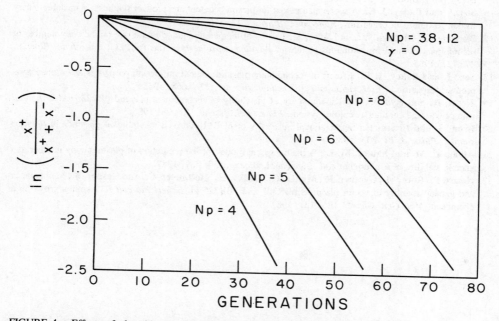

FIGURE 4. Effects of plasmid segregation in chemostat cultures. All replication and translation constants were fixed to yield a significant dynamic growth rate differential. The molecular weight of the plasmid was then varied, which altered the plasmid copy number without influencing the growth rate differential. In this way the effects of segregation become clear. As the copy number (Np) is lowered, the probability of yielding a plasmid-free cell upon division is raised; hence, the culture becomes more unstable.

REFERENCES

1. **Tsunekawa, H., Tateishi, M., Imanaka, T., and Aiba, S.,** TnA-directed deletion of the trp operon from RSF2124-trp in *Escherichia coli, J. Gen. Microbiol.,* 127, 93, 1981.
2. **Imanaka, T., Tsunekawa, H., and Aiba, S.,** Phenotypic stability of trp operon recombinant plasmids in *Escherichia coli, J. Gen. Microbiol.,* 118, 253, 1980.
3. **Nordstrom, U. M., Engberg, B., and Nordstrom, K.,** Competition for DNA polymerase III between the chromosome and the R-factor R1, *Mol. Gen. Genet.,* 135, 185, 1974.
4. **Godwin, D. and Slater, J. H.,** The influence of the growth environment on the stability of a drug resistance plasmid in *Escherichia coli* K12, *J. Gen. Microbiol.,* 111, 201, 1979.
5. **Kim, S. H. and Ryu, D. Y.,** Instability kinetics of *trp* operon plasmid ColE1-*trp* in recombinant *Escherichia coli* MV12 [pVH5] and MV12trp[pVH5], *Biotechnol. Bioeng.,* 26, 497, 1984.
6. **Nakazawa, T.,** TOL plasmid in *Pseudomonas aeruginosa* PAO: thermosensitivity of self-maintenance and inhibition of host cell growth, *J. Bacteriol.,* 133(2), 527, 1978.
7. **Pierce, J. and Gutteridge, S.,** Large-scale preparation of ribulosebisphosphate carboxylase from a recombinant system in *Escherichia coli* characterized by extreme plasmid instability, *Appl. Environ. Microbiol.,* 49(3), 1094, 1985.
8. **Bentley, W. and Kompala, D. S.,** A novel structured kinetic modeling approach for the analysis of plasmid instability in recombinant bacterial cultures, *Biotechnol. Bioeng.,* 33, 49, 1989.
9. **Fredrickson, A. G.,** Formulation of structured growth models, *Biotechnol. Bioeng.,* 18, 1481, 1976.
10. **Williams, F. M.,** A model of cell growth dynamics, *J. Theor. Biol.,* 15, 190, 1967.
11. **Bailey, J. E. and Ollis, D. F.,** *Biochemical Engineering Fundamentals,* 2nd ed., McGraw-Hill, New York, 1986, 453.
12. **Nordstrom, K.,** Chairman's introduction: replication, incompatability, and partition, in *Plasmids in Bacteria,* Helinski, D. R., Cohen, S. N., Clewell, D. B., Jackson, D. A., and Hollaender, A., Eds., Plenum Press, New York, 1985.
13. **Lewin, B.,** *Genes III,* John Wiley & Sons, New York, 1987, 309.
14. **Primrose, S. B., Derbyshire, P., Jones, I. M., Robinson, A., and Ellwood, D. C.,** The application of continuous culture to the study of plasmid stability, in *Continuous Culture 8: Biotechnology, Medicine, and the Environment,* Dean, A. C. R., Ellwood, D. C., and Evans, C. G. T., Eds., Ellis Horwood, Chichester, U.K., 1984.
15. **Cooper, S. and Helmstetter, C. E.,** Chromosome replication and the division cycle of *Escherichia coli* B/r, *J. Mol. Biol.,* 31, 519, 1968.
16. **Seo, J. and Bailey, J. E.,** A segregated model for plasmid content and product synthesis in unstable binary fission recombinant organisms, *Biotechnol. Bioeng.,* 27, 156, 1985.
17. **Koizumi, J., Monden, Y., and Aiba, S.,** Effects of temperature and dilution rate on the copy number of recombinant plasmid in continuous culture of *Bacillus stearothermophilus* (pLP11), *Biotechnol. Bioeng.,* 27, 721, 1985.
18. **Seo, J. and Bailey, J. E.,** Effects of recombinant plasmid content on growth properties and cloned gene product formation in *Escherichia coli, Biotechnol. Bioeng.,* 27, 1668, 1985.
19. **Siegel, R. and Ryu, D. Y.,** Kinetic study of instability of recombinant plasmid pPlc23trpA1 in *E. coli* using two-stage continuous culture system, *Biotechnol. Bioeng.,* 27, 28, 1985.
20. **Bron, S. and Luxen, E.,** Segregational instability of pUB110-derived recombinant plasmids in *Bacillus subtilis, Plasmid,* 14, 235, 1985.
21. **Ataai, M. M. and Shuler, M. L.,** A mathematical model for the prediction of plasmid copy number and genetic stability in *Escherichia coli, Biotechnol. Bioeng.,* 30, 389, 1987.
22. **Noack, P., Roth, M., Geuther, R., Muller, G., Undisz, K., Hoffmeier, C., and Gaspar, S.,** Maintenance and genetic stability of vector plasmids pBR322 and pBR325 in *Escherichia coli* K12 strains grown in a chemostat, *Mol. Gen. Genet.,* 184, 121, 1981.

Chapter 11

PRELIMINARY ANALYSIS OF A HOLLOW FIBER EXTRACTIVE FERMENTOR FOR THE PRODUCTION OF ETHANOL

Ronald L. Fournier, Ramesh Kagalanu, and Samer F. Naser

TABLE OF CONTENTS

I. INTRODUCTION

Extractive fermentation is receiving considerable attention for fermentations that are product inhibited. Numerous approaches have been described in the literature.[1-12] Potential advantages of extractive fermentation include increased specific productivity, reduced costs of downstream separation and purification steps, ability to process high concentrations of substrate in the feed, and increased volumetric productivity.

The use of hollow fiber membranes in extractive fermentation has several advantages. These include very high membrane surface areas per unit volume of fermentor, elimination of emulsions due to physical separation of the solvent and aqueous phases, protection of the cells from physical toxicity of the extracting solvent,[12] independent variation of aqueous and solvent flow rates, as well as the amount of surface area within the fermentor.

In this chapter we report on efforts to develop a hollow fiber extractive fermentor (HFEF) for the production of ethanol. We present results of the testing of a mathematical model of the HFEF[13] and a preliminary economic analysis of a full-scale facility incorporating the HFEF for the production of ethanol.[14]

II. MATHEMATICAL MODEL OF A HOLLOW FIBER EXTRACTIVE FERMENTOR

As illustrated in Figure 1, an HFEF consists of a cylindrical shell literally containing thousands of hollow fiber membranes. The cells and aqueous nutrient phase flow continuously through the shell space, whereas the solvent flows concurrently through the lumen of the hollow fibers. As the product is produced on the shell side it diffuses through the membrane to the solvent phase, where it is continuously removed. This reduces the concentration of the product in the shell-side fermentation phase, resulting in a decrease in product inhibition and an increase in volumetric productivity.

The inset in Figure 1 illustrates the vicinity of the hollow fiber membrane wall. For this analysis, the membrane is assumed to be hydrophobic and its pores completely wetted by the solvent phase. Bulk flow of the solvent through the membrane wall can be stopped by maintaining the pressure on the shell side several pounds per square inch above that of the solvent side.[12] The aqueous phase is a three-phase mixture consisting of cells, nearly insoluble gas generated or added to the fermentation process, and water containing the fermentation substrate, product, and traces of dissolved solvent.

The model of the HFEF is based on the above physical description and provides for the axial changes in the concentrations of substrate, product, solvent, water, and cells, the variation of the aqueous- and solvent-phase velocities due to reaction and mass transfer, and the axial change in the gas velocity and gas holdup. Analysis of the HFEF is based on the following assumptions:

1. The aqueous and solvent phases are at equilibrium at their interface, which is assumed to be located at the outer surface of the hollow fiber membrane wall.
2. Distribution of component i between the phases in equilibrium at the interface is provided by a component distribution coefficient.
3. The solvent is biocompatible.
4. Mass transfer processes can be described by simple film-type mass transfer coefficients.
5. The aqueous and solvent phases are in plug flow.
6. Cells flow at the same velocity as the aqueous phase.
7. Gas is assumed to be insoluble.
8. Stripping of the product by the gas is neglected.

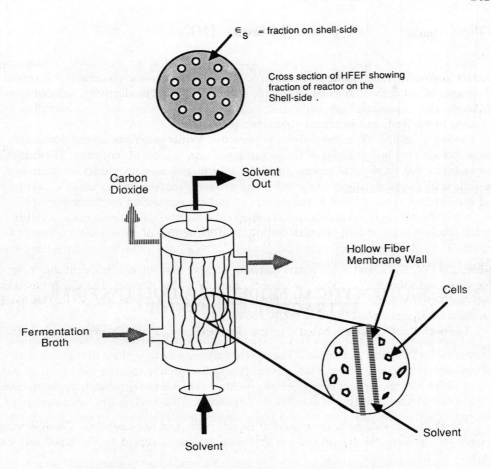

FIGURE 1. Schematic diagram of a hollow fiber extractive fermentor.

In the following discussion, the HFEF model will be formulated for the special case of ethanol production.[13] Other fermentations can be handled by straightforward modifications.

The specific cell production rate for a strain of *Saccharomyces cerevisiae* is provided by the following expression presented by Maiorella et al.:[15]

$$\mu = 0.46 \left[\frac{C_{GW}}{0.315 + C_{GW}} \right] \left[1 - \frac{C_{EW}}{87.5} \right]^{0.36} \tag{1}$$

The second term in brackets accounts for the inhibitory effect of ethanol, the maximum ethanol concentration being 87.5 g/l.

The component mass balances for the aqueous and solvent phases are as follows:

Aqueous phase mass balances

$$\frac{d}{dz} [\epsilon_{L} v_{W} C_{iW}] = - \left[\frac{2(1 - \epsilon_{S})}{\epsilon_{S} R} K_{iS} \right] (m_{i} C_{iW} - C_{iS})$$

$$+ (1 - \epsilon_{G}) X \mu Y_{i} \tag{2}$$

Cells

$$\frac{d}{dz}[v_W(1 - \epsilon_G)X] = (1 - \epsilon_G)\mu X \tag{3}$$

CO_2

$$\frac{d}{dz}[v_G] = \frac{(1 - \epsilon_G)}{44P/RT}\mu \frac{Y_{CO_2/G}}{Y_{X/G}}X \tag{4}$$

Solvent phase mass balances

$$\frac{d}{dz}[v_S C_{iS}] = \left[\frac{2RK_{iS}}{R'^2}\right](m_i C_{iW} - C_{iS}) \tag{5}$$

where i = W, E, G, and S for water, ethanol, glucose, and solvent, respectively. $Y_i = 0$ for water and solvent, $Y_{E/G}/Y_{X/G}$ for ethanol, and $1/Y_{X/G}$ for glucose. The C_{iW} is based on cell-free liquid volume, whereas the cell density X (in grams dry weight per liter) is based on the total aqueous phase volume (cells and liquid).

The overall solvent-phase-based mass transfer coefficient for the ith component is defined as

$$\frac{1}{K_{iS}} = \frac{m_i}{k_{iW}} + \frac{R}{D_{ei}}\ln\frac{R}{R'} + \frac{R}{R'}\frac{1}{k_{iS}} \tag{6}$$

The shell-side volume is comprised of liquid, cells, and insoluble gas. The following expressions provide the fraction of the shell-side volume occupied by the liquid and the cells:

$$\epsilon_L = (1 - \epsilon_G)(1 - X\phi) \tag{7}$$

$$\epsilon_C = (1 - \epsilon_G)X\phi \tag{8}$$

where ϕ is the specific volume of the dry cells. ϵ_G is the gas holdup and is determined by a suitable expression.[13]

The total (shell and hollow fiber volume) ethanol volumetric productivity can be determined by an ethanol balance from the fermentor entrance to the axial position z. It is assumed that no ethanol enters the fermentor.

$$P_E = \frac{\epsilon_L \epsilon_S v_W C_{EW} + (1 - \epsilon_S)(R'/R)^2 v_S C_{ES}}{z} \tag{9}$$

The above set of equations comprises the mathematical model for ethanol production in the HFEF.

III. HFEF SIMULATION RESULTS AND DISCUSSION

To illustrate the solution of the HFEF model and its potential for increased volumetric productivity, the following example was defined. The HFEF was assumed to contain hollow fibers with a total length of 250 cm. The fibers were microporous, with an inner diameter of 400 μm, a wall thickness of 25 μm, and a porosity of 40%. The solvent used for the

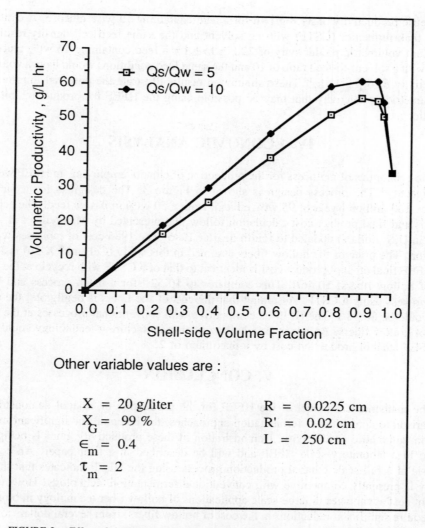

FIGURE 2.   Effect of shell-side volume fraction on volumetric productivity for Q ratios of 5 and 10, as calculated from the HFEF mathematical model.

simulation was tri-n-butylphosphate (TBP), since it has been recognized in several recent papers as a good solvent for ethanol extraction[11,12] and is not toxic to yeast in the dissolved form.[12]

Figure 2 illustrates the predicted relationship between the HFEF volumetric productivity, P_E, and the shell-side volume fraction, ϵ_s. Q_s/Q_w is the volumetric solvent-to-aqueous-feed ratio. The feed to the fermentor contains 40 wt% glucose, and the entering cell density is 20 g/l. Glucose conversion is 99%.

The model predicts the existence of an optimum volume fraction of hollow fibers in the fermentor, which maximizes the volumetric productivity. This optimum is a result of a classic trade-off between the volume fraction of the fermentor required for fermentation and that required for efficient removal of the ethanol product to minimize product inhibition. For $Q_s/Q_w = 10$, the maximum volumetric productivity occurs when ϵ_s is about 0.95 — that is, 61.6 g/l · h.

It is interesting to compare the optimum productivities presented in Figure 2 with those possible by conventional fermentation approaches. Fermentation in the absence of solvent in a plug flow fermentor, for the same entering feed cell density, results in a maximum volumetric productivity of 34.8 g/l · h for a feed containing 13 wt% glucose. A continuous stirred tank fermentor (CSTF) with no solvent and the same feed cell density results in a maximum volumetric productivity of 22.8 g/l · h for a feed containing 16 wt% glucose. A CSTF with a solvent-to-feed ratio of 10 and the same feed conditions would have a volumetric productivity of 6.8 g/l · h.[16] These simulation results illustrate the potential improvements in volumetric productivity that may be possible using the HFEF for product-inhibited fermentations.

IV. ECONOMIC ANALYSIS

The economics of a process for the production of ethanol employing an HFEF were also investigated.[14] The process design is shown in Figure 3. The design is for a facility to produce 100 million l/year of 95 wt% ethanol from a 50 wt% molasses feed. The economic analysis and final product cost calculation follow that presented by Maiorella et al.,[15] with costs (in U.S. dollars) updated to fourth quarter 1986. The 1986 cost of molasses was $70/short ton. The price of the hollow fibers assumed in this study (Celgard® X-20) is $4/ft².[17] The HFEF final ethanol product cost is identical to that of a CSTF with recycle at the current cost of hollow fibers, $0.46/l. This compares to $0.52/l for a batch process and $0.50/l for a process using a CSTF. Assuming that the cost of the fibers is negligible, the ethanol cost would be $0.38/l. The HFEF is competitive with conventional processes at the current price of hollow fibers. An advance in hollow fiber manufacturing technology could reduce the HFEF ethanol production cost by a maximum of 25%.

V. CONCLUSION

The mathematical model of the HFEF for the production of ethanol demonstrates, in comparison to conventional fermentation approaches, the potential for significant improvement in volumetric productivity. Demonstration of these productivity gains is being investigated in a laboratory-scale HFEF and will be described in a later paper. An economic analysis of a full-scale ethanol production process using the HFEF indicates that this technology is presently competitive with conventional fermentation technology. However, significant cost advantages in large-scale applications of hollow fiber technology in bioreactors will require significant reductions in the cost of hollow fibers, from several dollars to several cents per square foot.

LIST OF SYMBOLS

C_{iS}	Concentration of i in the solvent phase
C_{iw}	Concentration of i in the aqueous phase
D_{bi}	Component i diffusion coefficient in solvent
D_{ei}	Component i effective diffusivity, $= \epsilon_m D_{bi}/\tau_m$
ϵ_C	Shell-side cell fraction
ϵ_G	Shell-side gas holdup
ϵ_L	Shell-side liquid fraction
ϵ_m	Membrane porosity
ϵ_S	Fraction of fermenter volume on shell side
k_{iS}, k_{iw}	Individual component mass transfer coefficients for the solvent and aqueous phases
K_{iS}	Overall mass transfer coefficient for component i, defined by Equation 6
L	Length of HFEF; length of a hollow fiber
m_i	Equilibrium distribution coefficient

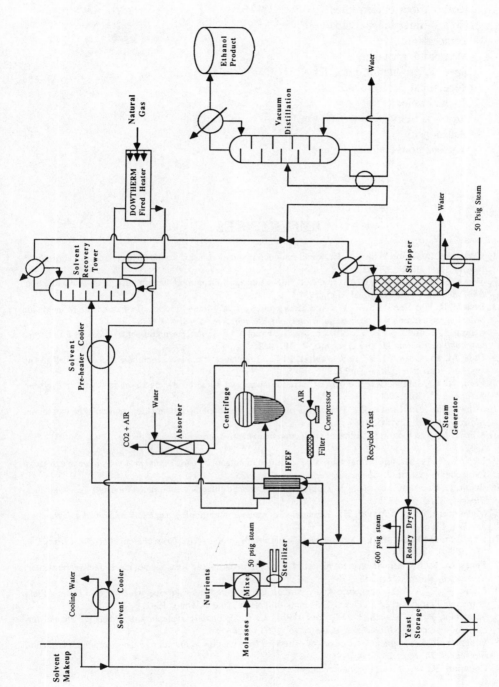

FIGURE 3.   Flowsheet of the process using an HFEF to produce ethanol.

P	Pressure
P_E	Total volumetric productivity, defined by Equation 9
ϕ	Specific volume of dry cells
R	Hollow fiber outer radius
R'	Hollow fiber inner radius
T	Temperature
τ_m	Membrane tortuosity
μ	Specific production rate, 1/h
v_G	Superficial gas velocity
v_S	Solvent velocity
v_W	Aqueous velocity
X	Cell density
X_G	Glucose conversion

REFERENCES

1. **Cysewske, G. R. and Wilke, C. R.,** Rapid ethanol fermentations using vacuum and cell recycle, *Biotechnol. Bioeng.,* 19, 1125, 1977.
2. **Wang, H. Y., Robinson, F. M., and Lee, S. S.,** Enhanced alcohol production through on-line extraction, *Biotechnol. Bioeng. Symp.,* 11, 555, 1981.
3. **Lencki, R. W., Robinson, C. W., and Moo-Young, M.,** On-line extraction of ethanol from fermentation broths using hydrophobic absorbents, *Biotechnol. Bioeng. Symp.,* 13, 617, 1983.
4. **Chung, I. S. and Lee, Y. Y.,** Effect of *in situ* ethanol removal on fermentation of D-xylose by *Pachysolen tannophilus, Enzyme Microb. Technol.,* 7, 217, 1985.
5. **Dale, M. C., Okos, M. R., and Wankat, P. C.,** Simultaneous product separation. I. Reactor design and analysis, *Biotechnol. Bioeng.,* 27, 932, 1985.
6. **Finn, R. K.,** Inhibitory cell products: their formation and some new methods of removal, *J. Ferment. Technol.,* 44, 305, 1966.
7. **Minier, M. and Goma, G.,** Production of ethanol by coupling fermentation and solvent extraction, *Biotechnol. Lett.,* 3, 405, 1981.
8. **Minier, M. and Goma, G.,** Ethanol production by extractive fermentation, *Biotechnol. Bioeng.,* 24, 1565, 1982.
9. **Ishii, S., Taya, M., and Kobayashi, T.,** Production of butanol by *Clostridium acetobutylicum* in extractive fermentation system, *J. Chem. Eng. Jpn.,* 18, 125, 1985.
10. **Brink, L. E. S. and Tramper, J.,** Optimization of organic solvent in multiphase biocatalysis, *Biotechnol. Bioeng.,* 27, 1258, 1985.
11. **Matsumura, M. and Markl, H.,** Elimination of ethanol inhibition by perstraction, *Biotechnol. Bioeng.,* 28, 534, 1986.
12. **Cho, T. and Shuler, M. L.,** Multimembrane bioreactor for extractive fermentation, *Biotechnol. Prog.,* 2, 53, 1986.
13. **Fournier, R. L.,** A mathematical model of a microporous hollow fiber membrane extractive fermentor, *Biotechnol. Bioeng.,* 31, 235, 1988.
14. **Naser, S. F.,** A Technoeconomic Analysis of an Ethanol Production Process which Uses a Hollow Fiber Extractive Fermentor, M.S. thesis, University of Toledo, Toledo, Ohio, 1987.
15. **Maiorella, B. L., Blanch, H. W., and Wilke, C. R.,** Economic evaluation of alternative ethanol fermentation processes, *Biotechnol. Bioeng.,* 26, 1003, 1984.
16. **Fournier, R. L.,** Mathematical model of extractive fermentation: application to the production of ethanol, *Biotechnol. Bioeng.,* 28, 1206, 1986.
17. **Callahan, R. W.,** personal communication.

Chapter 12

SIMULATION RESULTS OF APPLYING A ROBUST MULTIVARIABLE CONTROL DESIGN METHODOLOGY TO A GENERALIZED BACTERIAL GROWTH SYSTEM

Tse-Wei Wang, Charles F. Moore, and J. Douglas Birdwell

TABLE OF CONTENTS

I. INTRODUCTION

The application of computers in bioreactor control has made the application of more advanced control technology feasible. Research development in this area has been slow. Several serious problems have slowed the progress of development in this area: (1) the lack of accurate models for the growth dynamics of living cells, (2) variations in the growth-associated parameters during the growth processes, (3) lack of reliable on-line sensors for growth monitoring, and (4) the highly nonlinear and interactive nature of most growth systems. Any realistic control scheme must take these factors into consideration. In addition to controlling the basic operating variables such as pH, temperature, and dissolved oxygen level, a higher level of control must exist in order to operate the bioreactor in an optimum manner, such as maximizing productivity and maintaining the long-term stability of the bioreactor operation — both important goals in industrial settings. Due to the lack of on-line sensors, on-line estimating schemes would have to be implemented to arrive at the state estimates. Due to imprecision in the modeling process and the interacting nature of the system dynamics, controller design schemes that guarantee overall system stability in the presence of model uncertainty appear very attractive for this purpose. During the last 15 years, theoretical development of several robust multivariable control design methodologies has emerged in the control communities, such as linear quadratic Gaussian/loop transfer recovery (LQG/LTR), H^{∞} (H-infinity minimization), and μ (structured singular value) synthesis approaches. Robustness, as used here, means the ability to guarantee closed-loop stability in the presence of model uncertainty. Of these methods, LQG/LTR was chosen for initial investigation because a full set of reliable computer software exists for its implementation.[1] This design methodology, however, has some limitations associated with it. For instance, the methodology requires the open-loop plant to be minimum phase (no zeros in the closed right-half plane). In addition, the LQG/LTR design is based on *linear* system theory and assumes the perturbations to the system to be linear. As far as the authors are aware, results reported here are among the first attempts to apply any type of robust controller design algorithm to a bioreactor system.

Recently, several groups have reported the use of a Kalman filter (KF)-based estimation algorithm to estimate, on-line, the state variables of a bioreactor system.[2-4] Among them, Stephanopoulos and San[4] proposed an algorithm to perform reliable on-line estimation of states and some growth-associated parameters of a bioreactor system via an extended KF. This method appears very promising in that it circumvents the necessity of on-line sensors which have yet to be developed. The work of Stephanopoulos and San serves as the basis for studies undertaken in the research reported here. The results presented in this chapter are based on the thesis research done by one of us,[5] and preliminary results have been presented elsewhere.[6-8]

II. PROBLEM STATEMENT

Given the above considerations, the task is to design a controller for a nonlinear bioreactor such that the controller maintains long-term overall system stability and keeps system operation as close to a chosen nominal setpoint as possible, in spite of external disturbances and internal model uncertainty. The emphasis is on maintaining the overall system stability. All other requirements of the system performance are to be accommodated only after the stability requirement is first satisfied. This stability requirement is of paramount importance in continuous operation mode.

III. BACKGROUND

A. BIOREACTOR DESCRIPTION

The generic biosystem under consideration is a continuously operated bioreactor with one substrate in the feed and no product formation. This simple case is considered in order to avoid complexity in control design brought about by an overly complex model. The bacterial biomass consumes the substrate to produce more biomass, and the biomass is harvested as the desired end product. The state variables of interest are the concentrations of biomass (bacterial cell mass), b, and substrate, s, in the reactor as a function of time. These two values, if precisely known, fully describe the state of the bioreactor system at any time; however, the specific growth rate, μ (defined as the rate of bacterial concentration growth per unit of biomass concentration), is also of interest. Therefore, μ is treated as a third state variable. The values of the three states will be estimated by the KF of the LQG controller.

Variables that can be used as manipulated variables to effect changes in the state variables are D, the dilution rate (D = F/V, where F is the flow rate and V is the volume of the bioreactor), and s_f, the substrate concentration in the feed.

The choice of suitable measurement variables is difficult. Stephanopoulos and San[4] proposed an on-line method, based on the technique of elemental balance,[9,10] for calculating the total rate of growth, R, and the growth yield of biomass with respect to substrate, Y_s, from feasible physical measurements of the concentrations and flow rates of the inlet and outlet oxygen and carbon dioxide gases. They derived a set of equations that related R and Y_s to the previously mentioned gas flow rates. The calculated values of R will serve as the measurement values in the state-space representation of our model.

B. LQG/LTR CONTROL METHODOLOGY OVERVIEW

The LQG/LTR control design is a robust multivariable design based on linear system theory.[11-13] Given a set of performance constraints that the closed-loop system must satisfy and the bounds on the overall uncertainty of the model, the LQG/LTR methodology yields a controller that will simultaneously satisfy these performance constraints and stability requirements. The LQG controller contains two subsystems, a KF block cascaded with a linear quadratic regulator (LQR) block in the feedback path. Figure 1 shows the block diagram of a closed-loop system containing an LQG controller, as well as the LQG controller's internal structure.

For this bioreactor system, there are two inputs and only one output. In this case, as required by the LQG/LTR algorithm, the filter must be designed first to meet the performance and robustness constraints. Then, in the LTR step, a suitable regulator gain will be calculated such that the robustness property with respect to perturbations modeled at the output node is "recovered" in the overall closed-loop system from the first-phase design.

IV. FORMULATION OF THE PROBLEM

A. SYSTEM MODELING

An unstructured, unsegregated modeling approach was used in this research, similar to that of Stephanopoulos and San,[4] with one important difference, however. In our approach, the empirical Monod relationship between the substrate and the specific growth rate was introduced in the μ dynamic equation. The μ dynamics were modeled as a first-order process with Gaussian white noise present. The reason for this change was to make the three-state model system completely controllable and completely observable. The state and measurement equations are shown below.

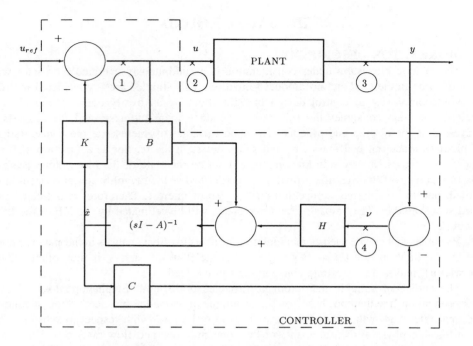

FIGURE 1. Internal LQG structure; H = filter gain; K = regulator gain; y = output; A, B, and C denote system matrices; nodes 2 and 3 denote the input and output nodes, respectively; nodes 1 and 4 denote nodes internal to the controller.

State dynamics equations:

$$\dot{b} = (\mu - D)b \tag{1}$$

$$\dot{s} = D(s_f - s) - \mu b/Y_s + \xi_1(t) \tag{2}$$

$$\dot{\mu} = m[(\mu_{max}s)/(k_s + s) - \mu] + \xi_2(t) \tag{3}$$

Measurement equations:

$$R = \mu b + \eta(t) \tag{4}$$

where b = biomass concentration; s = substrate concentration; μ = specific growth rate (per unit time); Y_s = biomass yield with respect to substrate; D = dilution rate (per unit time); s_f = substrate concentration in the feed; μ_{max}, k_s = constants of the Monod growth model equation, $\mu_{max}s/(k_s + s)$; $\xi_1(t)$, $\xi_2(t)$, $\eta(t)$ are independent, white noise processes; and m is a proportionality constant.

LQG/LTR design method requires a nominal linear description of the system. Therefore, the nonlinear model was first linearized around a chosen nominal setpoint. In this research, the nominal setpoint is chosen such that the biomass productivity, P (P = Db), is maximized at a given feed substrate concentration, s_f. The choice of a feasible setpoint is critical. It should be chosen such that the system would still be able to achieve it after disturbances such as parameter variations have taken place. For this system, the following values have been chosen as the nominal setpoint:

$$D = 0.42 \ (h^{-1}) \qquad k_s = 0.05 \ (g/l) \qquad s = 0.2625 \ (g/l)$$

$$s_f = 9.0 \ (g/l) \qquad Y_s = 0.5 \qquad \mu = 0.42 \ (h^{-1})$$

$$\mu_{max} = 0.5 \ (h^{-1}) \qquad b = 4.36875 \ (g/l) \qquad m = 3 \ (h^{-1}) \qquad (5)$$

The linearized model has the standard form

$$\dot{x} = Ax + Bu + w(t) \qquad (6)$$

$$y = Cx + v(t) \qquad (7)$$

where $x = [\Delta b \ \Delta s \ \Delta \mu]'$ is the state vector (where "'" means transpose and "Δ" means deviation from the nominal value); $u = [D \ s_f]'$, the control input; $y = [\Delta R]'$, the measurement vector; $w(t)$, $v(t)$ = independent white noise vectors; and A, B, C = linearized system matrices. Note that in this modeling approach there are two inputs, D and s_f, one output, R, and three states, b, s, and μ.

To implement the LQG/LTR control design methodology, the expert software system called *CASCADE*[14] was used. The inputs to the expert system are the linearized system descriptions (Equations 6 and 7), the specification of performance constraints (e.g., zero tracking error), and the uncertainty bound (see below). The outputs from the expert system are the KF and the regulator gains, H and K, respectively, and an analysis of the expected closed-loop performance.

B. DEVELOPMENT OF THE UNCERTAINTY BOUND

The development of an uncertainty bound is a key step in any robust control design methodology. The determination of the bound relies heavily on the designer's understanding of the plant dynamics, especially its largest probable deviation from nominal plant dynamics. For this research, unstructured, multiplicative uncertainty, modeled at the output node, was considered. The uncertainty bound was calculated by varying the growth parameters, μ_{max}, Y_s, and k_s, and the steady-state values of b, s, and μ by as much as $\pm 15\%$ from the corresponding nominal values, and then solving for the uncertainty matrix, ΔL, as the solution to

$$G(s) = [I + \Delta L(s)]G_0(s) \qquad (8)$$

where G and G_0 are the perturbed and nominal open-loop transfer functions of the linearized dynamics, respectively; ΔL represents the multiplicative system uncertainty matrix; and s $\in$ C (the set of complex numbers) is the Laplace variable. Let s (the Laplace variable) be a point $j\omega$ (j denotes the imaginary number $\sqrt{-1}$, and ω denotes the frequency in radians per unit time) on the imaginary axis. Then, at each frequency ω, the largest of the maximum singular values of all the possible $\Delta L(j\omega)$ was chosen. Thus, a plot of the largest maximum singular value as a function of frequency ω could be obtained. This datum would be an input to the *CASCADE* expert system.

Difficulty was encountered in calculating a representative uncertainty bound for this system, in that different results were obtained when two different methods were used to calculate the uncertainty bound. Both a gradient search method and a search along the boundary of the parameter set were implemented. The results were different at high frequencies; it appears that several local maxima exist. The maximum of the two solutions, as a function of frequency, was used; however, there is no guarantee that these values represent global maxima over the parameter set. It was very difficult to develop an appropriate bound on $\Delta L(j\omega)$. Further work needs to be done to resolve this difficulty.

V. RESULTS

A. ROBUSTNESS REQUIREMENT

The goal of the controller is to maintain *stable* bioreactor operation at values near the chosen setpoint in the presence of external perturbations and internal model uncertainty. The controller is designed using loop shaping, with LQR and KF blocks as parameterized models which satisfy frequency-dependent performance and robustness constraints. In this research, robustness at the plant output node is required; therefore, a KF which meets the performance constraints is designed first. A 26-dB reduction in output tracking error (with respect to the reference output setpoint) is specified for a very low frequency. This corresponds to a 5% allowable error in steady state. The design method cannot satisfy this constraint simultaneously with the robustness constraint imposed on the system. This is due to the characteristics of the open-loop plant. Therefore, the system is augmented to include an integrator at the output node to guarantee zero steady-state error. The bounds imposed on the singular values of the return difference and inverse return difference transfer functions are infinite loop gain at zero frequency and a maximum crossover frequency of 1.0 rad/h derived from the uncertainty bound singular value plot. The maximum and minimum singular values are balanced at a low but nonzero frequency of 0.01 rad/h. Balancing at a nonzero low frequency is required due to the augmented integrator. In addition, balancing at high frequency yields a KF which, although it satisfies the design specifications, does not update the estimates of three of the four states. This is not desirable. The balancing technique for nonzero, finite frequency was developed by Birdwell and Laub.[15] The filter and regulator gains completely define the controller structure. The expert system *CASCADE* produced as its output the specifications of the KF and regulator gains.

The design produced a filter gain H and a regulator gain K as shown below:

$$H = \begin{bmatrix} -0.2841 \\ 0.1177 \\ -0.02496 \\ -0.9987 \end{bmatrix} \tag{9}$$

$$K = \begin{bmatrix} 0.875 & 1.717 & 7.701 & 4.357 \\ -2.596 & -3.865 & -15.426 & -8.745 \end{bmatrix} \tag{10}$$

The associated singular value plots of the return difference and inverse return difference transfer functions are shown in Figures 2 and 3. Since there is only one output, the loop return ratio at the plant output node, L(s) (defined as the transfer function X_{out}/X_{in}, where X_{out} and X_{in} denote the signals immediately to the left and right of the break, respectively, if the system loop is broken at the plant output node, node 3 in Figure 1), is a scalar. Therefore, there is only one singular value, which is simply the absolute value of L. It can be seen that the performance and robustness constraints are satisfied. A less than 5% tracking error is insured, as can be seen from the singular value plot in Figure 2, in that the gain of the return difference function at the low-frequency end is above 26 dB (0.05^{-1}). The robustness requirement is satisfied because the singular value plot of the inverse return difference function does lie above that of the uncertainty bound (as shown in Figure 3). This implies that the closed-loop system is guaranteed to remain stable when uncertainty, modeled at the plant output node, up to the bound, exists in the true plant dynamics.

B. SIMULATION RESULTS

During simulation, the state propagation, measurements generation, Kalman filtering,

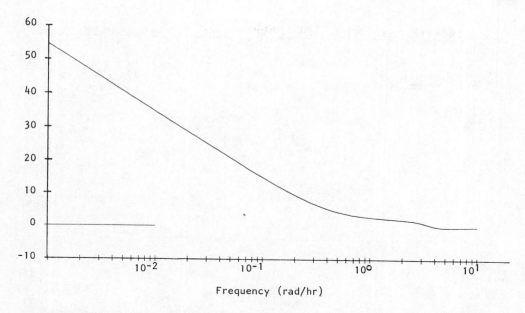

FIGURE 2. Singular value plot of the return difference transfer function; L denotes the loop transfer function. The y-axis values are I + ΔL(jω).

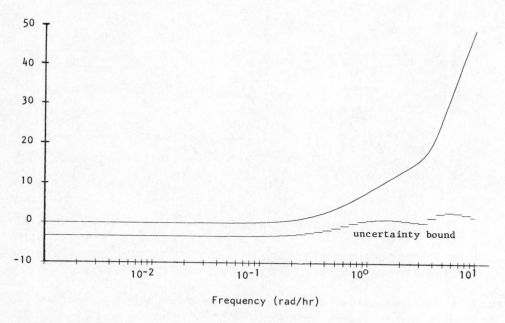

FIGURE 3. Singular value plots of the inverse return difference transfer function and the uncertainty bounds. L denotes the loop transfer function; the lower curve denotes the maximum singular value plot of the system uncertainty bound. The y-axis values are [I + ΔL(jω)]$^{-1}$.

and control input generation were all carried out in continuous time, while noise simulations were carried out in discrete time with a sampling interval of 0.01 h.

The loop was closed and simulated for nominal parameter and state values (values shown in Equation 5). Figure 4 shows the simulation results. It can be seen that the response tracks the nominal values. (Note the finely divided vertical scales in this figure.) Next, the convergence property of the linear KF was tested. This simulation was carried out for the closed-

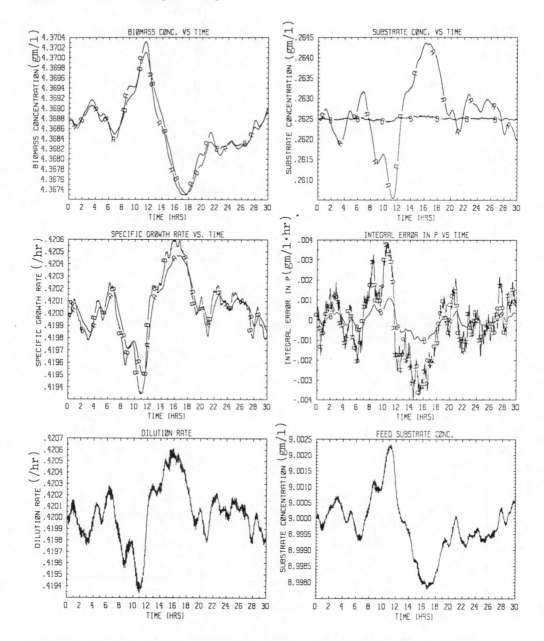

FIGURE 4.   Closed-loop simulation using nominal parameter and state values. -A-A- is for the "true" state values; -B-B- is for "estimated" values. Note the finely scaled ordinate axis.

loop plant. Results show that the estimates from the filter do converge to the state values, even when "incorrect" initial values were used to initialize the filter estimates, as shown in Figure 5.

Finally, the robustness properties of the closed-loop system were tested by the introduction of parameter and state perturbations during simulation. Figure 6 shows the result of varying μ_{max} and Y_s by $+15\%$, k_s by -15% at t = 1 h, and state b by -10% at t = 4 h. The step change in b simulates the case where some biomass has died as a result of an environmental upset such as toxicity in the feed. It is noted that stability is indeed maintained for the remaining time interval. Note that a positive change is introduced in μ_{max} and Y_s,

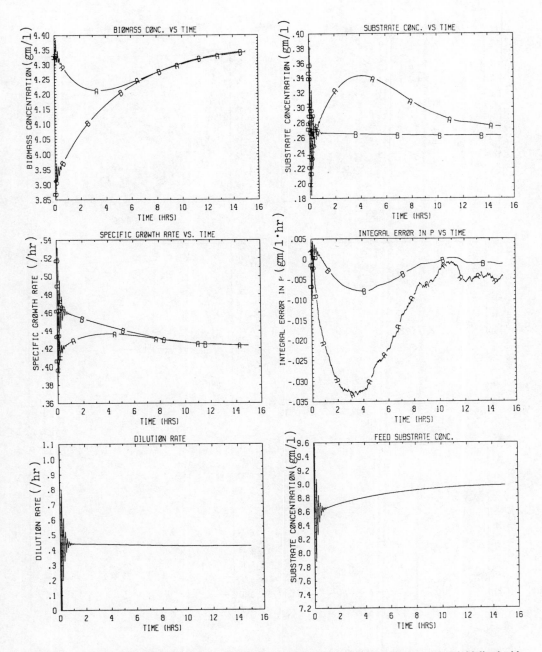

FIGURE 5. Convergence properties of the estimator in closed-loop simulation; the filter has been initialized with "incorrect" values. -A-A- is for the "true" state values; -B-B- is for "estimated" values.

although in practice the system can only deteriorate. The reason is that the nominal values chosen for μ_{max} and Y_s are at the low end of the spectrum of all possible parameter variations. Therefore, any variation introduced during simulation must be above the corresponding nominal parameter value in order for the values to stay within the spectrum of possible variations.

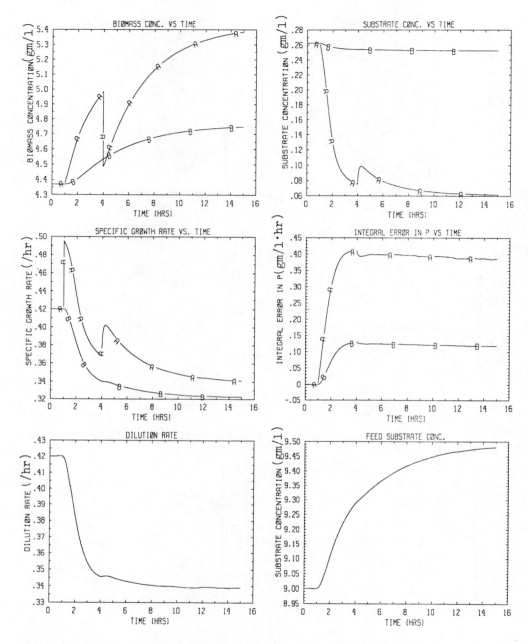

FIGURE 6.   Maintenance of stability in closed-loop simulation of the LQG control system with parameter and state variations. μ_{max}, Y_s, and k_s are changed by $+15\%$, $+15\%$, and -15%, respectively, at t = 1, and the b state is perturbed by -10% at t = 4; -A-A- is for the "true" state values; -B-B- is for "estimated" values.

VI. DISCUSSION AND FUTURE WORK

This chapter has presented the design of a robust multivariable controller for a nonlinear bacterial growth system, using the LQG/LTR design methodology. Simulation results show that the closed-loop system remains stable in the simultaneous presence of parameter variations and step changes of the states. However, several clarifications are in order concerning its application in real-time settings.

First, it is assumed that low-level feedback control loops are in place to control basic operating conditions at the desired setpoint. The basic operating conditions include temperature, pH, agitation rate, gas flow rates, reactor volume, and feed flow rate. These controls can be implemented by local proportional-integral-derivative (PID) controllers, with the setpoint dictated via the results of calculations by the main supervisory program stored in the computer. The proposed LQG controller is to be cascaded onto the low-level feedback loops.

Second, the design philosophy of LQG/LTR is based on linear systems theory. The robustness properties are with respect to linear perturbations. Bacterial growth system dynamics are highly nonlinear. When setpoint control strategy is employed and the nonlinear system is linearized around a nominal setpoint, the resultant controller is expected to work properly only when the system does not deviate appreciably from the nominal setpoint. Simulation results show that the estimated and the "true" state values exhibit a steady-state offset after a parameter variation is introduced. This offset is attributed to the KF being a linear filter; the "true" state values are derived from the nonlinear dynamic equations. An extended KF (for nonlinear systems) which utilizes current state estimates to calculate an adaptive KF gain should track the states more accurately. Initial open-loop simulations using an extended KF show that the estimation is indeed closer than that given by the linear KF. When a disturbance is introduced to the system, the estimates converge very closely, after some transient dynamics, to the "true" state values (values generated by simulating the original nonlinear state equations with nominal values, Equations 1 through 4). The inclusion of an extended KF in the LQG controller block in place of the linear KF has also been carried out.[16] Simulation results show that the tracking ability of the extended KF is somewhat improved; however, the resultant robustness property depends on the noise covariance matrices chosen for the state and measurement noise vectors. The proper choice of the noise covariances so as to render a robust design is not clear at this point.

It is further noted that the offset for the s state is very pronounced, possibly because s, although observable, is not used by the KF to update the estimates. This can be seen from the fact that the measurement is μb and does not depend directly on s. Therefore, s is not updated by the filter, although $\hat{s}$ (the substrate concentration estimate) is fed into the regulator in generating the control input (u = Kx).

In conclusion, the results presented here represent our first attempt to apply a robust controller design methodology to a bioreactor system. In general, the bioreactor operation is very sensitive to environmental perturbations such as variations in temperature and pH. As the results showed, the closed-loop system did remain stable, even in the simultaneous presence of several perturbations. A considerable amount of work in this direction remains to be done. However, this preliminary result is encouraging, providing an impetus for future research in this area. Successful demonstrations of multivariable controller designs such as the one presented here should have a strong impact on industrial biotechnological processes, where economics and automation are prime considerations in plant operations.

REFERENCES

1. **Laub, A.,** Efficient multivariable frequency response computations, *IEEE Trans. Autom. Control*, AC-26(2), 407, 1981.
2. **Staniskis, J. and Levisanskas, D.,** An adaptive control algorithm for fed-batch culture, *Biotechnol. Bioeng.*, 26, 419, 1984.
3. **Gallegos, J. and Gallegos, J.,** Estimation and control techniques for continuous culture fermentation processes, *Biotechnol. Bioeng.*, 26, 442, 1984.

4. **Stephanopoulos, G. and San, K.-Y.,** Studies on on-line bioreactor identification. I. Theory, *Biotechnol. Bioeng.,* 26, 1176, 1984.

5. **Wang, T.-W.,** Feasibility of Applying a Robust Multivariable Controller to a Nonlinear Bacterial Growth System, Ph.D. thesis, University of Tennessee, Knoxville, June 1986.

6. **Wang, T.-W., Moore, C. F., and Birdwell, J. D.,** Feasibility of control of bacterial growth systems, in Proc. 18th Southeastern Symp. on System Theory, Knoxville, TN, April 7 to 8, 1986.

7. **Wang, T.-W., Moore, C. F., and Birdwell, J. D.,** Applying modern control techniques to a nonlinear bacterial growth system, in Proc. 1986 American Control Conference, Seattle, WA, June 1986, 2062.

8. **Wang, T.-W., Moore, C. F., and Birdwell, J. D.,** Control of nonlinear bioreactor system using a robust multivariable control design methodology, in 1987 Annu. American Institute of Chemical Engineers Spring Natl. Meet., Houston, TX, March 29 to April 2, 1987.

9. **Jefferis, R. and Humphrey, A.,** in Proc. 4th Conf. Global Impacts of Applied Microbiology (GIAM-IV), São Paulo, Brazil, July 23 to 28, 1979, 767.

10. **Wang, H. Y., Cooney, C., and Wang, D. I. C.,** Computer control of baker's yeast production, *Biotechnol. Bioeng.,* 21, 975, 1979.

11. **Doyle, J. and Stein, G.,** Multivariable feedback design: concepts for a classical/modern synthesis, *IEEE Trans. Autom. Control,* AC-26(1), 4, 1981.

12. **Lehtomaki, N. A., Sandell, N. R., Jr., and Athans, M.,** Robustness results in linear-quadratic Gaussian based multivariable control designs, *IEEE Trans. Autom. Control,* AC-26(1), 75, 1981.

13. **Stein, G. and Athans, M.,** The LQG/LTR procedure for multivariable feedback control design, *IEEE Trans. Autom. Control,* AC-32(2), 105, 1987.

14. **Birdwell, J. D., Cockett, J. R. B., Heller, R., Rochell, R. W., Laub, A. J., Athans, M., and Hatfield, L.,** Expert systems techniques in a computer-based control system analysis and design environment, in Proc. 3rd International Federation of Automatic Control Symp. Computer Aided Design in Control and Engineering Systems, Lyngby, Denmark, August, 1985.

15. **Birdwell, J. D. and Laub, A.,** Balanced singular values for LQG/LTR design, in Proc. 1986 American Control Conference, Seattle, WA, June 1986, 409.

16. **Lien, C. Y. and Wang, T. W.,** Application of extended Kalman filter in the feedback control of nonlinear bacterial growth systems, in IEEE Southwest Conference, Knoxville, TN, April 1988, 123.

Chapter 13

RepA mRNA, A POTENTIAL RATE-LIMITING FACTOR IN REPLICATION OF R1 PLASMIDS

M. M. Ataai and J. Jachowicz

TABLE OF CONTENTS

I. INTRODUCTION

Plasmids R1, R6-5, and R100 belong to the FII incompatibility group and have similar modes of replication. They all need *Escherichia coli* enzymes in addition to a plasmid-encoded protein (RepA protein). A summary of the replication process is as follows.[1-5]

All of the genes required for replication are located in a 2.5-kb (kilobase) region[6] which also contains the origin of replication. The repA gene is expressed from two different promoters, repA and copB. CopB polypeptide represses the expression of repA from the repA promoter, and in wild-type plasmids the repA gene is expressed constitutively, mainly from the copB promoter.[7] CopA RNA, the primary inhibitor of replication initiation, is expressed constitutively upstream of the origin and is transcribed in the opposite direction from repA.[5,8] The CopA RNA appears to inhibit replication through interaction with CopT, the target sequence, which is part of the RepA mRNA, rather than one of the DNA strands.[9] Expression of the copB gene is also constitutive.[10] When a critical concentration of CopB protein is reached, the expression of repA from the repA promoter is almost fully repressed, and repA is expressed only from the copB promoter.[7,10-11] It has been shown that the RepA protein is absolutely required for *in vitro* replication of plasmid R1, but it is not clear whether or not it is the rate-limiting factor.[12] It is known that the rate of translation of RepA mRNA to form the RepA polypeptide is controlled by CopA RNA. The interaction of CopA RNA with RepA mRNA affects RepA mRNA in such a way that effective translation of RepA mRNA is inhibited. However, due to transcriptional polarity (premature termination of RepA mRNA in the absence of translation), if the RepA mRNA is the primer for replication, it is possible that RepA mRNA also acts as the rate-limiting factor in the replication process.[9]

Despite substantial information reported on replication control of R1, the criterion for initiation of plasmid replication is not well understood. However, there are three components described in the literature which could potentially serve as the rate-limiting factor: RepA mRNA, CopA RNA, and RepA protein. In the first case, RepA mRNA, as the rate-limiting factor for initiation of replication, has to extend to the replication origin, and it serves as a primer RNA. For the second case, because CopA RNA is the rate-limiting component, when its concentration falls below a critical value, a round of plasmid replication will be initiated. Finally, for RepA protein being the rate-limiting factor, it is assumed that when the amount of RepA protein exceeds a critical value, a round of replication is initiated.

The purpose of the work described in this chapter is to design and simulate various computer experiments to assess the plausibility of current hypotheses for replication initiation. The model formulated in this work includes the above-mentioned interactions among the control elements. Other efforts to model replication process of plasmids include the work by Lee and Bailey for λdv replication[13-15] and that of Ataai and Shuler for ColE1 plasmids.[16-17] The model for λdv replication required the knowledge of 12 parameters, of which 9 could be obtained independently and the other 3 were adjusted. The model for ColE1 plasmids requires no adjustable parameter. The model developed for R1 makes use of one or two adjustable parameters, depending on the criterion employed for replication initiation.

The modeling of the replication process of R1 plasmids is a challenging task, since the sequence of molecular events leading to replication initiation is not as well understood as that of either λdv or the ColE1 plasmid. Thus, formulation of a model for R1 replication may serve as an effective tool for assessing the extent of insight that a model of this nature can provide in a situation where, despite availability of a large body of information on the overall replication process, the mechanism for the key event of replication (i.e., initiation) is not known.

In the rest of this chapter, the formulation of the model is first presented; then, a description of how the model can be used to test various initiation mechanisms is provided.

II. MODEL FORMULATION

The rate of change in the number of CopB mRNA molecules of a single cell per unit time can be written as

$$d/dt(\text{CopB mRNA}) = K_{TC(copB)}PL - k_{d(CopB\ mRNA)}(\text{CopB mRNA}) \tag{1}$$

where (in Equation 1 and other equations) the symbols $K_{TC(i)}$, $K_{TL(i)}$, and $k_{d(i)}$ represent the average rates of transcription, translation, and degradation of component i, respectively. PL is the number of plasmid molecules per cell at time t (i.e., gene dosage). The first term in Equation 1 is the rate of formation of CopB mRNA (i.e., the product of the rate of transcription from the copB gene at time t and the gene dosage). The second term is the rate of degradation of CopB mRNA (i.e., the product of the degradation rate and the number of CopB mRNA molecules per cell). The rate equation for the CopB protein can be written as

$$d/dt(\text{CopB protein}) = K_{TL(CopB\ protein)}(\text{CopB mRNA}) - k_{d(CopB\ protein)}(\text{CopB protein}) \tag{2}$$

Similarly, the rate equations for CopA RNA, RepA mRNA, and RepA protein are written as

$$d/dt(\text{CopA RNA}) = (1 - f)K_{TC(copA)}PL - k_{d(CopA\ RNA)}(\text{CopA RNA})$$
$$- (k_2/VC)(\text{RepA mRNA})(\text{CopA RNA}) \tag{3}$$

$$d/dt(\text{RepA mRNA}) = K_{TC(repAp)}\{[k_i]/[k_i + (\text{CopB protein})]\}PL$$
$$+ K_{TC(copBp)}PL - k_{d(RepA\ mRNA)}(\text{RepA mRNA}) \tag{4}$$
$$- (k_2/VC)(\text{RepA mRNA})(\text{CopA RNA})$$

$$d/dt(\text{RepA protein}) = K_{TL(RepA\ protein)}(\text{RepA mRNA}) - k_{d(RepA\ protein)}(\text{RepA protein}) \tag{5}$$

$K_{TC(repAp)}$ and $K_{TC(copBp)}$ are the average transcription rates of RepA mRNA from the repA and copB promoters, respectively. $K_{TC(repAp)}$ is multiplied by a feedback inhibition factor to indicate the repression of transcription from the repA promoter due to the interaction of the CopB protein with the operator region of that promoter. Use of such a feedback term will cause complete derepression of the repA promoter if the copB gene is defective. The constant k_2 is the second-order binding constant between CopA RNA and the copT sequence of RepA mRNA. Apparently, the interaction between CopA RNA and the copT sequence is one to one.[5] The inhibition constant (k_i) is chosen so as to cause about 90% repression of the repA promoter if a wild-type level of CopB protein is present.[5] The rate of formation of CopA RNA (Equation 3) is multiplied by $(1 - f)$ to indicate the formation of active CopA RNA molecules.[18] As will be discussed later, in copB plasmid mutants a fraction (f) of CopA RNA molecules are elongated and are not active. For all other cases, except for copB mutants, the value of f is equal to zero. The values of the model parameters are given in Table 1. The procedure for determining the values of the model parameters is given in Reference 19. The model for plasmid replication was included in the single-cell model of *E. coli* developed by Shuler and co-workers.[20] The growth-rate dependency of the transcription and translation rates of the plasmid genes was assumed to be similar to that of an average *E. coli* gene, as it is predicted from the single-cell model.

TABLE 1
Values Used for Model Parameters in Equations 1—5

$K_{TC(copB)} = 30.0$	$K_{TC(repAp)} = 53.0$
$k_{d(CopB\ mRNA)} = 27.0$	$k_i = 222.0$ molecules
$K_{TL(CopB\ protein)} = 1200.0$	$K_{TC(copBp)} = 30.0$
$k_{d(CopB\ protein)} = 0.07$	$k_{d(RepA\ mRNA)} = 27.0$
$K_{TC(copA)} = 300.0$	$K_{TL(RepA\ protein)} = 1200.0$
$k_{d(CopA\ RNA)} = 27.0$	$k_{d(RepA\ protein)} = 0.07$
$f = 0.4$	$k_2 = 1.15 \times 10^{-11}$ ml/molecule-h

Note: Transcription rate constants are in transcripts/h-promoter; translation rate constants are in proteins/mRNA-h; degradation rate constants are in h^{-1}

III. RESULTS AND DISCUSSION

We have examined the hypotheses for RepA mRNA and CopA RNA as the rate-limiting substances for initiation of replication. To perform a simulation in which RepA mRNA is the rate-limiting factor, it was assumed that a round of replication will initiate when a RepA mRNA transcript extends to the replication origin before it binds a CopA RNA molecule. The mathematical formulation of this criterion is similar to that described previously for ColE1 plasmids.[16] In this case, there is only one adjustable parameter — a kinetic constant for the rate of binding between RepA mRNA and CopA RNA molecules. The value of the binding constant, k_2, was chosen such that the predicted plasmid copy number at a growth rate of 0.7 h^{-1} would be equal to the experimentally reported value. The value of k_2 found in this manner is 1.15×10^{-11} ml/molecule-h, which is close to the value reported for the second-order binding constant between the two RNA species of ColE1 plasmids.[21]

To test the second hypothesis, it was assumed that the level of plasmid-encoded inhibitor, CopA RNA, controls the rate of replication. Replication can take place only when the concentration of the inhibitor falls below a critical value. In this case, there are two adjustable parameters: the binding constant and a critical value for CopA RNA concentration. The value for the binding constant in this simulation is the same value as the one used for RepA mRNA as the rate-limiting step. The critical concentration of CopA RNA (5.41×10^{13} molecules per milliliter) was chosen so that the predicted plasmid copy number at a growth rate of 0.7 h^{-1} would equal that determined experimentally.

Four types of simulations were performed for every replication initiation criterion. First, the model prediction of plasmid copy number as a function of growth rate was compared with experimental values. The results are shown in Table 2. The agreement between the experiment and model predictions appears to be satisfactory when either CopA RNA or RepA mRNA is the rate-limiting element.

Second, we investigated the effect of copA gene dosage on replication initiation. It has been found experimentally that an increase in copA gene dosage proportionately reduces the plasmid copy number.[5] The model predictions are very consistent with experimentally measured values in this regard.

Third, we simulated the effect of mutations in the copB gene on plasmid copy number. It has been observed that mini-R1 plasmids lacking the promoter-proximal part of the copB gene are copy mutants having a six- to eightfold increased copy number relative to R1.[5] It has also been shown that in copB mutant plasmids some of the CopA RNA molecules are larger (i.e., not terminated properly) than in a wild-type plasmid. These elongated CopA RNAs have little or no inhibition activity.[15] The simulations were performed using an elongation factor, f, which accounts for the percentage of inactive CopA RNA molecules.[18] The results are presented in Table 3. Seven and fivefold increases in plasmid copy number

TABLE 2
The Effect of Growth Rate on Plasmid Copy Number and Comparison of Two Versions of the Model with Experimental Results

Cell growth rate (h^{-1})	Copy number experiment[a]	Copy number RepA mRNA rate limiting	Copy number CopA RNA rate limiting
0.1	26.0	30.9	35.0
0.2	16.0	15.7	14.6
0.4	6.0	5.9	4.6
0.6	3.4	3.5	3.2
0.7	3.1	3.1	3.1
0.8	2.8	2.6	2.4
0.9	2.3	2.2	2.0

[a] Experimental values are from Reference 23.

TABLE 3
The Effect of copB Mutation on Plasmid Copy Number

Elongation factor (f)	Copy number RepA mRNA rate limiting	Copy number CopA RNA rate limiting
0.4	23.06	15.1

TABLE 4
The Effect of Binding Rate between RepA mRNA and CopA RNA on Predicted Plasmid Copy Number, Using a Growth Rate of 0.7 h^{-1}

Ratio of binding constants (relative to wild-type value)	Copy number RepA mRNA rate limiting	Copy number CopA RNA rate limiting
0.1	30.1	2.9
0.2	15.2	2.9
1.0[a]	3.1	3.1
5.0	0.7	3.1

[a] Wild type.

relative to wild-type level were predicted by the model for RepA mRNA and CopA RNA as the rate-limiting element, respectively.

Fourth, we investigated the sensitivity of the predicted plasmid copy number to changes in the value of the binding constant. It has been shown experimentally that single base mutations in the DNA region coding for CopA RNA change the plasmid copy number.[4] This change in copy number may be due to the effect of mutation on the binding rate between the two RNAs. Model calculations indicate that the plasmid copy number changes in proportion to alterations in the binding constant, provided that RepA mRNA is used as the rate-limiting factor for replication initiation (see Table 4). If, on the other hand, CopA RNA is used as the rate-limiting component, the plasmid copy number is insensitive to variation in the value of the binding constant, as shown in Table 4. Based on this observation, we have concluded that the hypothesis for CopA RNA as the rate-limiting element in the replication process may not be a plausible one. The physical basis for the lack of sensitivity of predicted plasmid copy number to alterations in the value of k_2 (the binding constant) is the much higher transcription rate of CopA RNA than RepA mRNA. Hence, the intracellular concentration of CopA RNA is not sensitive to its rate of complex formation with RepA mRNA, which has a very low intracellular concentration. Therefore, the inhibitor-dilution model proposed by Pritchard[22] may not be applicable to the replication of R1 plasmids and, possibly, ColE1 plasmids. Also, in ColE1 plasmids the transcription rate of RNA I (the inhibitor RNA) is about seven times higher than that of RNA II. RNA II binds to RNA I and regulates the replication initiation frequency.

IV. SUMMARY AND CONCLUSIONS

In summary, a model of the replication process of R1 plasmids was developed. The model simulates the effect of cell growth rate, copA gene dosage, copB mutation, and the CopA RNA to RepA mRNA binding constant on plasmid copy number. Agreement between the experimental data and model predictions was found to be satisfactory using RepA mRNA as the rate-limiting factor, but a hypothesis based on CopA RNA as the rate-limiting element does not appear to be plausible. The possibility of the RepA protein being the rate-determining step in the replication initiation process is currently under investigation.

ACKNOWLEDGMENT

This work was supported by Public Health Service Grant No. R15GM37443 from the National Institute of General Medical Sciences of the National Institutes of Health.

REFERENCES

1. **Molin, S., Stougaard, P., Uhlin, B. E., Gustafsson, P., and Nordstrom, K.,** Clustering of genes involved in replication, incompatibility, and stable maintenance of the resistance plasmid R1drd-19, *J. Bacteriol.,* 138, 70, 1979.
2. **Dong, X., Womble, D. D., Luckow, V. A., and Rownd, R. H.,** Regulation of the transcription of the repAi gene in the replication control region of Inc FII plasmid NRI by gene dosage or repA2 transcription repressor protein, *J. Bacteriol.,* 161, 544, 1985.
3. **Rosen, J., Ryder, T., Inokuchi, H., Ohtsubo, H., and Ohtsubo, E.,** Genes and sites involved in replication incompatibility of an R100 plasmid derivative based on nucleotide sequence analysis, *Mol. Gen. Genet.,* 179, 527, 1980.
4. **Givskov, M. and Molin, S.,** Copy mutants of plasmid R1: effects of base pair substitutions in the copA gene on the replication control system, *Mol. Gen. Genet.,* 194, 286, 1984.
5. **Nordstrom, K., Molin, S., and Light, J.,** Control of replication of bacterial plasmids: genetics, molecular biology, and physiology of the plasmid R1 system, *Plasmid,* 12, 71, 1984.
6. **Kollek, R., Oertel, W., and Goebel, W.,** Isolation and characterization of the minimal fragment required for autonomous replication of a copy mutant (pKN102) of the antibiotic resistance factor R1, *Mol. Gen. Genet.,* 162, 51, 1978.
7. **Light, J. and Molin, S.,** The sites of action of the two copy number control function of plasmid R1, *Mol. Gen. Genet.,* 187, 486, 1982.
8. **Stougaard, P., Molin, S., and Nordstrom, K.,** RNA molecules involved in copy number control and incompatibility of plasmid R1, *Proc. Natl. Acad. Sci. U.S.A.,* 78, 6008, 1981.
9. **Light, J. and Molin, S.,** Post-transcriptional control of expression of the repA gene of plasmid R1 mediated by a small RNA molecule, *EMBO J.,* 2, 93, 1983.
10. **Light, J. and Molin, S.,** Expression of a copy number control gene (copB) of plasmid R1 is constitutive and growth rate dependent, *J. Bacteriol.,* 151, 1129, 1982.

11. **Light, J. and Molin, S.,** Replication control functions of plasmid R1 act as inhibitor of expression of a gene required for replication, *Mol. Gen. Genet.,* 184, 56, 1981.
12. **Masai, H., Kaziro, Y., and Arai, K. I.,** Definition of oriR, the minimum DNA segment essential for initiation of R1 plasmid replication *in vitro, Proc. Natl. Acad. Sci. U.S.A.,* 80, 6814, 1983.
13. **Lee, S. B. and Bailey, J. E.,** A mathematical model for λdv plasmid replication: analysis of wild type plasmid, *Plasmid,* 11, 151, 1984.
14. **Lee, S. B. and Bailey, J. E.,** A mathematical model for λdv plasmid replication: analysis of a copy number mutant, *Plasmid,* 11, 166, 1984.
15. **Lee, S. B. and Bailey, J. E.,** Analysis of growth rate effects on productivity of recombinant *Escherichia coli* population using molecular mechanism models, *Biotechnol. Bioeng.,* 26, 66, 1984.
16. **Ataai, M. M. and Shuler, M. L.,** Mathematical model for the control of ColE1 type plasmid replication, *Plasmid,* 16, 204, 1986.
17. **Ataai, M. M. and Shuler, M. L.,** A mathematical model for prediction of plasmid copy number and genetic stability in *Escherichia coli, Biotechnol. Bioeng.,* 30, 389, 1987.
18. **Stougaard, P., Light, J., and Molin, S.,** Convergent transcription interferes with expression of the copy number control gene, copA, from plasmid R1, *EMBO J.,* 1, 323, 1982.
19. **Jachowicz, J.,** A Mathematical Model for Replication of R1 Plasmid, M.S. thesis, Polytechnic University, Brooklyn, NY, 1988.
20. **Domach, M. M., Leung, S. K., Cahn, R. E., Cocks, G. G., and Shuler, M. L.,** Computer model for glucose-limited growth of a single cell of *Escherichia coli* B/r A, *Biotechnol. Bioeng.,* 26, 203, 1984.
21. **Tomizawa, J.,** Control of ColE1 plasmid replication: the process of binding of RNA I to the primer transcript, *Cell,* 38, 861, 1984.
22. **Pritchard, R. H.,** Control of DNA synthesis in bacteria, *Heredity,* 23, 472, 1968.
23. **Engberg, B. and Nordstrom, K.,** Replication of the R-factor R1 in *Escherichia coli* K-12 at different growth rates, *J. Bacteriol.,* 123, 179, 1975.

Chapter 14

BACILLUS SUBTILIS AS A HOST OF RECOMBINANT VECTORS: THE PROTEASE PROBLEM

M. M. Ataai, W. J. Strohm, and J. W. Jeong

TABLE OF CONTENTS

I. INTRODUCTION

Escherichia coli and *Bacillus subtilis* are the most widely used bacteria for hosting cloned genes, primarily because a vast amount of information is available about the genetics of both systems. Vectors for cloning foreign genes and procedures for their insertion into these hosts have been developed successfully.[1-3] Despite in-depth study at the molecular level, *E. coli* is not a familiar industrial microorganism due to several of its inherent properties. As a Gram-negative bacterium, *E. coli* has a toxic lipopolysaccharide outer membrane[3] which is hard to remove, but it must be removed in pharmaceutical applications. The characteristic double-walled membrane of *E. coli* usually restricts proteins produced either to intracellular or periplasmic space.[4-8] Thus, the recovery of the product would require either removal of the outer membrane, making the cell leaky (i.e., making the membrane permeable to allow the release of periplasmic proteins), or cell lysis.[9] Mutations that lead to a leaky phenotype also adversely affect cell viability and, hence, complicate their routine cultivation.[10] Owing to these major hindrances, the toxic lipopolysaccharide cell wall and the inability to excrete protein products into the extracellular fluid, there exists a great interest in developing cloning and expression systems utilizing organisms other than *E. coli*.

B. subtilis, a Gram-positive bacterium, offers several advantages as a host of foreign genes. It has a simple, nontoxic cell wall and a single membrane, thus allowing secretion of protein products into the culture fluid and facilitating protein purification. *B. subtilis* is nonpathogenic to humans and is able to grow on either minimal or a complex medium. These are substantial benefits, and, thus, a good deal of energy is being expended in developing plasmids useful for gene cloning in *B. subtilis*.

The use of *B. subtilis* as an efficient host for the production of proteins depends on several factors. First, a good understanding of its genetics and transcription/translation mechanisms is required. These processors are being clarified through the compilation of numerous data sets. Second, a good understanding of the protein secretion process might reveal factors which could be manipulated to allow the secretion of normally intracellular proteins. Finally, the major obstacle is extracellular proteolytic enzymes (proteases) which degrade the excreted foreign proteins. *B. subtilis* transformants carrying TEM β-lactamase,[11] interferons,[12,13] and rat proinsulin[14] have been designed successfully, but the protein products are rapidly degraded by the action of proteases. The purpose of this chapter is to address the protease problem in relation to *B. subtilis* as a host of recombinant vectors for protein production.

Several types of proteolytic enzymes have been identified and isolated from bacilli. The level of extracellular proteases is observed to increase rapidly when the cells reach their postexponential phase of growth. The best-characterized extracellular proteases excreted by bacilli belong in one of two classes: alkaline (serine active site) or zinc-containing neutral proteases.[15,16] Low-protease *B. subtilis* mutants having deletions in genes coding for the two major proteases have been constructed. One such mutant is *B. subtilis* strain DB104. The protease activity in the supernatant of batch-grown cells of DB104 is about 4% of the wild-type level.[17] This 4% activity was postulated to be a combination of intracellular proteases (released through cell lysis) and esterases which may have low, but demonstrable, proteolytic activity. It was this low, yet substantial, level of protease activity which we attempted to minimize. To our knowledge, all of the low-protease mutants constructed so far have some residual protease activity when cultivated in a batch reactor. Apparently, the use of the term ''protease-deficient mutant'' instead of ''low-protease mutant'', along with the lack of appreciation for variation in degree of susceptibility of different proteins to protease degradation, have led some researchers not directly involved with the protease problem to believe that this problem has been overcome.[18]

It is important to note that the degree of reduction in extracellular protease activity required to avoid the degradation of an excreted protein depends on the susceptibility of the protein to protease action. In this chapter we describe the effect of dilution rate in a chemostat on biosynthesis of extracellular proteases.

II. MATERIALS AND METHODS

Bacterial strains *B. subtilis* 168 and DB104 (a low-protease mutant deficient in both alkaline and neutral proteases), kindly supplied to us by Dr. Roy H. Doi (University of California, Davis), were used. The medium contained 8.0 g nutrient broth, 2.0 g glucose per liter, and the salt concentrations given in Reference 19. The protease assay was performed using azocasein as a substrate as described by Millet.[19] One of the steps in Millet's procedure which involves addition of NaOH to the assay solution was omitted.[23] A unit of protease activity was defined as the amount of enzyme which gives ΔA/time (min) $\times$ 1000 = 1.0, where ΔA is the change in absorbance using the azocasein assay described above. This unit of activity would correspond to the activity of a solution containing 2.0×10^{-4} mg alkaline protease per milliliter. The proteolytic activity was also monitored by measuring the degradation rate of TEM β-lactamase.[20] β-Lactamase (approximate activity = 590 units per milliliter) was kindly supplied to us by Mr. Jeffrey Chalmers of Cornell University. This β-lactamase was obtained through cultivation of *E. coli* cells containing plasmid pKK.[10]

The continuous culture experiments were performed using Multigen F-1000 (New Brunswick Scientific, New Brunswick, NJ) with a working volume of 330 ml. The culture medium was introduced through an inlet on the top of the vessel by means of a Master Flex pump (Cole Parmer, Chicago, IL). The temperature was maintained at 37 ± 0.2°C.

III. RESULTS AND DISCUSSION

It has been observed that biosynthesis of extracellular protease begins at late log phase of growth in a batch culture. From this observation one may conclude that operating a chemostat at a high dilution rate should repress protease biosynthesis to a great extent.

Figure 1 shows the culture optical density, pH, and specific protease activity vs. dilution rate in a glucose-limiting chemostat. The specific protease activity is the total units of protease activity per OD_{660} = 1.0. The first notable part of this plot is the dramatic inverse relationship between dilution rate and the extracellular protease levels. The specific protease activity in batch culture of *B. subtilis* 168 is 35 units. The specific activity at a dilution rate greater than 0.3 h^{-1}, on an average, is 0.6 units. This level of activity is 58 times lower than the batch value. Indeed, the level of activity at a moderate range of chemostat dilution rates is lower than the value found in batch culture of the low-protease mutant, DB104. The specific protease activity found in batch culture of DB104 is 1.4 units. Also, it appears that the pH of the culture decreases with an increase in optical density. Since the protease assay was performed at a constant pH, the measured value of protease activity reflects the true rate of synthesis.

Figure 2 shows the specific protease activity as a function of dilution rate in a chemostat culture of *B. subtilis* DB104. Again an inverse relation between the protease activity and dilution rate is observed. Also note that, similar to Figure 1, above some critical dilution rate (0.3 h^{-1}) protease synthesis is not significantly growth-rate dependent. The level of protease activity in chemostat culture of DB104 at a moderate to high range of dilution rate is about 0.25 units, which is six times lower than the corresponding value in batch culture (i.e., 1.4 units).

IV. SUMMARY AND CONCLUSIONS

In summary, protease activity substantially lower than that found in batch cultures can

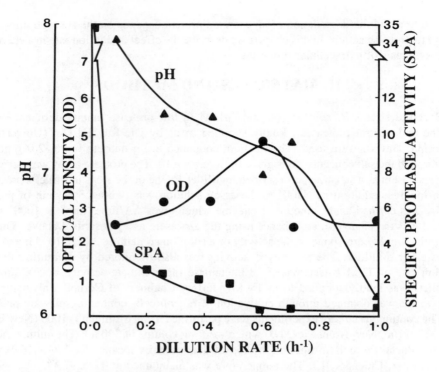

FIGURE 1.   Optical density, pH, and specific protease activity as a function of dilution rate in a continuous culture of *B. subtilis* 168. Symbols ▲, ●, and ■ denote pH, OD_{600}, and specific protease activity, respectively.

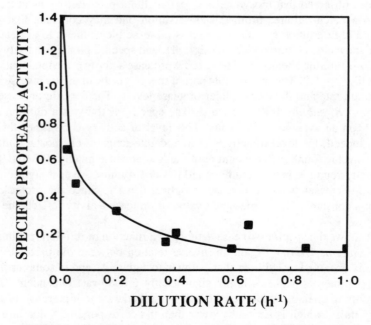

FIGURE 2.   Specific protease activity as a function of dilution rate in a chemostat culture of *B. subtilis* DB104.

be obtained in a continuous culture at a moderate to high range of dilution rates (58 times lower for the wild-type cell and 6 times lower for the low-protease mutant). As stated above, the degree of reduction in protease level required for protection of an excreted foreign protein depends on the protein itself. We have found TEM β-lactamase to be fairly stable when incubated in a solution containing 0.3 units of protease activity (about 100-fold less than the wild-type level in a batch culture). However, growth hormones in the same solution of protease are very unstable (half-life of about 15 min).[21] Therefore, stability of some proteins in the culture of a low-protease mutant does not indicate that a similar observation should be expected for other proteins.

It should be noted that the repression of the protease gene may also repress the promoter used for the expression and secretion of the foreign gene and its product, particularly when one considers that many of the promoters of the genes for the secreted proteins from *B. subtilis* are expressed during the stationary phase in a batch culture, when most of the protease genes are also expressed. However, this problem can be simply circumvented by using vegetative promoters which are expressed during the exponential phase of growth. For example, Wong et al.[22] have developed an efficient *B. subtilis* secretory system by employing a secretion vector containing the gene for TEM β-lactamase fused to a mutated subtilisin promoter which is expressed only during the growth phase.

Current research in molecular biology is focused on isolating and purifying unidentified proteases contributing to the residual protease activity found in cultures of low-protease mutants. Having purified new components with proteolytic activity, one can attempt to delete the genes coding for these new proteases. Although such an approach is a potentially useful one, the deletion of the genes could adversely affect cell physiology and/or growth. Our approach, partially described in this manuscript, clearly shows the possibility of a substantial reduction in protease biosynthesis by cultivating cells in a chemostat operated at a high growth rate. We have very recently shown that reduction of the protease level below the values reported in this manuscript is possible through other manipulation.

Experimental measurements of the effects of pH, dissolved oxygen concentration, the addition of inhibitors, and the nature of nutrient limitation on protease biosynthesis in a continuous culture of DB104, along with the results of a washout chemostat experiment which indicated the existence of a log-phase protease, are to be discussed elsewhere.

ACKNOWLEDGMENT

We are thankful to Professor Roy H. Doi of the Biochemistry Department at the University of California, Davis, for many helpful discussions.

REFERENCES

1. **Maniatis, T., Frisch, E. F., and Sambrook, J.,** *Molecular Cloning: A Laboratory Manual,* Cold Spring Harbor Laboratory, Cold Spring Harbor, NY, 1982.
2. **Dubnau, D.,** Genetic transformation in *Bacillus subtilis,* in *The Molecular Biology of the Bacilli,* Vol. 1, Dubnau, D., Ed., Academic Press, New York, 1982, 148.
3. **Cryczan, T. J.,** Molecular Cloning in *Bacillus subtilis,* in *The Molecular Biology of Bacilli,* Vol. 1, Dubnau, D., Ed., Academic Press, New York, 1982, 307.
4. **Gray, O. and Chang, S.,** Molecular cloning and expression of *Bacillus licheniformis* β-lactamase gene in *Escherichia coli* and *Bacillus subtilis, J. Bacteriol.,* 145, 422, 1981.
5. **Brammer, W. J., Muir, S., and McMorris, A.,** Molecular cloning of the gene for the β-lactamase of *Bacillus licheniformis* and its expression in *Escherichia coli, Mol. Gen. Genet.,* 178, 217, 1980.
6. **Clement, J. M., Perrin, D., and Hedgpeth, J.,** Analysis of receptor and β-lactamase synthesis and export using cloned genes in a Mini cell system, *Mol. Gen. Genet.,* 185, 302, 1982.

7. **Robson, L. M. and Chambliss, G. H.,** Cloning of the *Bacillus subtilis* DLG β-1,4-glucanase gene and its expression in *Escherichia coli* and *B. subtilis, J. Bacteriol.,* 165, 612, 1986.

8. **Cornelis, P., Dignette, C., and Willemot, K.,** Cloning and expression of a *Bacillus coagulans* amylase gene in *Escherichia coli, Mol. Gen. Genet.,* 186, 507, 1982.

9. **Anderson, J. J., Wilson, J. M., and Oxender, D. L.,** Defective transport and other phenotypes of a periplasmic leaky mutants of *Escherichia coli* K-12, *J. Bacteriol.,* 140, 351, 1979.

10. **Georgiou, G., Chalmers, J. J., Shuler, M. L., and Wilson, D. B.,** Continuous protein production form *E. coli* capable of selective protein excretion: a feasibility study, *Biotechnol. Prog.,* 1, 75, 1985.

11. **Yamane, K., Otozai, K., Ohmura, K., Nakayama, A., Yamazaki, H., Yamasaki, M., and Tamura, G.,** Secretion vector of *Bacillus subtilis* constructed from the *Bacillus subtilis* α-amylase promoter and signal peptide coding region, in *Genetics and Biotechnology of Bacilli,* Ganesan, A. T. and Hoch, J. A., Eds., Academic Press, New York, 1984, 181.

12. **Sibakov, M., Lehtovaara, P., Petterson, R., Lundstrom, K., Kalkkinen, N., Ulmanen, I., Takkinen, K., Kaarianinen, L., Palva, I., and Sarvas, M.,** Secretion of foreign gene products by the aid of a bacillus secretion vector, in *Genetics and Biotechnology of Bacilli,* Ganesan, A. T. and Hoch, J. A., Eds., Academic Press, New York, 1984, 153.

13. **Palva, I., Sarvas, M., Lehtovaara, P., Sibakov, M., and Kaarianinen, L.,** Secretion of *Escherichia coli* β-lactamase from *Bacillus subtilis* by the aid of β-amylase signal sequence, *Proc. Natl. Acad. Sci. U.S.A.,* 79, 5582, 1982.

14. **Mosbach, K., Birmbaum, S., Hardy, K., Davies, J., and Bulow, L.,** Formation of proinsulin by immobilized *Bacillus subtilis, Nature (London),* 302, 543, 1983.

15. **Leighton, T. J. and Doi, R. H.,** The relationship of serine protease activity to RNA polymerase modification and sporulation in *Bacillus subtilis, J. Mol. Biol.,* 76, 103, 1973.

16. **Keay, L., Moser, P. W., and Wildi, B. S.,** Proteases of the genus *Bacillus.* II. Alkaline proteases, *Biotechnol. Bioeng.,* 12, 213, 1970.

17. **Kawamura, F. and Doi, R. H.,** Construction of a *Bacillus subtilis* double mutant deficient in extracellular alkaline and neutral proteases, *J. Bacteriol.,* 160, 442, 1984.

18. **Holland, I. B., Mackman, N., and Nicaud, J.-M.,** Secretion of proteins from bacteria, *Bio/Technology,* 4, 427, 1986.

19. **Millet, J.,** Characterization of proteinases excreted by *Bacillus subtilis* Marburg during sporulation, *J. Appl. Bacteriol.,* 33, 207, 1970.

20. **Citri, N., Garber, N., and Sela, M.,** The effect of urea and guanidine hydrochloride on activity and optical rotation of Penicillinase, *J. Biol. Chem.,* 235, 3454, 1960.

21. **Doi, R. H.,** personal communication.

22. **Wong, S., Kawamura, F., and Doi, R. H.,** Use of the *Bacillus subtilis* Subtilisin signal peptide for efficient secretion of TEM β-lactamase during growth, *J. Bacteriol.,* 168, 1005, 1986.

23. **Hageman, J.H.,** personal communication.

Part 3. Biosensors

The term "biosensors" has at least two meanings: sensors of biological activity and measuring devices that make use of a biological process. Both of these meanings are deliberately used in this section. For example, in one chapter, Karube and Tamiya describe sensors that utilize enzymatic catalysis as a means of converting a substrate (analyte) and producing an electrical signal. On the other hand, Callis discusses the use of optical methods and Lauks et al. describe the application of field-effect transistor methods ("ChemFET") for measuring biochemical analytes.

Chapter 15

INTEGRATED BIOSENSORS

Isao Karube and Eiichi Tamiya

TABLE OF CONTENTS

I. INTRODUCTION

The determination of organic compounds is required in clinical, industrial, and environmental analyses. Most analyses of organic compounds can be performed by spectrophotometric methods utilizing specific enzyme-catalyzed reactions. However, these methods often require long reaction times and complicated procedures. On the other hand, sensors employing immobilized biocatalysts have definite advantages. Miniaturization of the enzyme sensors is a prerequisite for medical application. This has been achieved using semiconductor fabrication technology combined with enzyme immobilization techniques to produce highly selective microbiosensors.

A multifunctional biosensor is also desirable so that the determination of several compounds can be performed at the same time. The multifunctional biosensor can be constructed by the integration of microbiosensors.

In this chapter, we describe ion-sensitive field effect transistors (ISFETs) and microelectrodes prepared by silicon fabrication technology and employed as microbiosensor transducers. Microbiosensors for urea, adenosine triphosphate (ATP), glucose, and glutamate are described.

II. UREA FET SENSOR

The ISFET was first reported by Bergveld in 1970,[1] and Matsuo and Wise[2] improved the ISFET properties by utilizing silicon nitride (Si_3N_4) as the gate insulator and reported its use as a pH sensor. ISFET consists of a source, drain, and gate. Current between the source and drain is dependent on gate potential, which is selective related to H^+ concentration. In 1980, Caras and Janata[3] demonstrated that an immobilized penicillinase layer over the gate insulator of the ISFET could be used as a penicillin sensor. We have reported an enzyme-FET sensor.[4] Urea determinations are based on spectrophotometry, and they involve complicated and delicate procedures. Therefore, the development of an inexpensive and miniaturized sensor that is highly selective and sensitive is extremely desirable. Realization of these goals can be achieved using the ISFET transducer.

Fabrication of the ISFET uses basically the same procedures as are employed for the metal-insulator-semiconductor field effect transistor (MISFET), as reported by Matsuo and Esashi.[5] The structure of the ISFET is shown in Figure 1. The gate insulator of the ISFET is composed of two layers, the lower being thermally grown silicon dioxide (SiO_2) and the upper being Si_3N_4, which is sensitive to H^+ ions and also has a barrier effect on ion penetration. The thicknesses of the SiO_2 and Si_3N_4 layers are approximately 0.1 μm. The sensor system consists of two ISFETs; one ISFET is covered with a cross-linked polyvinylbutyral membrane containing amino groups, onto which urease has been immobilized through a Schiff base linkage (urea ISFET), and the other ISFET is only covered with a cross-linked polyvinylbutyral resin membrane reference field effect transistor (REFFET). The polyvinylbutyral membrane was spread onto the gate insulator of the ISFETs by a dropping method. A polymer solution was dropped onto the gate insulator of the two ISFETs and then immersed in a 5% glutaraldehyde solution at room temperature for approximately 1 d to advance the cross-linking reaction. Urease was immobilized on the ISFET by immersing the ISFET in a 5 mg/ml urease solution at 4°C for approximately 1 d.

Measurements of urea concentration were performed in a differential mode by comparing the difference in gate output voltages of the urea FET gate and the reference gate. A schematic diagram of the circuit is shown in Figure 2. An Ag/AgCl reference electrode was placed directly in solution with the urea FET and the REFFET. A change in solution pH affects the gate insulator surface potential, with a concomitant proportional change in the gate output voltage. The differential gate output voltage reached a steady state approximately 2 min after injection of urea.

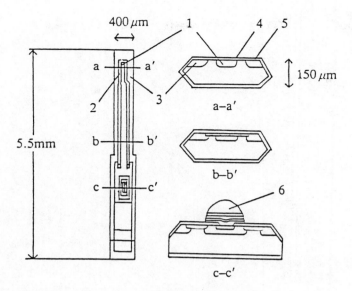

FIGURE 1. Structure of ISFET: 1, drain; 2, gate; 3, source; 4, Si$_3$N$_4$; 5, SiO$_2$; 6, solder. (From Turner, A. P. F., Karube, I., and Wilson, G. S., Eds., *Biosensors: Fundamentals and Applications,* Oxford University Press, New York, 1987, 472. With permission.)

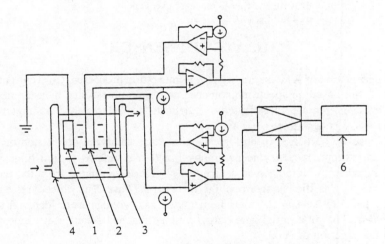

FIGURE 2. Circuit diagram of measuring system: 1, Ag/AgCl reference electrode; 2, urease FET; 3, REFFET; 4, cell; 5, differential amplifier; 6, recorder. (From Turner, A. P. F., Karube, I., and Wilson, G. S., Eds., *Biosensors: Fundamentals and Applications,* Oxford University Press, New York, 1987, 473. With permission.)

The initial rate of change of the differential gate output voltage after injection was plotted against the logarithm of the urea concentration. Figure 3 is a calibration curve of the urea sensor system. A linear relationship was obtained between the initial rate of voltage change and the logarithm of urea concentration over the range 1.3 to 16.7 mM urea. An examination of the selectivity of the urea sensor system showed that it did not respond to glucose, creatinine, and albumin.

The stability of the urea sensor system was also examined. The urea FET was stored at 4°C between measurements and exhibited a response to 16.7 mM urea for at least 2 weeks.

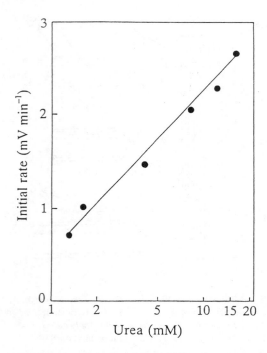

FIGURE 3.   Urea calibration curve. Experiments were performed at 37°C, pH 7.0.

III. ATP FET SENSOR

The determination of ATP is important in fermentation processes. Conventional methods of ATP assay are based on spectrophotometric and bioluminescence measurements. These methods, however, require complicated and delicate procedures, and a simpler and more inexpensive assay is desirable.

H^+-ATPase (EC 3.6.1.3) in biological membranes catalyzes production or hydrolysis of ATP. Furthermore, the enzyme has many functions, such as proton transport, which could be utilized for a biomolecular device.[6] H^+-ATPase, prepared from the thermophilic bacterium PS3, is classified as thermophilic F_1 (TF_1) ATPase. The procedures employed in constructing the ATP sensor and measurements of gate voltage were identical to those of the urea sensor. The differential gate output voltage reached steady state approximately 4 to 5 min after injection of ATP.

The initial rate of change of the differential gate output voltage after injection of ATP was plotted against the logarithm of the ATP concentration. Figure 4 is a calibration curve of the ATP sensor system. A linear relationship was obtained between the initial rate of the voltage change and the logarithm of ATP concentration over the range 0.2 to 1.0 mM ATP.

Slight responses were obtained when 1 mM glucose, urea, and creatinine were applied to the system. The response of the system to 1 mM ATP was retained for 18 d.[7]

IV. MICRO-GLUCOSE SENSOR

The determination of glucose is important in food and fermentation industries, and the development of biodevices would be of considerable help in routine laboratory work. The development of individual miniaturized enzyme-based sensors is required for preparation of integrated biosensors. Therefore, a micro-hydrogen peroxide (H_2O_2) electrode has been developed utilizing currently available integrated circuit (IC) technology. The structure of

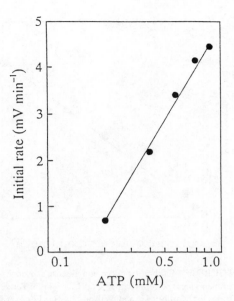

FIGURE 4. ATP calibration curve. Experiments were performed at 40°C, pH 7.0.

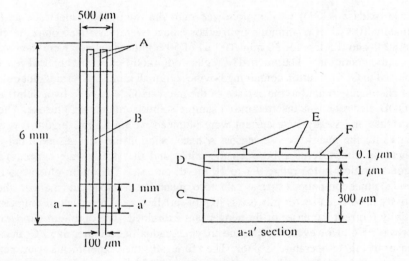

FIGURE 5. Schematic diagram of a microelectrode. A and E, Au; B, Ta$_2$O$_5$; C, Si; D, SiO$_2$; F, Si$_3$N$_4$. (From Turner, A. P. F., Karube, I., and Wilson, G. S., Eds., *Biosensors: Fundamentals and Applications*, Oxford University Press, New York, 1987, 476. With permission.)

micro-H$_2$O$_2$ sensor, which is 500 μm wide and 6 mm long, is shown in Figure 5. Au electrodes were deposited on the surface of the silicon nitride chip using vapor deposition methods. The Au electrodes were partially insulated by coating with Ta$_2$O$_5$.

The characteristics of the micro-H$_2$O$_2$ electrode were examined by cyclic voltammetry during immersion in H$_2$O$_2$ at various concentrations. A peak current was observed due to oxidation of H$_2$O$_2$ when a voltage of approximately 1.1 V was applied to the Au electrodes. A linear relationship was observed between the H$_2$O$_2$ concentration (1 μ*M* to 1 m*M*) and the steady-state current. This electrode could be used as the H$_2$O$_2$ electrode. The electrode was then utilized as a transducer in a micro-glucose sensor.

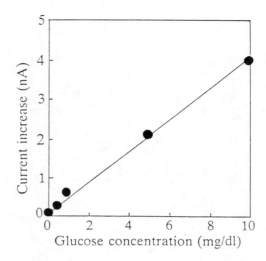

FIGURE 6.   Calibration curve of a glucose sensor. Experiments were performed at 37°C, pH 7.0. (From Turner, A. P. F., Karube, I., and Wilson, G. S., Eds., *Biosensors: Fundamentals and Applications,* Oxford University Press, New York, 1987, 476. With permission.)

Glucose oxidase (GOD) was immobilized onto the micro-H_2O_2 electrode as follows. Approximately 100 μl of α-aminopropyltriethoxysilane were vaporized onto the electrode surface at 80°C and 0.5 torr for 30 min; 100 μl of 50% glutaraldehyde were then vaporized under the same conditions. The micro-H_2O_2 electrode modified with chemical reagent was then immersed in GOD solution containing bovine serum albumin (BSA) and glutaraldehyde. GOD was chemically bound to the surface of the micro-H_2O_2 electrode by a Schiff linkage.

The GOD electrode was inserted into a sample solution containing glucose. The output current increase and steady-state current were obtained within 5 min. Figure 6 is the calibration curve for the micro-glucose sensor. A linear relationship was observed between the current increase (the difference between the initial and the steady-state currents) and the glucose concentration in the range 0.1 to 10 mg/dl glucose. The micro-glucose sensor did not respond to other compounds such as galactose, mannose, fructose, and maltose; therefore, the selectivity of this sensor for glucose is highly satisfactory. The effect of temperature on the sensitivity (current increase) of the sensor was examined. The optimum temperature for the sensor was 55°C. However, because the enzyme gradually denatures at 55°C, the stability of the sensor at this temperature is poor. Therefore, all other experiments were performed at 37°C. Continuous operation of the sensor in 10 mg/dl glucose produced a constant current output for more than 15 d and 150 assays. Thus, it is known that this microsensor can be used for the determination of glucose for more than 2 weeks.[8]

V. GLUTAMATE SENSOR

The determination of L-glutamate is very important because large amounts of L-glutamate are produced by fermentation for use as a food seasoning. Various glutamate sensors consisting of immobilized enzyme and an electrochemical device have been developed for the fermentation and food industries. Glutamate oxidase catalyzes the oxidation of glutamate. Oxygen is consumed by the reaction; therefore, an oxygen sensor or a hydrogen peroxide sensor can be employed as the transducer for a glutamate sensor.

A micro-oxygen sensor was developed by modifying the micro-H_2O_2 electrode. Figure 7 is a schematic diagram of the micro-oxygen sensor. It consists of a gas-permeable Teflon®

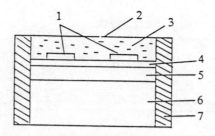

FIGURE 7. Schematic diagram of an oxygen electrode. 1, Au; 2, Teflon® membrane; 3, 0.1 M KOH; 4, Si_3N_4; 5, SiO_2; 6, Si; 7, silicon rubber.

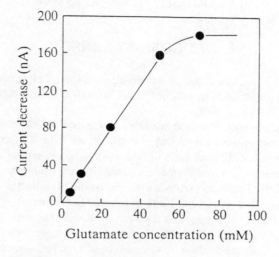

FIGURE 8. Calibration curve of a micro-glutamate sensor. Experiments were performed at 40°C, pH 7.5.

membrane, two micro-Au electrodes, and 0.1 M KOH electrolyte solution. The characteristics of the micro-oxygen electrode were examined by cyclic voltammetry at various concentrations of dissolved oxygen (an oxygen/nitrogen mixture was sparged through the sample solution). A peak current due to reduction of oxygen was observed when a voltage of approximately 1.1 V was applied to the Au electrodes. A linear relationship was observed between the oxygen concentration and the peak current obtained from the cyclic voltammogram. These results indicate that the micro-oxygen sensor can be used for the determination of dissolved oxygen. The micro-oxygen sensors were employed as a transducer in a micro-glutamate sensor.

Glutamate oxidase (*Streptomyces* sp., 25 U/mg protein) was immobilized on a cellulose triacetate membrane containing glutaraldehyde and 1,8-diamino-4-aminomethyloctane(triamine). The glutamate oxidase membrane was placed on the Teflon® membrane of the micro-oxygen sensor and covered with a nylon net. The steady state of current decrease (the difference between the initial and final currents) was obtained when the sensor was immersed in a sample solution containing glutamate.

Figure 8 shows the relationship between the current decrease and the glutamic acid concentration. When the current decrease at 5 min was used as the measure of the determination, a linear relationship was observed between the current decrease and the glutamic acid concentration in the 5 to 50 mM range. The optimum temperature for the sensor was

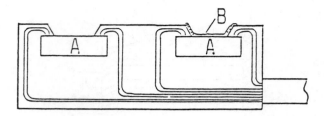

FIGURE 9.   Schematic diagram of glucose IC sensor. A, IC thermal sensors; B, immobilized enzyme membrane.

approximately 40°C, but gradual denaturation of the enzyme occurred at this temperature; therefore, all other experiments were performed at 30°C. The selectivity of the sensor for glutamic acid was found to be satisfactory. Hence, its application to food analysis and glutamate production is very promising.

VI. GLUCOSE-IC SENSOR

Enzymatic reactions are associated with enthalpy change. Calorimetry is applicable to biosensor systems. Mosbach and Danielsson[8,9] have written extensive reports on the enzyme thermistor system.

A micro-glucose-calorimetric sensor has been developed in our laboratory. IC temperature-sensing devices are suitable for the calorimetric sensor. The IC thermal sensors (Seiko Instruments and Electronics, Tokyo) used in this work were composed of three Darlington connected n-p-n transistors, in which the collector of each transistor was connected in common, the base of the first was connected to the common collector, and the emitter of the third transistor was connected to a C-MOS constant-current circuit. The total dimensions of the sensor were $1 \times 1 \times 0.3$ mm, and the surface was coated with an Si_3N_4 layer (0.95 μm). The case was then encapsulated with insulating epoxy resin, except for the surface of the IC thermal sensor. Glucose oxidase was immobilized on a cellulose triacetate membrane containing glutaraldehyde and triamine. The membrane was formed by casting a solution containing these compounds on the IC sensor (Figure 9). The measurement system consisted of digital voltmeter, power supply, and personal computer (Figure 10). Temperature control was performed with a stirred thermostatic bath connected via the computer to a third IC thermal sensor. The normal operational output voltage of the IC thermal sensors was approximately 1.2 V at 30°C. The time required for the enthalpy change to reach stationary state is dependent on the mass diffusion of glucose and the enzyme reaction velocity. The stationary response time is 2 min at 2 mM glucose and 4 min at 100 mM glucose. The calibration curve for glucose concentration vs. the glucose-calorimetric IC sensor output voltage is shown in Figure 11. Glucose concentrations between 2 and 100 mM can be determined by the system. The minimum detectable concentration of 2 mM and the poor linearity of our system can be greatly improved by coupling enzymatic reactions. Further studies are directed toward improving the sensitivity of the IC sensor.

The use of semiconductor fabrication technology enables the miniaturization and integration of biosensors using only minute amounts of sample solutions and enzymes. Apart from glucose determination in blood, diagnostics of some organic substrates (e.g., urine) generally require detection or determination of ten organic compounds at a time. Therefore, integrated microbiosensors play an important role in practical clinical tests.

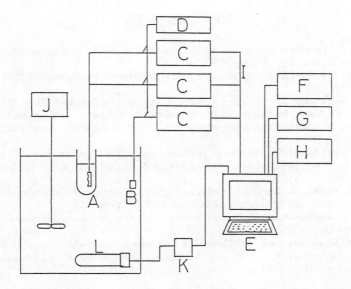

FIGURE 10. Schematic diagram of glucose-IC sensor system. A, glu-cose-IC sensor; B, IC thermal sensor; C, digital voltmeter; D, power supply; E, personal computer; F, floppy disk unit; G, printer; H, recorder; I, general-purpose interface bus; J, stirrer; K, relay; L, heater.

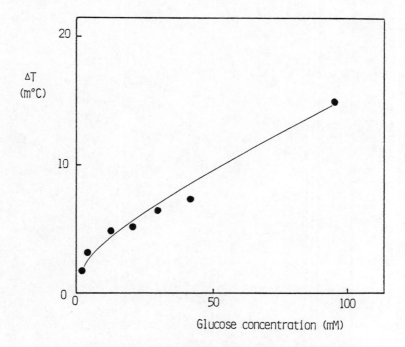

FIGURE 11. Calibration curve for determination of glucose.

REFERENCES

1. **Bergveld, P.,** Development of an ion-sensitive solid-state device for neurophysiological measurements, *IEEE Trans. Biomed. Eng.,* 17, 70, 1970.
2. **Matsuo, T. and Wise, K. D.,** An integrated field-effect electrode for biopotential, *IEEE Trans. Biomed. Eng.,* 21, 485, 1974.
3. **Caras, S. and Janata, J.,** Field effect transistor sensitive to penicillin, *Anal. Chem.,* 52, 1935, 1980.
4. **Miyahara, Y., Matsu, F., Moriizumi, T., Matsuoka, H., Karube, I., and Suzuki, S.,** Micro enzyme sensor using semiconductor and enzyme-immobilization, in *Proc. Int. Meet. Chemical Sensors,* Kodansha-Elsevier, Tokyo, 1983, 501.
5. **Matsuo, T. and Esashi, M.,** Methods of ISFET fabrication, *Sensors Actuators,* 1, 77, 1981.
6. **Kagawa, Y.,** A new model of proton motive ATP synthesis: acid-base cluster hypothesis, *J. Biochem.,* 95, 295, 1984.
7. **Gotoh, M., Tamiya, E., Karube, I., and Kagawa, Y.,** A microsensor for adenosine-5'-triphosphate pH-sensitive field effect transistors, *Anal. Chim. Acta,* 187, 287, 1986.
8. **Mosbach, K. and Danielsson, B.,** Thermal bioanalyzers in flow streams-enzyme thermistor devices, *Anal. Chem.,* 53, 83A, 1981.
9. **Danielsson, B. and Mosbach, K.,** Theory and application of calorimetric sensors, in *Biosensors: Fundamentals and Applications,* Turner, A. P. F., Karube, I., and Wilson, G. S., Eds., Oxford University Press, New York, 1987, 575.

Chapter 16

SENSING TECHNIQUES FOR BIOPROCESS MONITORING AND CONTROL

James B. Callis

TABLE OF CONTENTS

I. INTRODUCTION

The importance of control in bioprocessing is universally acknowledged, and at present there is no lack of (1) detailed models for growth of cellular populations which incorporate dependence upon external variables, (2) high-speed digital computers, and (3) sophisticated control and optimization algorithms. What is lacking, however, are suitable sensing techniques that can provide the quality and quantity of information about cells to facilitate implementation of the control strategies now possible. At the very least, one would like to have measurements that bear a close correspondence to the mean state vector of the population contained within a bioreactor. Even better control could be achieved if information could be obtained regarding the mean physiological status of the organisms and the relevant distribution functions. At the same time, information is required about the milieu, such as the concentration of products and nutrients.

As long ago as 1974 Humphrey[1] recognized the need for "gateway" sensors that could give the relevant information on the physiological status of a culture, and since that time there has been a growing effort to provide such sensing capabilities.

The above considerations are the proper concern of process analytical chemistry, an emerging subdiscipline of analytical chemistry, the mission of which is to supply quantitative and qualitative information about a chemical or biochemical process. The field has received considerable attention recently with the advent of chemical and biochemical microsensors, and a flurry of activity in this area has begun.[2,3] However, it must be realized that there is much more to the practice of process analytical chemistry than improved chemical sensing.[4] The issues of sampling,[5] extraction of information from the sensor data, and integration of the information into process control must be given equal consideration.

The purpose of this chapter is to review the status of the field of process analytical chemistry as it relates to bioprocessing and, hopefully, to convince the reader that important technological, methodological, and chemometric advances have greatly improved the capabilities of optical spectroscopy for multicomponent analysis and that optical spectroscopy is a most viable approach for biomonitoring.

II. MODES OF PROCESS ANALYTICAL CHEMISTRY

As a conceptual framework, we have found it useful to identify five modes of process analytical chemistry: off-line, at-line, on-line, in-line, and noninvasive (see Table 1 and Figure 1).[6] The first two modes, off-line and at-line, are distinguished by the requirement for manual removal of the samples and transport to a measuring instrument. In off-line analysis the sample is analyzed in a well-equipped central laboratory, whereas in at-line analysis a dedicated instrument is located at the reactor site.

A. ON-LINE

In the on-line mode, an automated sampling system is used to extract the sample, condition it, and present it to an analytical instrument for measurement. It is possible to subdivide on-line modes into two categories: (1) intermittent methods that require injection of a portion of the sample stream into the instrument and (2) continuous methods that permit the sample to flow continuously through the instrument.

An obvious example of an intermittent on-line process analyzer is a liquid chromatograph. The considerations in interfacing this device to a bioreactor are provided by Lenz et al.[7] Yet another example of the intermittent instrument is flow injection analysis (FIA). Here, a well-defined volume of sample is injected by means of a valve into a carrier liquid stream. As the sample zone moves through the channel, it disperses into and reacts with the carrier stream on its way toward a flow cell containing a detector. FIA may be conveniently thought of as an on-line alternative to manual wet chemical analysis.[8] The major advantages of the

TABLE 1
The Five Modes of Process Analytical Chemistry

Mode	Name	Current example	Future example
First	Off-line	Gas chromatography-mass spectrometry (GC-MS)	Capillary electrophoresis with tandem MS detection
Second	At-line	Colorimeter	Flow injection analyzer
Third	On-line (intermittent)	Capillary GC	Microbore liquid chromatography (LC)
Third	On-line (continuous)	Correlation IR	FTIR using reflectance cell
Fourth	In-line	Conductivity sensor	Micro UV-Vis with fiber optic probe
Fifth	Noninvasive	Diffuse reflectance NIR	Multifrequency microwave radar

FIA approach are its high throughput, reliability, and excellent precision. FIA is an appealing paradigm for development of an automated universal microlaboratory that integrates the functions of sample acquisition, cleanup, concentration, separation, and sensing/identification.

Optical spectroscopy encompassing the ultraviolet-visible (UV-Vis), near infrared (NIR), and mid-infrared (IR) can form the basis for continuous on-line process analyzers. The state of the art in IR instrumentation for the laboratory is, of course, the Fourier transform IR (FTIR), and recently such instruments have become rugged enough to envisage their use on-line. A particular problem for IR is the extremely strong absorption by water, which limits the pathlengths to micrometer dimensions.

For a liquid stream, attenuated total reflectance offers a way to eliminate ultranarrow pathlength cells and their attendant difficulties with blockage from particulate matter in the sample and interference fringes. In addition, this approach is very well suited to analysis of multiphase bioprocesses. The cylindrical geometry cell provides a simple method for implementing attenuated total reflectance technology in a form readily adapted to bioprocess applications.[9]

Phillips and colleagues[10] have provided a demonstration of the potential of FTIR spectroscopy with an attenuated total reflectance cell to monitor substrate and product concentrations in broths. Their system monitors analyte concentrations in the physiological ranges, despite the presence of water, microorganisms, and major and minor trace media components. While the method is not yet either reliable or accurate enough for routine use, further work is clearly justified.

In the UV-Vis arena, great strides in technology have made scanning versions of these instruments far more rugged and reliable. Three key concepts are (1) the adaptation of single-beam techniques that use stored baselines rather than mechanically sampled reference and sample compartments, (2) the use of concave holographic gratings that yield acceptable stray light rejection in a single monochromator and eliminate the need for collimating mirrors, and (3) the introduction of photodiode arrays that eliminate mechanical scanning of the grating.[11] To a classical analytical chemist, the use of UV-Vis spectroscopy on highly scattering suspensions of yeast to make quantitative measurements would seem impossible.

Nevertheless, Chance and his colleagues have been successfully characterizing the redox states of cytochromes in suspensions of yeast, mitochondria, and photosynthetic bacteria for a number of years.[12] Second-derivative spectroscopy has proved especially useful in this regard, since it is relatively insensitive to baseline offsets.[13]

B. IN-LINE

The major disadvantage of on-line analysis is the need to construct a separate analytical line that properly samples the main stream and presents it to the instrument at a suitable temperature and pressure. This has led to the fourth mode: in-line process analytical chem-

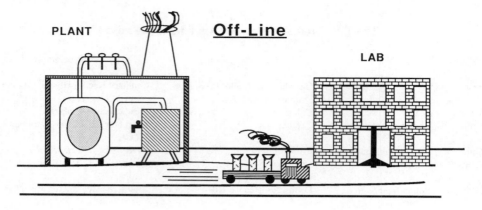

A

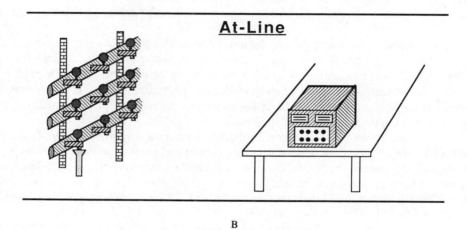

B

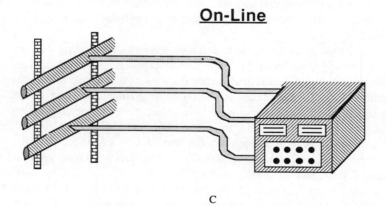

C

FIGURE 1.   The five modes of process analytical chemistry: A, off-line; B, at-line; C, on-line; D, in-line; E, noninvasive.

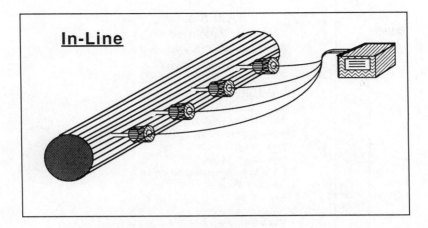

FIGURE 1D.

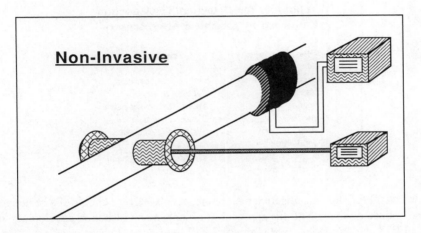

FIGURE 1E.

istry. Here, chemical analysis is done *in situ*, directly inside the process line, using a probe that is chemically sensitive. In its ideal form, the device might resemble a typical industrial temperature probe. Such an implementation would obviously be attractive to process engineers.

The key to scaling down chemical sensors is widely held to be their fabrication using tools that originate in the microelectronics revolution: microlithography and micromachining.[2] Virtually all of the microsensors fabricated to date use a two-step detection scheme (see Tables 2 and 3). In the first stage, chemical selectivity is obtained by means of a chemical transformation or by physisorption or chemisorption to a chemically selective surface. In the second stage, some physical consequence of the chemically selective stage, such as release of heat, change of optical absorption, etc., is converted to an electrical signal by means of a suitable microtransducer.

The most noted of the microsensors are the ChemFET and the ISFET.[14] The latter was first reported by Bergveld[14a] over 15 years ago and arose from a desire to scale down common pH or ion-specific electrode technology. It is therefore important to note that such devices share all of the advantages and disadvantages of potentiometric electrochemical detectors. It is now widely appreciated that to develop long-lived, reliable versions of these devices will require considerable development in encapsulation and packaging technology.

TABLE 2
Methods Employed in
Microsensors to Obtain
Chemical Selectivity

Chemical transformations
   Electrochemical
   Enzyme catalyzed
Physisorption to a surface
   From gas phase
   From liquid phase
Binding to an immobilized receptor
   Immunoadsorption
   Ion exchange
Membrane transport
   Electrolyte separation

TABLE 3
Techniques for Detection/Transduction
Employed by Chemical Microsensors

Physical property	Transducer(s)
Chemical potential	Field effect transistor
Optical	Photodiode
Heat	Thermistor, pyroelectric
Mass	Piezoelectric balance
Mass	Surface acoustic wave
Conduction	Dielectrometer
Surface resistance	Chemiresistor, Taguchi
Current flow	Ampometer

In the ChemFET, the ion-selective membrane is replaced by a chemically selective layer. Janata has pioneered the use of this approach for biochemical detection developing enzyme-based (EnFET) and antibody-based (ImmunoFET) field effect transistors. One of the major issues in the design of these microsensors is response optimization. The interested reader is referred to a very sobering paper by Janata's group that shows the multitude of considerations involved.[15]

A second class of microsensors takes advantage of the electro-optical revolution: fiber optic sensors.[3] Here, probe light is generated remotely and is conveyed to the sensor end by a light guide, where it interacts with a chemical probe by absorption, scattering, fluorescence, or Raman emission. The light, which has been encoded with chemical information, is then returned by the same or a second light guide to an electro-optical transducer that creates the desired electrical signal. Such an optical technique is very versatile. One has the entire range of chemistries developed over the years as "spot tests" to provide inspiration. Many of these can be immobilized onto glass or polymer beads and encapsulated into microcuvettes to form all sorts of "optrodes".

Thus far, we have confined our thinking about chemical microsensors to the one sensor/one analyte concept. However, much more is possible, and we find it useful to think in terms of a hierarchy of complexity of devices that the microelectronics paradigm provides, with the simple, single sensors as the first level, sensor arrays as the second, optical spectrometers as the third, and the integrated chemical microlaboratory as the fourth level of complexity.

The second level in the hierarchy is the sensor array. Here, the ability to use microlithography to fabricate multiple sensors on one chip is an obvious way to avoid the cumbersomeness of multiple probes for multiple analytes. The straightforward approach to the implementation of a multifunction sensor is to devote one sensor to each analyte of interest. However, this method does not address the well-known problem of interferences in the chemical selectivity step. An alternative strategy is to use a set of relatively unselective sensors, each one of which has a different response profile to all of the analytes of interest.[16] In this case, the array generates a response pattern that may be analyzed by various powerful methods of multivariate statistics and pattern recognition. The combination of sensor array and data processing constitutes a robust analytical system that can yield simultaneous multicomponent analyses with the capability of recognizing and correcting for interferences and drift.[17] The same multivariate methods may also be used to optimize specific multicomponent analysis problems by selecting the minimum number of sensors or sensing channels (e.g., wavelengths) to yield optimal performance.[18]

The third member of the microsensor hierarchy is the ultraminiature optical spectrometer. One possible approach employs the highly successful optical imaging device, which is a microfabricated array of photosensors. Previously, these devices have been used as electronic photographic plates in conjunction with conventional spectrographs. Alternatively, extremely small "solid-state" spectrometers have been constructed using tiny diffraction gratings and graded index of refraction lenses as collimators.[19] Unfortunately, diffraction gratings do not scale down very well because, due to the fixed range of useful wavelengths, fewer and fewer grating lines are sampled and lower resolution results. A different approach is multibeam interferometry, which is widely known as the basis of interference filters. Here, wavelength analysis takes place over the dimensions of micrometers (rather than centimeters). One suitable implementation of this concept is the linear interference wedge filter, in which the wavelength of maximum transmission varies continuously along one axis. A prototype compact planar spectrometer using a wedge filter and linear array detector was found to give quite acceptable results[20] in terms of sensitivity, linearity, and resolution.

The remaining member of the hierarchy is the chemical microlaboratory. The goal is to integrate all of the functions of a chemical analysis laboratory in a unit no larger than a credit card. The key to this development is the integration of microconduits, microvalves, and microsensors by various types of hybrid circuit fabrication techniques.[21] The electronic and optical sensors described above should last much longer when incorporated into a microlaboratory because of extensive sample conditioning and intermittent use.

C. NONINVASIVE

The final mode of bioprocess analytical chemistry uses noninvasive methods and represents the ultimate in desirability. Because the probe does not physically contact the sample, the sampling problem is greatly alleviated. This mode obviously has a great deal in common with remote sensing and nondestructive testing. This point has been emphasized by Clark and co-workers[22] in their review of biosensing for bioreactor monitoring and control.

NIR spectroscopy from 700 to 1100 nm has much to offer[23] in this regard. First, the extinction coefficients in this region are very small and allow pathlengths of 0.5 to 20 cm for clear solutions. For the most part, absorptions in this region arise from the second and third overtones of CH, OH, and NH stretches together with combination bands from other vibrations. Since these are highly forbidden transitions, most materials are very transparent (exceptions are metals and graphite-containing composites). At these optical distances, a thin layer of adsorbed material in the windows does not fatally degrade the results. Also, quantitative measurements of highly scattering materials can be made because scattering coefficients are much greater than absorption coefficients. As a consequence, both diffuse log inverse reflectance and diffuse absorbance measurements show good linear correlation with analyte concentration. Diffuse reflectance is particularly useful because only a single window into the process need be established. In addition, the high degree of light scattering results in the sampling of a large volume. Also, the hardware is very inexpensive and employs readily available fiber optic components, conventional monochromators, tungsten lamps, and silicon detectors. Finally, the spectra are characterized by an extremely high signal-to-noise ratio. Thus, very subtle shifts in the spectra, which might go undetected by the unaided eye, can become the basis for highly successful analytical procedures developed by computer learning methods.

Another noninvasive technology to consider is ultrasound. The major use of ultrasound has been for outline imaging of flaws, employing reflections from index of refraction boundaries. However, a recent paper makes a strong case for further research into the mechanisms for ultrasound absorption, which is based, at least in part, on chemical composition.[24]

III. THE ROLE OF CHEMOMETRICS

We have now completed an outline of some of the possibilities for chemical sensing in the service of process analysis. It remains to describe how to analyze the data from the array of instruments and sensors that have been placed on-line. Our approach is empirical and involves the use of multivariate statistical calibration.[25] In this method, the computer is provided with a training set consisting of both the spectrum of each of the samples and the corresponding set of properties measured by an independent reference method. It is important that the training set consist of representative samples whose variation spans that of the set of all samples which might reasonably be encountered by the method. Once the data set is acquired, the computer searches for correlations between the spectra and the sought-after property. Eventually, a transformation vector is produced that, when multiplied by the spectral vector, yields an estimate for the property. Once the derived transformation vector has been validated with a double-blind study of a second set of samples for which spectra and properties have both been measured, the method is ready for use.

As examples of this, consider the results obtained by Phillips and co-workers,[10] in which fermentation broths were analyzed simultaneously for lumped media, glucose, ethanol, and glycerol using FTIR spectroscopy in the 2000 to 1000 cm^{-1} range. Using multivariate analysis, quantitation was obtained despite severely overlapping spectra. Equally impressive is the routine analysis of wheat for starch, moisture, and protein content.[22] This method uses NIR reflectance spectroscopy and multivariate statistics; its precision and accuracy are excellent, and the measurement requires only the grinding of the wheat to a uniform texture followed by a 10-s spectral measurement.

Such measurements need not be confined to the determination of chemical composition. In the case of gasoline, for example, a study of 43 samples of unleaded gasoline of known octane number was made using NIR spectroscopy over fiber optics. By simple step-forward multilinear regression, a linear equation involving only absorptions at three wavelengths was found that correlated with the reference octane number with an error of 0.3 octane, the known precision of the octane engine.[26] Further studies involving a wider variety of samples will clearly be required to verify the method and to assure its robustness, but the effort is justified because the spectroscopic test takes only 20 s, and the instrumentation costs far less and requires considerably less maintenance than a knock engine.

Another advantage of on-line spectroscopy is that a number of properties, both chemical and physical, can be measured simultaneously from one spectrum. In the case of gasoline, we have been able to correlate the spectrum to both research and motor octane, atmospheric pressure ionization (API) density, Reid vapor pressure, distillation points and total aromatics, olefins, and aliphatics — eight tests in 20 s.

One of the most intriguing aspects of multivariate calibration is the capability of generating a correlation spectrum that is associated with a particular property.[27] This may prove to be an extremely valuable tool in materials research: for example, the correlation spectrum of tensile strength might reveal the presence of specific supramolecular structures or unique intermolecular bonds that determine the specific property. Such knowledge could lead to new materials with significantly improved properties. Thus, multivariate calibration could also become a tool for discovery.

IV. APPLICABILITY TO BIOPROCESS ANALYSIS

How can these advances be applied to bioprocess analysis? Let us start with the observation that optical spectroscopies contain considerable information about the physiological status of a cell population. In the visible range, one obtains data on the oxidation states of cytochromes in the mitochondria. In the NIR and IR, one obtains information about bulk constituents such as protein, substrates, products, media, etc. Thus, information directly related to the physiological status of a cell culture is contained within these spectra. Unfortunately, the information is often present in an obscure form, and we have no "first principles" theory for obtaining it. However, multivariate statistics offers an approach. All that is needed is an independent method for measuring the desired set of properties. Provided that the signal-to-noise ratio is high enough, the information can be retrieved. If such highly relevant parameters as total cell mass, percent viable cells, percent synthesizing DNA, etc. can be obtained on a routine basis, better control is sure to result.

ACKNOWLEDGMENT

This work was supported by the Center for Process Analytical Chemistry from grants provided by the National Science Foundation (#ISI-8415075), the 44 sponsors, and the University of Washington.

REFERENCES

1. **Humphrey, A. E.,** Current developments in fermentation, *Chem. Eng.*, 81, 98, 1974.
2. **Wohltjen, H.,** Chemical microsensors and microinstrumentation, *Anal. Chem.*, 56, 87A, 1984.
3. **Seitz, R.,** Chemical sensors based on fiber optics, *Anal. Chem.*, 56, 16A, 1984.
4. **Hirschfeld, T., Callis, J. B., and Kowalski, B. R.,** Chemical sensing in process analysis, *Science*, 226, 312, 1984.
5. **Cornish, D. C., Jepson, G., and Smurthwaite, M. J.,** *Sampling Systems for Process Analyzers*, Butterworths, London, 1981.
6. **Callis, J. B., Illman, D. L., and Kowalski, B. R.,** Process analytical chemistry, *Anal. Chem.*, 59, 625A, 1987.
7. **Lenz, R., Boelcke, C., Peckman, U., and Reuss, M.,** A new automatic sampling device for the determination of filtration characteristics and the coupling of an HPLC to fermenters, in *Modeling and Control of Biotechnological Processes*, Johnson, A., Ed., Pergamon Press, Oxford, 1986.
8. **Ruzicka, J. and Hansen, E. H. H.,** *Flow Injection Analysis*, John Wiley & Sons, New York, 1981.
9. **Rein, A. J. and Wilks, P.,** Cylindrical internal reflection cell, *Am. Lab.*, 14(10), 152, 1982.
10. **Alberti, J. C., Phillips, J. A., Fink, D. J., and Wacasz, F. M.,** Off-line monitoring of fermentation samples by FTIR/ATR: a feasibility study for real-time process control, *Biotechnol. Bioeng. Symp.*, 15, 689, 1985.
11. **Miller, J. C., George, S. A., and Willis, B. G.,** Multichannel detection in high performance liquid chromatography, *Science*, 218, 241, 1982.
12. **Chance, B. and Hess, B.,** On the control of metabolism in ascites tumor cell suspensions, *Ann. N.Y. Acad. Sci.*, 63, 1008, 1956.
13. **Nagel, B., Bayer, C., Iske, U., and Glombitza, F.,** UV-Vis diffuse reflectance spectroscopy in fermentation process control, *Am. Biotechnol. Lab.*, July/August, 12, 1986.
14. **Janata, J. and Huber, R. J., Eds.,** *Solid State Chemical Sensors*, Academic Press, New York, 1985.

14a. **Bergveld, P.,** Development of an ion-sensitive solid-state device for neurophysiological measurements, *IEEE Trans. Biomed. Eng.,* 17, 70, 1970.

15. **Caras, S. D., Janata, J., Saupe, D., and Schmidt, K.,** pH-based enzyme potentiometric sensors. I. Theory, *Anal. Chem.,* 57, 1917, 1985.

16. **Hirschfeld, T.,** Remote and *in-situ* analysis, *Adv. Instrum.,* 40, 305, 1985.

17. **Kalivas, J. H. and Kowalski, B. R.,** Generalized standard addition method for multicomponent instrument characterization and elimination of interferences in inductively coupled plasma spectroscopy, *Anal. Chem.,* 53, 2207, 1981.

18. **Carey, W. P., Beebe, K. R., Kowalski, B. R., Illman, D. L., and Hirschfeld, T.,** Selection of adsorbates for chemical sensor arrays by pattern recognition, *Anal. Chem.,* 58, 149, 1986.

19. **Fuh, M. S. and Burgess, L.,** Wavelength division multiplexor for fiber optic sensor readout, *Anal. Chem.,* 59, 1780, 1987.

20. **Pfeffer, J. C., Skoropinski, D. B., and Callis, J. B.,** Simple, compact visible absorption spectrophotometer, *Anal. Chem.,* 56, 2973, 1984.

21. **Ruzicka, J.,** Flow injection analysis: from test tube to integrated microconduits, *Anal. Chem.,* 55, 1040A, 1983.

22. **Clark, D. J., Calder, M. R., Carr, R. J. G., Blake-Coleman, B. C., Moody, S. C., and Collinge, T. A.,** The development and application of biosensing devices for bioreactor monitoring and control, *Biosensors,* 1, 213, 1985.

23. **Norris, K. H.,** Instrumental techniques for measuring quality of agricultural crops, *NATO Adv. Study Ser. A,* 46, 471, 1983.

24. **Slutsky, L. J.,** On the possible importance of chemical relaxation in acoustic absorption in biological systems, *IEEE Trans. Sonics Ultrasonics,* UFFC-33, 147, 1986.

25. **Martens, H. and Naes, T.,** Multivariate calibration. I. Concepts and distinctions, *Trends Anal. Chem.,* 3, 204, 1984; *NATO ASI Sev. Ser. C.,* 138, 147, 1984.

26. **Kelly, J. J., Barlow, C. H., Jinguji, T. M., and Callis, J. B.,** Prediction of gasoline octane numbers by near-infrared spectroscopy in the spectral range 660—1215 nm, *Anal. Chem.,* 61, 313, 1989.

27. **Honigs, D. E., Hieftje, G. M., and Hirschfeld, T.,** A new method for obtaining individual component spectra from those of complex mixtures, *Appl. Spectrosc.,* 38, 317, 1984.

Chapter 17

MASS FABRICATED ION, GAS, AND ENZYME MULTISPECIES SENSORS USING INTEGRATED CIRCUIT TECHNOLOGY

I. R. Lauks, H. J. Wieck, N. J. Smit, S. Cozzette, G. Davis, and S. Piznik

TABLE OF CONTENTS

I. INTRODUCTION

Significant growth in the development and application of electrochemical sensors has occurred in the last 20 years. Research activity has largely focused on two areas: the improvement of membranes, particularly for use in biological fluids, and the establishment of the fluids and delivery systems under which the electrodes optimally perform. More recently, attention has been given to developing sensors broader in scope than traditional ion-selective electrodes (ISEs) (e.g., enzyme-based electrodes) for various metabolites, as well as higher sensitivity immunosensors.[2] With the basic chemistry and structure of electrochemical sensors being well understood, the technology is poised for further developments in mass fabricated devices, and a broader class of applications and configurations of particular interest is the fabrication of multispecies sensors to perform panels of tests.

The approach of using semiconductor chip technology to microfabricate electrochemical sensors dates back to the 1970s. However, the specifics of how to implement this technology are only now beginning to be addressed. Initially, it was thought that the best approach was to develop sensing structures that were integral to the electronics — hence, the proliferation of research programs involving ion-sensitive field effect transistors (ISFETs) or ChemFETs.[3-4] Unfortunately, the ChemFET is based upon sensing a charge buildup in a chemically sensitive membrane placed over the gate insulator of a field effect transistor (FET), which places the sensing and transduction functions in close proximity. This has led to new and stringent encapsulation requirements. While some of these programs have had success in terms of working structures, many have fallen by the wayside. An alternative design separates the electrode and electronic functions, thereby retaining the advantage of the macroelectrodes while allowing mass fabrication. An additional problem which was not adequately addressed was the issue of mass fabrication, including the issue of automated encapsulation and membrane deposition.

II. MANUFACTURING APPROACH

Our approach was to focus almost exclusively on the manufacturing technology by developing a method for planar fabrication of several classes of electrochemical sensors. First, we focused on a design for the interface to the integrated electronics which permits standard silicon processing. Hence, conventional low-noise, high-input impedance operational amplifiers were employed to buffer the potentiometric signals into an analog multiplexer. The electronic section could then be fully passivated and encapsulated at the wafer level. Next, we turned our attention to chemically sensitive layers that permit thin film processing and form the basis for broad classes of sensors. One of the materials specially developed for this was amorphous iridium oxide, which behaves well as a pH electrode. Finally, the techniques of micropatterning these thin films at the wafer level have been developed so that we now are ready to mass fabricate sensors.

Figure 1 shows electronic layouts for two different combinations of sensors. One (Figure 1A) is a purely potentiometric interface, while Figure 1B shows the interface required for a blood gas chip. The latter contains a single-ended and differential voltage follower for the pH and CO_2 sensors, respectively, together with a potentiostat and a current-to-voltage converter for the O_2 sensor. Both interfaces employ a multiplexer so as to present a common interconnection to an instrument. These have been designed and implemented using standard complementary metal oxide semiconductor (CMOS) circuitry, thereby avoiding needless development effort on our part. The actual sensor-amplifier interface is shown in Figure 2. In order to reduce leakage currents to a minimum, the lines connecting the sensors to the amplifiers employ both a signal line and a guard line. This is bootstrapped to the amplifier

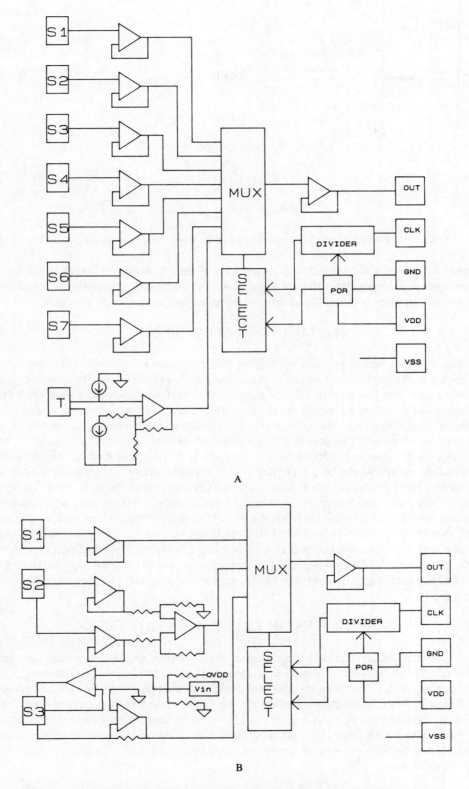

FIGURE 1.   Multispecies chip designs. (A) Seven-species potentiometric interface with temperature; (B) a single-ended potentiometric, a differential potentiometric, and an amperometric interface, designed for a blood gas and pH multispecies chip.

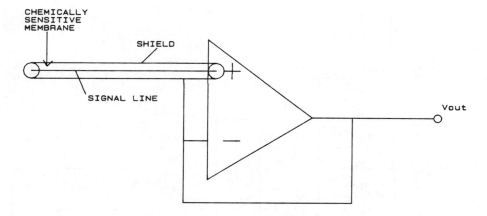

FIGURE 2.   Interface between the sensor and the preamplifier, showing the shielded signal line.

output in order to reduce signal line capacitance. Figure 3 shows the actual layout of the seven-species potentiometric chip of Figure 1A. The significant separation of the sensors from the electronics is chosen to simplify integration into an analytical system.

III. MULTISENSOR DESIGN

The thin film technology can be divided into three classes of sensor structures; these are shown in Figure 4. Figure 4a shows the iridium oxide pH sensor and the halide sensors, where a single film is deposited onto the metalized conducting signal line. Figure 4b depicts the next class of structures, which are the basic ISEs and gas sensors. In these, a solid-state structure is used as a base. On top of this is patterned a hydrogel containing various anions and cations of interest. This acts as a faradic interface between the solid-state structure and the selective membrane patterned on top. For the potassium and calcium electrode the selective membrane would be a PVC doped with the appropriate ionophore, while for the gas sensors a gas-conducting membrane, such as a silicone rubber, might be used. The third class of structures uses patterned hydrophilic conductive membranes into which specific enzymes have been immobilized. With this structure, various metabolites such as glucose, urea, creatinine, etc. can be measured. It is well known that for the amperometric-based enzyme sensors, redox species such as ascorbate and glutathione are significant interferers. With the ability to make matched pairs of sensors, it is possible to treat such interferers as a common mode signal if only one of the sensors is sensitive to the metabolite being determined.

IV. RESULTS OF EXPERIMENTAL TESTS

A previous paper discussed the response characteristics of microfabricated sensors.[5] Speed of response as well as linearity in aqueous buffered solution have been demonstrated. Figures 5 and 6 show correlation studies for microfabricated potassium and glucose sensors in biological fluids. Both the potentiometric potassium sensor and the enzyme-based amperometric glucose sensor exhibit good correlation when results using these electrodes are compared with results derived from a traditional clinical chemistry analyzer. Figure 5 also demonstrates the sensor-to-sensor reproducibilities obtainable using integrated circuit fabrication technology.

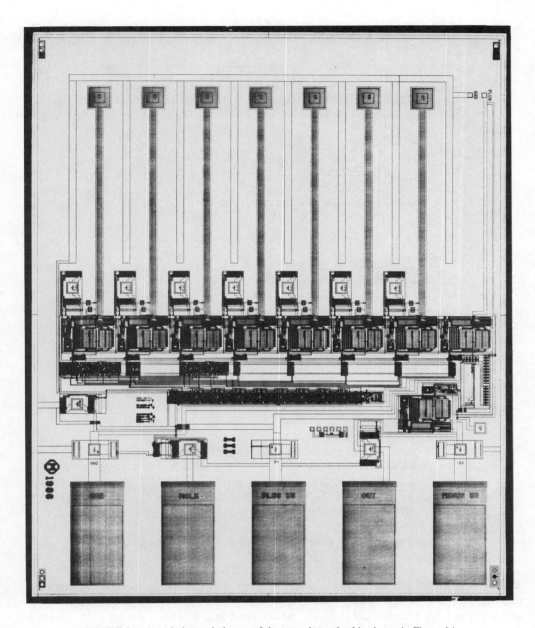

FIGURE 3.   Actual electronic layout of the potentiometric chip shown in Figure 1A.

V. APPLICATIONS

Given that these sensors can be completely fabricated with a wafer level process, it is possible to achieve very low multispecies chip manufacturing costs for high-volume applications where disposable or quasidisposable sensors make sense. In this light, it is interesting to note that almost all of the assays performed in a clinical laboratory take place in disposable cuvettes. Many of these assays require precise dilution of the sample, leading to expensive instrumentation. On the other hand, the sensors in a traditional ISE instrument are in direct contact with the sample. This places considerable burden on the operator to maintain the sensors in a fully functional state. Although a key advantage of ISEs is that no sample

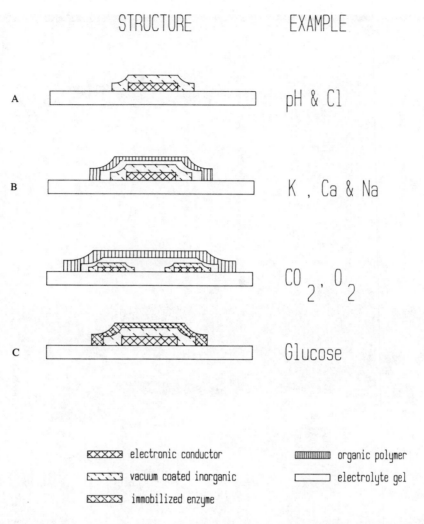

FIGURE 4.   Sensor structures. (a) Solid state; (b) selective membranes; (c) immobilized enzymes.

dilution is required, this advantage is offset by more complex fluidics used to maintain the sensors. If, however, the sensor price is low enough, one can envisage an instrument in which the sensors are disposable or are incorporated into a limited-use cartridge. Such a system would compete favorably with those in the clinical and physician's office labs, especially as whole blood can be used with no sample dilution required. An example of a disposable cartridge is the open-heart sensor pack developed by Diamond Sensors. Arden Medical has demonstrated disposable sensor cards for both the physician's office and the critical care settings in hospitals. All of these systems are much simpler to operate than the conventional ISEs because the sensors are disposable. While these companies have addressed the mass fabrication of the sensors, their technology limits them in the final cost of the devices. Consequently, prices for the disposable devices range from several dollars to several hundreds of dollars.

Miniaturized, low-cost sensors also find application in catheters for blood gas determinations in an arterial line or a syringe. The present means of performing blood gas analyses is to draw a sample, pack it in ice, and send it to the stat lab where the blood gas instrument

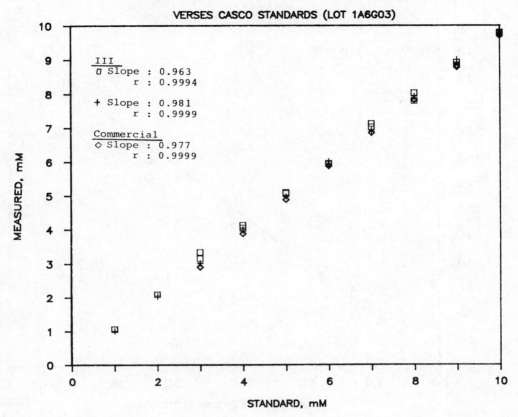

FIGURE 5.   Correlation of results obtained using microfabricated potassium electrodes.

resides. The sample is then manually warmed and run on the instrument. The data then have to be transferred back to the patient's site. This laborious procedure can be avoided completely if the sensors can be molded into a syringe or arterial line. One very obvious application for a miniaturized, multispecies Na, Cl chip is in a disposable cystic fibrosis screening test. The present method of transferring sweat from the patient's skin to the ISE analyzer via gauze and a test tube is both time consuming and technique dependent.

Table 1 summarizes the wide variety of medical applications which exist for multispecies chemical microsensors if, as it appears, they can be completely planar fabricated at the wafer level.

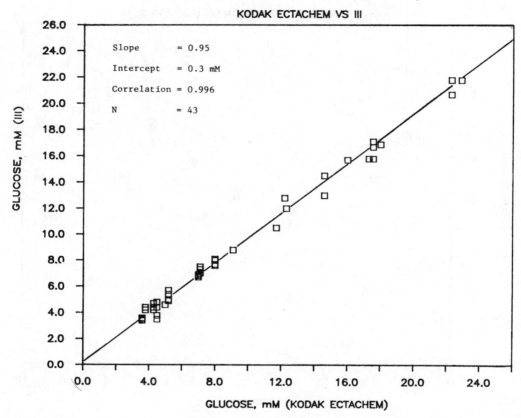

FIGURE 6.   Correlation of results obtained using microfabricated glucose sensors and a Kodak Ektachem® system.

TABLE 1
Medical Applications for Multispecies Chemical Microsensors

- Disposable electrodes (e.g., for cystic fibrosis screening)
- Clinical analyzers
- Stat analyzers
- Analyzers for the
    Physician's office
    Bedside
    Patient's home
- Clinical monitors
    Noninvasive:
        Transcutaneous
        Eyelid
        Oral
    Extracorporeal:
        Open-heart
        Dialysis
    In vivo, on-line:
        Critical care
        Anesthesiology
        Drug delivery
        Fetal scalp
        Esophageal and gastric (e.g., pH)

REFERENCES

1. **Czaban, J. D.,** Electrochemical sensors in clinical chemistry: yesterday, today, and tomorrow, *Anal. Chem.,* 57, 345, 1985.
2. **Meyerhoff, M. E. and Rechnitz, G. A.,** Electrode-based enzyme immunoassays using urease conjugated in conjunction with ammonia-sensing electrode, *Anal. Biochem.,* 95, 483, 1979.
3. **Janata, J. and Huber, R. J.,** Ion-sensitive field effect transistors, in *Ion-Selective Electrode Reviews,* Vol. 1, Thomas, J. D., Ed., Pergamon Press, Elmsford, NY, 1980.
4. **Sibbald, A., Whalley, P. D., and Covington, A. K.,** A miniature flow-through cell with a four-function chemfet integrated circuit for simultaneous measurements of potassium, hydrogen, calcium and sodium ions, *Anal. Chim. Acta,* 159, 47, 1984.
5. **Lauks, I., Groves, M., and Wieck, H.,** Mass fabricated ion, gas and enzyme multispecies sensors using IC technology, in Proc. Electrochemical Sensors for Biomedical Applications, Vol. 14, The Electrochemical Society, Pennington, NJ, 1986, 116.

Chapter 18

FLUX DETERMINATION IN CELLULAR BIOREACTION NETWORKS: APPLICATIONS TO LYSINE FERMENTATIONS

Joseph J. Vallino and Gregory Stephanopoulos

TABLE OF CONTENTS

I. INTRODUCTION

Mutation/selection or genetic engineering techniques employed to increase product yield in fermentations often only alleviate metabolic regulation occurring at or near the end of a biosynthetic chain that leads to that particular product. However, microorganisms have evolved biosynthetic pathways to produce energy and metabolites necessary for growth and replication, not for the overproduction of specific biocompounds. Consequently, not only does the regulation of the product of interest need to be removed, but also the main fueling reactions of the cell must be rerouted in such a way as to channel the main carbon flux into the biosynthetic pathway required for product synthesis.

To achieve, or at least approach, theoretical product yields in an optimal manner, one must know how carbon partitioning occurs between the fueling reactions, biomass, and product. This necessitates the development of means to determine the partitioning of carbon flux between the primary biosynthetic pathways. Once acquired, this information can be used to rechannel the carbon in the fueling reactions into the product pathway by amplification or attenuation of the enzymes associated with the limiting reactions. This type of global modification process has been termed metabolic engineering, yet there does not exist any simple way to determine the flux of carbon through the primary metabolic pathways.

Although the fluxes can be estimated, in theory, by constructing a set of differential equations to model the concentrations of the important metabolites in the network, these models rely on the regulatory and kinetic properties of individual enzymes in the network. These enzymes are poorly understood or not known at all for most microbes. Even though such models are useful, kinetic and regulatory information of individual enzymes *in vitro* does not necessarily indicate how the overall biosynthetic network functions as a whole *in vivo*. As will be shown, models this detailed are not necessary to estimate the fluxes in the network.

Experimentally, it is possible, to a limited extent, to calculate the carbon flux on the basis of radiocarbon labeling experiments; however, these experiments often disrupt the cellular environment, are difficult and expensive to conduct, and cannot be used in a fermentation environment due to the large volumes involved.

We have developed a methodology that determines the carbon flux through the primary biosynthetic pathways employed for product synthesis from measurements of extracellular metabolites only. This algorithm, presented in Section II, couples measurements of extracellular metabolites with the known metabolic pathways of the microbe of interest, and a pseudo-steady-state (PSS) approximation for intracellular metabolites is used to generate a bioreaction network equation. Once constructed, singularity and sensitivity analysis routines determine if the system is well posed. Often these algebraic systems have more equations than unknowns; consequently, redundancy analysis is used to check the consistency of measurements and the PSS approximations. Least-squares or quadratic programming techniques are employed to solve the equation and produce an estimate of the carbon flux.

The actual fermentation being studied is the production of lysine from glucose by *Corynebacterium glutamicum*. This process is a good example of mutation/selection techniques employed to construct production mutants free of feedback regulation by lysine and threonine on aspartate kinase. However, these fermentations have molar yields (moles lysine produced per mole glucose consumed) of approximately 30 to 40%,[1] while the theoretical maximum is about 75%[2] when $NADPH_2$ is assumed not to be equivalent to $NADH_2$ and about 86%[3] when $NADPH_2$ and $NADH_2$ are assumed to be equivalent. The difference in theoretical yields is caused by CO_2 evolution in the pentose phosphate pathway (PPP) when it is used to generate $NADPH_2$. Since essentially no other products are produced and biomass synthesis is minimal, glucose is simply being oxidized to carbon dioxide and water. Consequently, it might be possible to redirect this carbon flux into lysine. Literature data from

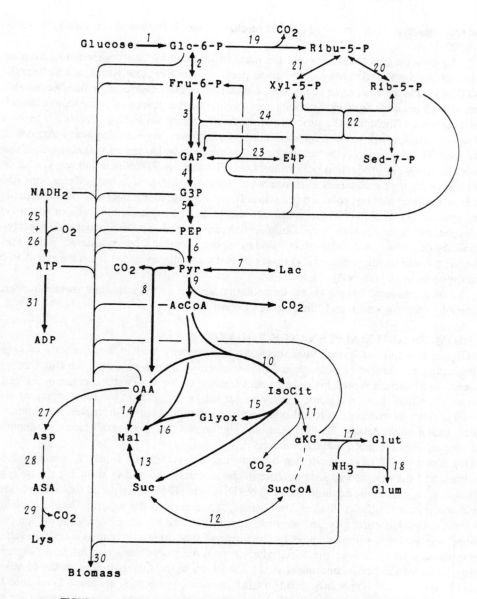

FIGURE 1.   Bioreaction network for lysine production by glutamic acid bacteria.

a lysine fermentation of *Brevibacterium* are used to illustrate the methodology, since *Brevibacterium* species that produce lysine or glutamate are basically identical to *C. glutamicum* for our purposes.

II. METHODOLOGY

Consider the simplified biosynthetic reaction network for lysine production by glutamic acid bacteria illustrated in Figure 1. The estimation of the carbon flux through each reaction in this network from extracellular measurements is the desired goal. Although this simplified network is used to represent cellular metabolism, it is quite obvious that not all biosynthetic reactions have been incorporated. There are thousands of such reactions; to include them all would be impractical. The first step in the development of the methodology is to extract those reactions that represent the major carbon fluxes. This information can usually be found in the literature.

To construct a simplified network, the main fueling and metabolite-generating bioreactions, such as the Embden-Meyerhof-Parnas pathway, the PPP, the Krebs or tricarboxylic acid (TCA) cycle, and the glyoxylate shunt, are assembled. In order to maintain observability of the overall network from extracellular measurements, the pathways that couple extracellular metabolites (including the product) to the fueling reactions must be included. Since a PSS approximation will be used for intracellular metabolites, regenerating reactions such as those for ATP and NAD via the respiratory chain are included to insure that no intracellular metabolite has a net production or consumption. Elaborate or ill-defined pathways, such as biomass synthesis or maintenance requirements, must be expressed as lumped reactions (the details of lumping will be explained in Section II.A). Finally, to minimize the dimensionality of the system, only metabolites that are involved at branch points in the biosynthetic pathways are considered in the network. For example, there are several metabolites between aspartyl-semialdehyde (ASA) and lysine (Lys); however, they need not be considered, since they comprise a nonbranching sequence of reactions that must all proceed at the same rate if PSS assumption is to be satisfied.

The reactions used for the lysine fermentation as well as the important metabolites are presented in the Appendix and illustrated in Figure 1.

A. BIOMASS AND MAINTENANCE EQUATIONS

Of the multitude of intracellular reactions, the majority lead to the synthesis of large biomolecules needed for cell growth and maintenance. Although essential, the flux through any one such reaction represents only a small fraction of the total carbon entering the cell. The sum total of these reactions, however, cannot be neglected, for they represent the pathways through which carbon and other essential nutrients are incorporated into biomolecules. In our methodology, all such reactions are lumped into one overall biomass equation and coupled into the main fueling reactions.

Ingraham et al.[4] have shown that biomass can be synthesized from 12 precursor metabolites, and the researchers have calculated the amounts required to form 1 g dry weight of biomass, as well as the amounts of ATP, $NADH_2$, and $NADPH_2$ required for biosynthesis. Their analysis is similar to that of Stouthamer.[5] Even though the reported biomass yields are for *Escherichia coli*, they are adequate approximations for *C. glutamicum* for the following two reasons. First, in most lysine fermentations biomass synthesis only requires approximately 10 to 20% of the initial glucose, and the growth rate is quite small or zero during lysine production. Consequently, the sensitivity of the flux estimates on the biomass yields is quite small, especially during lysine production. Second, we have been able to duplicate the biomass yields reported by Ingraham et al.[4] for threonine and methionine to within less than 5%. Therefore, the use of *E. coli* biomass yields for *C. glutamicum* is warranted; however, this will probably not be true for all microorganisms and/or fermentations. From the tables given by Ingraham, the lumped equation for biomass synthesis for *C. glutamicum* is shown in Table 1.

The ATP requirement in the lumped biomass equation is the theoretical amount and is equivalent to an ATP yield of 28 g of biomass synthesized per mole of ATP consumed. The lumped equation for biomass does not account for ATP consumption due to maintenance, futile cycles, transport costs, or the energy required to maintain concentration gradients across the cell wall. Consequently, an excess amount of ATP is usually produced by the fueling reactions. To maintain a PSS approximation for ATP, this excess is removed by incorporating the following equation into the network for the conversion of ATP into ADP:

$$ATP \rightarrow ADP \qquad (1)$$

TABLE 1
Lumped Biomass Equation from
Ingraham's Data[4]

Substrates (moles)	Products (moles)
GLC6P (0.0205)	
FRU6P (0.0071)	
RIB5P (0.0898)	
E4P (0.0361)	
GAP (0.0129)	
G3P (0.150)	
AKG (0.058)	
OAA (0.107)	
PEP (0.0519)	Biomass (1.0)
PYR (0.125)	ADP (3.89)
ACCOA (0.327)	NADP (1.37)
NH_3 (0.796)	NADH (0.312)
ATP (3.89)	
NADPH (1.37)	
NAD (0.312)	
LEU (0.043)[a]	
MET (0.015)[a]	
THR (0.024)[a]	
LYS (0.033)[b]	
GLUT (0.025)[b]	
GLUM (0.025)[b]	

Note: Mol wt of biomass = 100.

[a] Auxotroph.
[b] Pathway included.

Although the ATP balance is not necessary for flux determination, Equation 1 is informative since it indicates how much excess ATP, above the theoretical, is produced by the cells. For the lysine fermentation, a P/O ratio of 2 is used.

The reactions illustrated in Figure 1 and presented in the Appendix comprise the metabolism of the cell for the production of lysine by glutamic acid bacteria. A similar approach can be applied to any other organism for any product so long as the biochemistry is known. For protein products, a lumped equation similar to the biomass equation would probably have to be used.

Since it is desired to estimate the extent (i.e., the flux) of each reaction in the network from extracellular measurements only, a metabolite balance based on the biochemistry is used to construct an algebraic equation for the metabolism of the cell. The development of this equation, called the bioreaction network equation (BRNE), is described in the next section and follows a similar approach established for butyric acid bacteria by Papoutsakis[6,7] for the construction of the fermentation equation.

B. THE BIOREACTION NETWORK EQUATION

The extent or flux of each reaction listed in the Appendix is unknown and can be represented as x_i for the ith reaction in the network. The BRNE represents a metabolite balance and is constructed by determining the time rate of change of each metabolite in the network as a function of all the unknown flows, x_i, producing or consuming that metabolite. For example, in the lysine fermentation the rate of production of pyruvate based on the reactions given in the Appendix is

$$r_{pyr} = x_6 - x_7 + x_8 - x_9 - x_{29} - 0.13x_{30} \qquad (2)$$

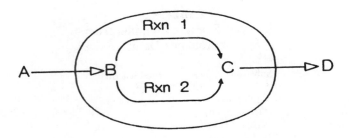

FIGURE 2.   Example of a singular reaction network.

The resulting set of equations for the rate of change of each metabolite in the network is placed in matrix form as follows:

$$\mathbf{Ax = r} \tag{3}$$

where $\mathbf{A}$ is the bioreaction network matrix reflecting the assumed biochemistry, $\mathbf{r}$ is the production rate vector for all the metabolites in the network, and $\mathbf{x}$ is the unknown flux vector to be determined.

The $\mathbf{A}$ matrix is determined by the biochemistry of the microbe and the fermentation, as described above, and has a dimension of $m \times n$, where m is the number of metabolites in the network, n is the number of reactions, and $m \geqslant n$. For the lysine fermentation, $m = 33$ and $n = 31$.

1. Singularities

A solution to the BRNE will exist provided the $\mathbf{A}$ matrix is nonsingular. Singularities in the $\mathbf{A}$ matrix may arise due to reaction dependence or network observability problems. Consider, for example, the simplified network depicted in Figure 2. In this case, Reactions 1 and 2 are indistinguishable from the extracellular measurements of A and D, and the resulting $\mathbf{A}$ matrix would be singular. Reactions in the network that produce such singularities must be either lumped together or removed; in the above example, both yield the same result.

When the complete TCA cycle and the glyoxylate shunt are incorporated into the network for lysine fermentation, a singularity arises indicating that the determination of carbon flux distribution between these two branches is not possible from available extracellular measurements. It has been shown, however, that α-ketoglutarate dehydrogenase (EC 1.2.4.2) of the TCA cycle is not active in glutamic acid bacteria[8] (shown as the dashed line in Figure 1). When this reaction is removed from the network, so is the singularity. Similarly, carboxylation reactions, such as the malate enzyme (EC 1.1.1.40), pyruvate carboxylase (EC 6.4.1.1), oxaloacetate decarboxylase (EC 4.1.1.3), and phosphoenolpyruvate carboxylase and carboxykinase (EC 4.1.1.31 and 4.1.1.32), must be lumped into one reaction (Reaction 8 in the Appendix). It should also be pointed out that matrix singularities are one of the main reasons that biosynthetic reactions are lumped into a single equation.

Once the bioreaction network matrix, $\mathbf{A}$, is assembled and reactions that produce singularities lumped together or removed, the production rate vector, $\mathbf{r}$, must be determined.

2. Production Rate Vector

Since biomass is treated as a product just like any other, each element of the $\mathbf{r}$ vector represents the time rate of change in total concentration of a metabolite in the network minus the intracellular component of that metabolite which has already been accounted for in the measurement of biomass, or

$$r_j = \frac{dC_j^T}{dt} - \mu C_j^I \qquad \text{and} \qquad r_{biomas} = \frac{dC_{biomass}}{dt} \qquad (4)$$

or

$$r_j = \frac{dC_j^E}{dt} + \frac{dC_j^I}{dt} - \mu C_j^I \qquad \text{and} \qquad r_{biomas} = \frac{dC_{biomass}}{dt} \qquad (5)$$

where C_j^T is the total concentration of metabolite j (total moles of metabolite j per total volume of broth) and is equal to $C_j^E + C_j^I$, where the superscripts E and I refer to extracellular and intracellular components, respectively. The subscript j refers to all metabolites listed in the Appendix except for biomass, and μ is the specific growth rate. Since it is desired to base the methodology on measurements of extracellular metabolites only, a PSS (i.e., balanced growth) approximation is used for all intracellular metabolites, given by

$$\frac{dC_j^I}{dt} = \mu C_j^I \qquad (6)$$

Consequently, after substitution of Equation 6 into Equation 5, the production rate vector, $\mathbf{r}$, simplifies to

$$r_i = 0 \qquad \text{for all intracellular metabolites}$$

$$r_k = \frac{dC_k^E}{dt} \qquad \begin{array}{l}\text{for all extracellular metabolites} \\ \text{and biomass}\end{array} \qquad (7)$$

that is, r_i is set to zero for all intracellular metabolites and r_k is determined by measuring the production rate of the extracellular component of metabolite k. Of course, dilution and sampling effects on concentration data must be accounted for when appropriate. The BRNE is now determined completely and can be solved.

Before presenting the solution, it is worth noting that the bioreaction network algorithm (BRNA) does not require information concerning intracellular control mechanisms or kinetic rate constants for reactions. The overall network is simply a metabolic balance governed by the particular biochemistry of the microbe. Consequently, the only uncertainties are the lumped equations and the PSS hypothesis, both of which are reasonable approximations. Furthermore, Equation 6, the PSS approximation, need not be held to rigorously. As long as r_j for all intracellular metabolites, given by Equation 5, is small compared to the fluxes producing or consuming metabolite j, the estimated fluxes will be a valid representation of the true state of the cellular metabolism.

3. Solution and Sensitivity Analysis

The estimated flux vector, $\bar{\mathbf{x}}$, is given by the least squares solution[9] to the BRNE 3:

$$\bar{\mathbf{x}} = (\mathbf{A}^\mathrm{T}\mathbf{A})^{-1}\mathbf{A}^\mathrm{T}\mathbf{r} \qquad (8)$$

A unique solution exists to this equation, since $\mathbf{A}$ is of full rank and $\mathbf{r}$ is defined. Even though a solution exists, it might be very sensitive to slight perturbations in the measurements, $\mathbf{r}$, or perturbations in the biochemistry, $\mathbf{A}$. As a first check, the condition number of the system is calculated. The condition number is defined as

$$C(\mathbf{A}) = \|\mathbf{A}\| \cdot \|\mathbf{A}^\#\| \qquad (9)$$

where $\| \ \|$ is any matrix norm and $\mathbf{A}^{\#}$ is the pseudoinverse, given as

$$\mathbf{A}^{\#} = (\mathbf{A}^\mathrm{T}\mathbf{A})^{-1}\mathbf{A}^\mathrm{T} \qquad (10)$$

If $C(\mathbf{A})$ is small (say <100, Euclidian norm), then the system is considered well posed; however, if $C(\mathbf{A})$ is large (say >1000, Euclidian norm), then there are possible sensitivity problems that should be investigated further (see Reference 9 for more details on the condition number). It should be noted that no measurements are needed for the determination of the condition number, indicating that modifications to the biochemistry or new networks can be tested for sensitivity problems without conducting experiments. The condition number for the lysine fermentation is 62 based on the Euclidian norm.

The sensitivity of the solution with respect to biochemistry and measurements is determined from the appropriate derivatives shown here:

$$\frac{\partial \bar{\mathbf{x}}}{\partial a_{ij}} = (\mathbf{A}^\mathrm{T}\mathbf{A})^{-1}\left[\frac{\partial \mathbf{A}^\mathrm{T}}{\partial a_{ji}} - \left(\frac{\partial \mathbf{A}^\mathrm{T}}{\partial a_{ji}}\mathbf{A} + \mathbf{A}^\mathrm{T}\frac{\partial \mathbf{A}}{\partial a_{ij}}\right)(\mathbf{A}^\mathrm{T}\mathbf{A})^{-1}\mathbf{A}^\mathrm{T}\right]\mathbf{r} \qquad (11)$$

$$\frac{\partial \bar{\mathbf{x}}}{\partial \mathbf{r}} = (\mathbf{A}^\mathrm{T}\mathbf{A})^{-1}\mathbf{A}^\mathrm{T} \qquad (12)$$

where $\dfrac{\partial \mathbf{A}}{\partial a_{ij}}$ represents an $m \times n$ matrix whose ijth element is a 1 and all others 0. These equations are similar to those presented by Boot[10] and can be used to isolate sensitivity problems. To remove such sensitivity problems, one must again either lump reactions together or remove them. Equation 12 allows one to determine which metabolites require accurate measurement, since the derivative indicates how sensitive the solution is to a particular measurement.

4. Redundancy Analysis

This analysis follows the original work of Romagnoli and Stephanopoulos[11] and its application to fermentation data discussed by Wang and Stephanopoulos.[12] When the row dimension of $\mathbf{A}$, m, is greater than the column dimension, n, and the matrix is of full rank, then there are more equations than unknowns, indicating an over-determined system and, hence, the least squares solution. The $m - n$ dependent equations can be used to check the consistency of the measurements, the validity of the PSS hypothesis, and possibly other assumptions.

Referring to the above papers for a complete description of the gross error identification methodology, the essentials of the algorithm can be summarized as follows. If the redundant equations are satisfied by the available measurements, the latter are consistent with the assumed bioreaction network structure and associated assumptions. If inconsistencies are detected, the redundant equations may be used to identify the source of gross error among the measurements and PSS assumptions. This is done by systematically eliminating, one at a time, the measurements or steady-state assumptions and using one of the redundant equations to determine the eliminated measurement. The remaining redundant equations are used to test the consistency of the resulting system of equations and measurements through the calculation of a statistically balanced consistency index.[12] A sharp decrease in the value of the consistency index resulting from the elimination of a measurement or assumption is a strong indication of the latter being a source of error, suggesting that it be dropped from further consideration.

To implement the consistency analysis described by Wang and Stephanopoulos,[12] the independent equations of the BRNE must be removed and the resulting set of redundant equations placed in the form $\mathbf{Br} = \mathbf{O}$ via the following procedure. Through Gauss elimination, the $\mathbf{A}$ matrix can be partitioned into an $n \times n$ upper diagonal matrix, $\mathbf{U}$, and an $m - n \times n$ zero matrix by multiplication of the appropriate $m \times m$ permutation matrix, $\mathbf{E}$. Equation 3 can then be expressed as

$$\mathbf{EAx} = \begin{bmatrix} \mathbf{U} \\ \overline{\mathbf{O}} \end{bmatrix} \mathbf{x} = \mathbf{Er} = \begin{bmatrix} \mathbf{E_1r} \\ \overline{\mathbf{E_2r}} \end{bmatrix}$$

or

$$\mathbf{O} = \mathbf{E_2r} \tag{13}$$

where $\mathbf{E_2}$ is the lower part of the permutation matrix whose dimension is $m - n \times m$. Equation 13 is now in the appropriate form.

The net result of the consistency analysis is the ability to detect and isolate errors in measurements, PSS approximations, and, to a limited extent, biochemistry.

Following the redundancy analysis and consistency tests, a final check of the PSS approximation can be performed utilizing the reaction fluxes obtained from the solution of Equation 3. These fluxes can be used to back calculate the apparent time rates of change of all intracellular metabolites. The latter can then be compared to the fluxes of all reactions producing or consuming the corresponding metabolite. PSS will be a valid approximation for a metabolite if its rate of change is found to be small compared to the fluxes of the reactions affecting that metabolite. If a PSS approximation is found to be invalid, further information is needed on that particular metabolite in the form of additional measurements or kinetic expressions. Such metabolites can also be removed from the network, thereby removing the PSS constraint, provided that this does not produce a singularity in $\mathbf{A}$.

5. Additional Information

In general, the more knowledge that can be incorporated into the flux determination algorithm, the more reliable the flux estimates will be. Thus far, we have not taken into account the directionality of reactions. It is well known that some reactions are completely reversible, while others are considered irreversible. Also, in rare cases a maximum rate might be known for a particular reaction. In either case, directionality of a reaction, or upper bounds on reactions rates, imposes inequality constraints on the BRNE of the type

$$\mathbf{Ax} = \mathbf{r} \qquad \text{subject to} \qquad \mathbf{Cx} \geqq \mathbf{b} \tag{14}$$

where $\mathbf{C}$ is the constraint matrix that dictates which reactions are constrained to be greater than or equal to $\mathbf{b}$, which is usually taken as the zero vector. This inequality constrained least squares (ICLS) problem is not amenable to solution by analytical methods and must be solved numerically. Liew[13] describes a technique for solving ICLS problems. Simply stated, a linearized equation is produced from the derivative of the quadratic least squares equation, $J = (\mathbf{Ax} - \mathbf{r})^T(\mathbf{Ax} - \mathbf{r})$, with respect to $\mathbf{x}$. The linearized equation is solved by a modification of the simplex method, which is but one of the many ways to solve such quadratic programming problems.[14] The solution to the ICLS problem is

$$\bar{\mathbf{x}} = (\mathbf{A}^T\mathbf{A})^{-1}\mathbf{A}^T\mathbf{r} + (\mathbf{A}^T\mathbf{A})^{-1}\mathbf{C}^T\mathbf{z} \tag{15}$$

where $\mathbf{z}$ is a numerically determined k-dimensional dual vector.

It is emphasized that the constraints should not be used to overcome an ill-posed system, but rather to fine tune the unconstrained solution to Equation 8. For example, if the solution to Equation 8 produces a slightly negative flow entering the PPP (i.e., ribulose-5-phosphate

TABLE 2
Production Rates of Extracellular Metabolites[15]

Time (h)	Production rate (mmol/h/l)				
	Glucose	Biomass	Lysine	Oxygen	CO_2
0—12	−2.32[a]	2.92	1.10	−12.5	13.4
12—24	−7.83[a]	3.66	2.28	−49.1	52.5
24—36	−14.3	2.58	1.53	−55.6	59.5
36—48	−16.1	1.25	2.38	−57.8	61.9

[a]   Glucose measurement not used due to inconsistency.

producing glucose-6-phosphate), then the above constraints can be used to set this flow to zero. Ideally, the constraints should not be violated for a perfectly posed system (i.e., $z = O$), but this is hard to achieve, and the constraints may add information that is not directly obtainable from the measurements.

For the lysine fermentation data, no constraints were violated; consequently, Equation 8 was used instead of Equation 15.

C. SUMMARY OF THE ALGORITHM

The following is a summary of the algorithm presented in the previous sections. For a particular microbe and fermentation, one constructs the primary biosynthetic pathways necessary for product synthesis and energy production, including those pathways necessary to generate the intermediate metabolites for the lumped biomass equation and those necessary to regenerate intracellular metabolites. These pathways can usually be found in the literature. The ones for the lysine fermentation are illustrated in Figure 1 and listed in the Appendix. These reactions are then placed in an equational form called the BRNE. The BRNE is then tested for singularity and sensitivity problems. Once a well-posed system is obtained (through lumping, etc.), production rates of all extracellular metabolites are measured experimentally as a function of time to determine r and are checked for consistency. This vector is then substituted into Equation 8 or 15, depending on whether system constraints are invoked, to obtain an estimate, $\bar{x}$, of the carbon flux through the primary pathways.

III. APPLICATIONS

The algorithm was tested with literature data reported by Erickson et al.[15] for a batch lysine fermentation of *Brevibacterium* (species not reported), since it has been well established[16] that glutamic-acid-producing strains of *Brevibacterium* and *Corynebacterium* have similar biochemistry and should be classified under *C. glutamicum*. Data on the production rates of biomass, lysine, and carbon dioxide and the consumption rates of glucose and oxygen were used from that paper. These data, reported every 12 h for the first 48 h, are summarized in Table 2. When these data were checked for consistency, it was found (see Reference 12) that the calculated performance index was outside the 90% confidence level for the 0- to 12-h and 12- to 24-h time periods. When the glucose measurement was removed, the index improved to within acceptable limits; therefore, the glucose measurement was not used for the first two time periods.

Upon substitution of the corrected rate data into Equation 8, the flux map through the lysine bioreaction network of Figure 1 was generated, with the results shown in Table 3. The reaction numbers refer to the reactions given in the Appendix and correspond to the numbers listed on Figure 1.

To compare the estimate of the flux network with experimental data, we have calculated the percentage of glucose entering the PPP from the flux estimate, assuming partial recycling

TABLE 3
Estimated Fluxes for the Bioreaction Network in
mmol/h/l

Reaction no.	Time periods (h)			
	0—12[a]	12—24[a]	24—36	36—48
1	5.39	13.0	14.2	15.8
2	1.11	5.62	8.00	8.60
3	3.70	10.2	10.9	11.9
4	8.60	22.5	23.3	25.6
5	8.14	22.0	22.9	25.4
6	7.96	21.8	22.8	25.3
7	0.00	0.00	0.090	0.135
8	0.621	11.7	15.4	14.7
9	7.04	30.7	35.7	36.7
10	3.31	14.9	17.4	18.2
11	0.407	0.318	0.297	0.366
12	−1.17	−2.42	−2.03	−3.00
13	2.79	14.6	17.2	17.7
14	5.53	29.3	34.7	35.8
15	2.83	14.6	17.2	17.9
16	2.79	14.6	17.4	18.0
17	2.56	4.96	4.17	6.17
18	0.0958	0.0721	0.106	0.145
19	4.22	7.27	5.88	6.92
20	1.62	2.69	2.27	2.58
21	2.60	4.58	3.47	4.12
22	1.35	2.36	1.87	2.22
23	1.35	2.36	1.66	1.90
24	1.25	2.23	1.44	1.67
25	11.1	41.8	47.2	49.2
26	1.40	7.31	8.59	8.84
27	1.22	2.38	1.93	2.95
28	1.17	2.42	2.03	3.00
29	1.15	2.44	2.02	2.94
30	2.91	3.67	2.72	1.45
31	40.7	184	213	226

[a] Glucose measurement not used.

of fructose-6-phosphate (F6P). It has been shown by radiolabeling experiments[17,18] that the best method for calculating the fraction of glucose entering the PPP is to assume partial recycling of F6P. This assumes that the fraction of F6P that is recycled back into the PPP is the same as the fraction of glucose that enters the PPP. From the network shown in Figure 1, the percentage of glucose entering the PPP, based on partial recycling, is given by

$$PPP\ (\%) = \frac{100(x_1 - x_2)}{x_1 + x_{23} + x_{24}} \qquad (16)$$

where x_i represents the flux in reaction i, depicted in Figure 1 and listed in the Appendix. Oishi and Aida[19] have measured, using radiorespirometry,[20] the percentage of glucose entering the PPP for *B. ammoniagenes* in the stationary phase (i.e., no growth) and under high biotin concentration (20 μg/l). The *PPP (%)* calculated from the flux estimates and Equation 16 along with the measurements of *PPP (%)* of Oishi and Aida are listed in Table 4.

The conditions under which Oishi and Aida conducted their experiments correspond to

TABLE 4
Estimated and Measured Glucose
Entering the Pentose Phosphate
Pathway

Percent Glucose Entering PPP

Time (h)	PPP (%)[a]	PPP (%)[b]
0—12	54[c]	—
12—24	42[c]	—
24—36	35	—
36—48	37	38[d]

[a] Calculated by partial recycle of F6P, Equation 16.
[b] Datum from Oishi and Aida.[19]
[c] Glucose measurement not used.
[d] Resting cells, with 20 μg/l biotin.

the conditions that prevail during the end of the fermentation, i.e., stationary phase and high biotin concentration. From Table 4 we can see that the estimated result of 35 to 37% glucose entering the PPP corresponds extremely well with the experimentally measured value of 38% for cells in stationary phase. This indicates that the estimated flux obtained from the BRNA does represent the true state of the metabolism in the cell for the PPP.

Furthermore, there are some other interesting features of this flux network. Reaction 31 indicates the amount of excess ATP produced over that quantity theoretically required. This flow is an order of magnitude greater than any other flux in the network and represents approximately 64, 82, 85, and 86% of ATP produced for the 0- to 12-h, 12- to 24-h, 24- to 36-h, and 36- to 48-h time periods, respectively. This finding indicates that most of the glucose consumed by the cells is simply oxidized to carbon dioxide and water, and it accounts for the large flux of carbon cycling through the modified TCA cycle (i.e., the TCA cycle plus the glyoxylate shunt and the decarboxylation reaction) and the low lysine yield (20% molar). Consequently, if this flux can be redirected to oxaloacetate (OAA), the lysine yield may be increased.

For glutamic acid production it has been proposed that Reaction 11 generates the required $NADPH_2$ for Reaction 17; however, this cannot be true for lysine production since, based on the flux network, Reaction 11 is very small compared to Reaction 17. Reaction 11 in this case is not large, since α-ketoglutaric acid is produced from glutamate in the lysine reactions (see Reactions 27 and 29); consequently, $NADPH_2$ needed for Reaction 17 must be generated via the PPP, which could be another limiting reaction in the production of lysine.

It should also be noted that the decarboxylation reaction (Reaction 8) is essential when the glyoxylate shunt is functioning, which is reasonable since Reaction 8 is used to complete the modified TCA cycle, as postulated by Shiio et al.[21]

IV. SUMMARY

The primary metabolic pathways of a cell have been represented by a linear equation, termed the BRNE. Several routines have been developed to check the singularity, sensitivity, and consistency of the model, and techniques have been described as means to correct these problems when they arise, such as lumping or removal of reactions and removal of statistically inconsistent measurements.

The biochemistry of lysine fermentation for glutamic acid bacteria has been constructed based on available literature data. With the carboxylation reactions lumped into one reaction (Reaction 8), α-ketoglutarate dehydrogenase in the TCA cycle removed, excess ATP removed via Reaction 31, and biomass synthesis represented as one reaction (Table 1), the bioreaction network matrix A is nonsingular and has a dimension of 33×31 and a condition number of 62. The corresponding biochemical reactions are listed in the Appendix.

The fluxes generated by the unconstrained solution, Equation 8, from Erickson's data appear reasonable, and the percentage of glucose entering the PPP calculated from the flux estimates correlates well with experimental data presented by Oishi and Aida.[19] It also appears that the low lysine yield of 20% molar might be caused by overproduction of ATP by the oxidative pathways of glutamic acid bacteria.

ACKNOWLEDGMENTS

We thank the Massachusetts Institute of Technology (MIT) Biotechnology Process Engineering Center and the National Science Foundation (PYI grant no. CBT-8514729) for their support of this research.

APPENDIX: BIOCHEMISTRY AND METABOLITES*

The following reactions are used to describe the metabolism of glutamic acid bacteria for lysine fermentation. Also given are the important metabolites considered. From this information the bioreaction network matrix, A, is constructed.

A. BIOCHEMICAL REACTIONS
Embden-Meyerhof-Parnas Pathway
1. GLC + ATP → GLC6P + ADP
2. GLC6P ↔ FRU6P
3. FRU6P + ATP → 2 GAP + ADP
4. GAP + ADP + NAD → NADH + G3P + ATP
5. G3P ↔ PEP + H_2O
6. PEP + ADP → ATP + PYR
7. PYR + NADH ↔ LAC + NAD

Carboxylation Reactions
8. OAA → PYR + CO_2

TCA Cycle
9. PYR + COA + NAD → ACCOA + CO_2 + NADH
10. ACCOA + OAA + H_2O ↔ ISOCIT + COA
11. ISOCIT + NADP ↔ AKG + NADPH + CO_2
12. SUCCOA + ADP ↔ SUC + COA + ATP
13. SUC + H_2O + FAD ↔ MAL + FADH
14. MAL + NAD ↔ OAA + NADH

Glyoxylate Shunt
15. ISOCIT ↔ SUC + GLYOX
16. ACCOA + GLYOX + H_2O → MAL + COA

---

* See Appendix Section B for abbreviation designations.

Glutamate and Glutamine Production
17. NH_3 + AKG + NADPH $\leftrightarrow$ GLUT + H_2O + NADP
18. GLUT + NH_3 + ATP $\rightarrow$ GLUM + ADP

Pentose Phosphate Cycle
19. GLC6P + H_2O + 2 NADP $\rightarrow$ RIBU5P + CO_2 + 2 NADPH
20. RIBU5P $\leftrightarrow$ RIB5P
21. RIBU5P $\leftrightarrow$ XYL5P
22. XYL5P + RIB5P $\leftrightarrow$ SED7P + GAP
23. SED7P + GAP $\leftrightarrow$ FRU6P + E4P
24. XYL5P + E4P $\leftrightarrow$ FRU6P + GAP

ATP Generation; P/O = 2
25. 2 NADH + O_2 + 4 ADP $\rightarrow$ 2 H_2O + 4 ATP + 2 NAD
26. 2 FADH + O_2 + 2 ADP $\rightarrow$ 2 H_2O + 2 ATP + 2 FAD

Aspartate Family
27. OAA + GLUT $\leftrightarrow$ ASP + AKG
28. ASP + NADPH + ATP $\rightarrow$ ASA + ADP + NADP
29. ASA + PYR + NADPH + SUCCOA + GLUT $\rightarrow$ SUC + AKG + CO_2 + H_2O
    + LYS + NADP + COA

Biomass Production; Assume Mol Wt (Biomass) = 100
30. 0.0205 GLC6P + 0.00709 FRU6P + 0.0898 RIB5P + 0.0361 E4P + 0.0129 GAP
    + 0.15 G3P + 0.0519 PEP + 0.125 PYR + 0.327 ACCOA + 0.058 AKG +
    0.107 OAA + 0.0326 LYS + 0.796 NH_3 + 0.025 GLUT + 0.025 GLUM + 3.89
    ATP + 1.37 NADPH + 0.312 NAD = BIOMASS + 3.89 ADP + 1.37 NADP
    + 0.312 NADH

Unaccounted for ATP Requirements
31. ATP $\rightarrow$ ADP

B. LIST OF IMPORTANT METABOLITES
Each metabolite listed below corresponds to one element of the **r** vector.

1. Acetyl coenzyme A (ACCOA)
2. α-Ketoglutarate (AKG)
3. Aspartate semialdehyde (ASA)
4. Aspartate (ASP)
5. Adenosine 5'-triphosphate (ATP)
6. BIOMASS
7. CO_2
8. Erythrose-4-phosphate (E4P)
9. Flavin adenine dinucleotide (FADH)
10. Fructose-6-phosphate (FRU6P)
11. 3-Phosphoglycerate (G3P)
12. Glyceraldehyde-3-phosphate (GAP)
13. Glucose (GLC)
14. Glucose-6-phosphate (GLC6P)
15. Glutamine (GLUM)
16. Glutamate (GLUT)
17. Glyoxylate (GLYOX)
18. Isocitrate (ISOCIT)
19. Lactate (LAC)
20. Lysine (LYS)
21. Malate (MAL)
22. NADH
23. NADPH
24. O_2
25. Oxalocetate (OAA)
26. Phosphoenolpyruvate (PEP)
27. Pyruvate (PYR)
28. Ribose-5-phosphate (RIB5P)
29. Rubulose-5-phosphate (RIBU5P)
30. Sedoheptalose-7-phosphate (SED7P)
31. Succinate (SUC)
32. Succinyl CoA (SUCCOA)
33. Xylulose-5-phosphate (XYL5P)

REFERENCES

1. **Kinoshita, S. and Nakayama, K.,** Amino acids, in *Economic Microbiology,* Vol. 2, Rose, A. H., Ed., Academic Press, London, 1978, 209.
2. **Simon, J., Engasser, J., and Germain, P.,** Substrate flux mapping in Corynebacteria for amino acids production, in *Third European Congress on Biotechnology, Munchen,* Vol. 2, Verlag Chemie, Weinheim, Federal Republic of Germany, 1984, 49.
3. **Shvinka, J., Viesturs, U., and Ruklisha, M.,** Yield regulation of lysine in *Brevibacterium flavum, Biotechnol. Bioeng.,* 22, 897, 1980.
4. **Ingraham, J. L., Maaloe, O., and Neidhardt, F. C.,** *Growth of the Bacterial Cell,* Sinauer Associates, Sunderland, MA, 1983, chap. 3.
5. **Stouthamer, A. H.,** A theoretical study on the amount of ATP required for synthesis of microbial cell material, *Antonie van Leeuwenhoek J. Microbiol. Serol.,* 39, 545, 1973.
6. **Papoutsakis, E. T.,** Equations and calculations for fermentations of butyric acid bacteria, *Biotechnol. Bioeng.,* 26, 174, 1984.
7. **Papoutsakis, E. T.,** Equations and calculations of product yields and preferred pathways for butanediol and mixed-acid fermentations, *Biotechnol. Bioeng.,* 27, 50, 1985.
8. **Shiio, I.,** Significance of α-ketoglutaric dehydrogenase on the glutamic acid formation in *Brevibacterium flavum, J. Biochem. (Tokyo),* 50, 164, 1961.
9. **Noble, B. and Daniel, J. W.,** *Applied Linear Algebra,* Prentice-Hall, Englewood Cliffs, NJ, 1977, 330.
10. **Boot, J. C. G.,** On sensitivity analysis in convex quadratic programming problems, *Oper. Res.,* 11, 771, 1963.
11. **Romagnoli, J. A. and Stephanopoulos, G.,** Rectification of process measurement data in the presence of gross errors, *Chem. Eng. Sci.,* 36, 1849, 1981.
12. **Wang, N. S. and Stephanopoulos, G. N.,** Applications of macroscopic balances to the identification of gross measurement errors, *Biotechnol. Bioeng.,* 25, 2177, 1983.
13. **Liew, C. K.,** Inequality constrained least-squares estimation, *J. Am. Stat. Assoc.,* 71, 746, 1976.
14. **Boot, J. C. G.,** Binding constraint procedures of quadratic programming, *Econometrica,* 31, 464, 1963.
15. **Erickson, L. E., Selga, S. E., and Viesturs, U. E.,** Application of mass and energy balance regularities to product formation, *Biotechnol. Bioeng.,* 20, 1623, 1978.
16. **Abe, S., Takayama, K., and Kinoshita, S.,** Taxonomical studies on glutamic acid-producing bacteria, *J. Gen. Appl. Microbiol.,* 13, 279, 1967.
17. **Dawes, E. A. and Holms, W. H.,** Metabolism of *Sarcina lutea.* II. Isotopic evaluation of the routes of glucose utilization, *Biochim. Biophys. Acta,* 29, 82, 1958.
18. **Shiio, I., Otsuka, S., and Tsunoda, T.,** Glutamic acid formation from glucose by bacteria. III. On the pathway of pyruvate formation in *Brevibacterium flavum* No. 2247, *J. Biochem. (Tokyo),* 47, 414, 1960.
19. **Oishi, K. and Aida, K.,** Studies on amino acid fermentation. XI. Effect of biotin on the Embden-Meyerhof-Parnas pathway and the hexose-monophosphate shunt in a glutamic acid-producing bacterium, *Brevibacterium ammoniagenes* 317-1, *Agric. Biol. Chem.,* 29, 83, 1965.
20. **Wang, C. H., Stern, I., Gilmour, C. M., Klungsoyr, S., Reed, D. J., Bialy, J. J., Christensen, B. E., and Cheldelin, V. H.,** Comparative study of glucose catabolism by the radiorespirometric method, *J. Bacteriol.,* 76, 207, 1958.
21. **Shiio, I., Otsuka, S., and Takahashi, M.,** Glutamic acid formation from glucose by bacteria. VI. Metabolism of the intermediates of the TCA cycle and of the glyoxylate bypass in *Brevibacterium flavum* No. 2247, *J. Biochem. (Tokyo),* 50, 34, 1961.

Part 4. Bioseparations I — Free Fluid Methods

Labile biological products and living cells present very special purification problems. The required downstream processes are themselves a considerable subject for research. Free fluid methods are preferred in cases where precipitation and adsorption result in loss of activity. Solid and semisolid purification media are seldom applicable to large-scale separation of particulates such as living cells and subcellular particles. The first three chapters deal with free electrophoresis, and they progress from basic testing to recent developments in scaleup to testing that utilizes the microgravity environment. Similarly, two-aqueous-phase systems are themselves the subject of research and are only gradually finding their way into routine processing. These systems can also be used in the presence of particulates and are much easier to scale up then electrophoresis methods, which require efficient heat rejection. The issues of monitoring and analysis of fractions are also addressed.

Chapter 19

ANALYTICAL CELL ELECTROPHORESIS AS A TOOL IN PREPARATIVE CELL ELECTROPHORESIS

Paul Todd, Jeffrey Kurdyla, Burton E. Sarnoff, and William Elsasser

TABLE OF CONTENTS

I. INTRODUCTION

A number of applications of analytical electrophoresis to problems in cell biology have been studied over the past two decades. Most applications have been to problems of hematology and immunology. In such studies, particular investigators[1-6] reported electrophoretic mobility values and/or distributions of particular cell subpopulations, typically from peripheral human blood. The type of cell in each analyzed subpopulation could not always be identified, as the cells were not separated and analyzed after electrophoresis, but sometimes previously separated subpopulations were used as electrophoretic mobility landmarks.

Conversely, preparative electrophoresis has successfully separated a variety of cell types, such as B and T murine lymphocytes,[7] graft-vs.-host bone marrow cells,[8] and functional cells of the kidney[9-12] and pituitary.[12,13] Continuous free-flow electrophoresis (CFE)[7-10,12] and density gradient electrophoresis[11,13,14] have been the preparative methods of choice. Such processes as particle sedimentation,[13] droplet sedimentation,[15] field-flow separation,[16] electrohydrodynamics,[17] and Poiseuille and electroosmotic flow profiles[18] are known to affect particle motion under most conditions of preparative free electrophoresis. Nevertheless, it is routinely assumed that migration distance in the free fluid chambers of these systems is proportional to electrophoretic mobility (EPM) and that separated cells are collected in fractions based on EPM and not other physical variables; however, this thesis has been subjected to very limited testing.

The purpose of this chapter is to review and present the results of experiments with living cells and test particles in which postseparation analytical electrophoresis (in which the above processes are not pertinent) was used to determine the EPM of separated subpopulations.

II. REVIEW

Sedimentation is one of the physical variables most likely to interfere with the purely electrokinetic separation of suspended particles.[12,13] Preparative cell electrophoresis experiments have been performed in the absence of sedimentation and buoyancy using orbital space flight as a means of eliminating inertial acceleration.[12,19-23] These are the only preparative cell electrophoresis experiments in which the EPM of collected subpopulations was measured, and only two experiments have been reported, both involving human embryonic kidney (HEK) cells from primary cultures.

In the work of Barlow et al.,[20] free zone electrophoresis was performed on frozen cell suspensions on the Apollo-Soyuz Test Project mission in 1975. A glycerol-containing dilute phosphate buffer was used. Fractions were collected by slicing the frozen electrophoresis column after return to earth. Each fraction yielded a small number of viable cells (Figure 1, top panel), which were subcultured twice to yield enough cells for investigations that included evaluation of secretory functions[20] and measurement of EPM in selected fractions by microscopic analytical electrophoresis using a cylindrical chamber and a Rank Brothers (Cambridge, U.K.) electrophoresis microscope. It can be seen in the lower three panels of Figure 1 that increasing EPM of the resulting subcultures correlated well with increasing migration distance. In the simplest interpretation, the findings shown in Figure 1 suggest that the EPM of progeny cells is correlated with that of their ancestors and, at least in the absence of sedimentation, free zone electrophoresis separates cells according to EPM.

In the work of Morrison et al., CFE of kidney cells was performed on U.S. Space Shuttle flight STS-8 along with other cell separation experiments.[12] A low-ionic-strength triethanolamine acetate buffer was used. Fractions were collected in flight as previously described,[12,22,23] and after return to earth aliquots of living cells were distributed into cultures for functional testing. After two subcultivations, adequate numbers of cells were available

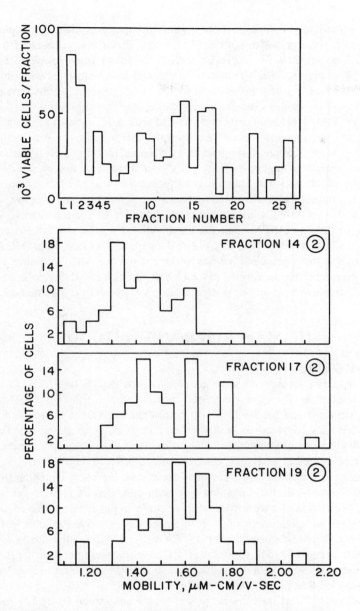

FIGURE 1.   Mobilities of kidney cell fractions separated by free zone electrophoresis. Top panel: electrophoretic migration pattern of cultured human embryonic kidney cells separated by free zone electrophoresis in low gravity on the Apollo-Soyuz Test Project mission; number of viable cells vs. fraction number. Migration was from left to right. Lower three panels: electrophoretic mobility distributions (number of cells vs. mobility) of cells that were present after two subcultivations of fractions 14, 17, and 19 from the distribution in the top panel. (Redrawn from Barlow, G. H., Lazer, S. L., Reuter, A., and Allen, R., in Bioprocessing in Space, NASA TM X-58191, Morrison, D. R., Ed., Lyndon B. Johnson Space Center, Houston, TX, January 1977, 125. With permission.)

for EPM determinations on some fractions by microscopic analytical electrophoresis using a Zeiss® ''Cytopherometer'' with platinized platinum electrodes purchased from Cam-Apparatus, Ltd., Cambridge, U.K. It can be seen in the lower three panels of Figure 2 that increasing EPM of the resulting subcultures correlated well with increasing migration distance, except that fraction 96, the lowest mobility fraction tested, contained cells with a broad range of EPMs. In the simplest interpretation, the findings shown in Figure 2 also suggest that the EPM of progeny cells is correlated with that of their ancestors and, at least in the absence of sedimentation, free zone electrophoresis separates cells according to EPM. The EPM histogram of fraction 96 suggests that the lowest mobility cells give rise to progeny cells that differentiate into higher mobility cells — a notion that deserves further testing.

The two reports to date that have provided postseparation EPM distributions of living cells (HEK cells in both cases) demonstrate that electrophoretic separation by two different methods and in two different buffers, but in the absence of inertial forces, do generally separate cells according to EPM, and the separated cells generally give rise to progeny whose EPM distributions reflect the EPM of the separated ancestral cells. Missing are data on cells separated in the presence of inertial forces and not permitted to multiply. Therefore, a series of experiments were performed in which test particles and living cells were separated by CFE, and individual fractions or pooled fractions were analyzed immediately for EPM distributions.

III. MATERIALS AND METHODS

A. TEST PARTICLES

Polystyrene latex microspheres were generously provided by Interfacial Dynamics Corporation, Portland, OR. Two lots were used in this study: lot 2-27-92 (297-nm-diameter carboxylated particles) and lot 10-36-23 (865-nm-diameter carboxylated particles). These particles have identical EPM values in the buffers tested, and both are provided as 8% solids suspended in distilled water without surfactant. For electrophoresis experiments, the particle suspensions were diluted directly into electrophoresis buffer.

Rabbit and goose erythrocytes were purchased from Hazelton Dutchland Laboratories (Denver, PA). Goose blood had been prepared with citrate as an anticoagulant, and rabbit blood was defibrinated. Cells were washed three times in phosphate-buffered saline (PBS), pH 7.4, and fixed for a minimum of 24 h in a solution of 1.5% formalin in PBS at 4°C. The formalin solution was replaced after 48 h with fresh 1.5% formalin in PBS, and the fixed cells were stored in 1.5% formalin at 4°C until measurements in CFE buffer were made. This treatment results in a negligible change in EPM at neutral pH and stabilizes the cells as test particles that can be used for several months.[24]

In preparation for EPM measurements, cells were suspended in freshly prepared CFE buffer and centrifuged at about $140 \times g$ for about 8 min in a Sorvall RT6000 benchtop centrifuge. The cells were then resuspended in CFE buffer and centrifuged twice more. The cells were finally resuspended in fresh CFE buffer and were subjected to measurement.

B. CULTURED HUMAN CELLS

Human embryonic kidney cell strain HEK-1593 (local designation), obtained from MA Bioproducts (Rockville, MD), was used at the fifth passage *in vitro*. Cells were cultivated at 37°C in a 5% CO_2, humidified atmosphere in medium ''MM1'', which consisted of one part Medium 199, one part ''Alpha'' modification of Eagle's MEM, and one part Dulbecco's modification of Eagle's MEM, to which were added (in grams per liter) 16.2 sodium bicarbonate, 1.2 bactopeptone, 0.02 folic acid, 0.072 *i*-inositol, and 0.001 nicotinic acid to form the basic medium. This recipe was generously provided by Dr. M. L. Lewis of Krug International, Houston, TX. To this medium were added 10% fetal bovine serum, 10 U/ml

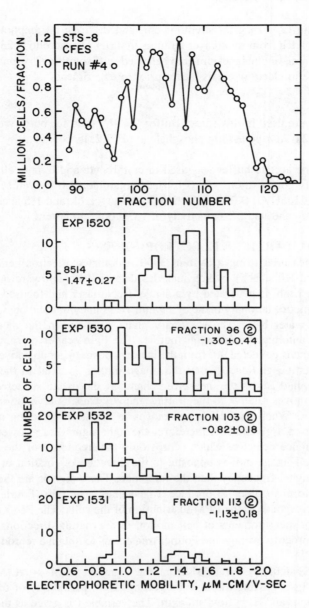

FIGURE 2. Mobilities of kidney cell fractions separated by continuous-flow electrophoresis. Top panel: electrophoretic migration pattern of cultured human embryonic kidney cells separated by continuous-flow electrophoresis in low gravity on the U.S. Space Shuttle STS-8 mission; number of viable cells vs. fraction number. Migration was from left to right. Second panel shows EPM distribution (number of cells vs. mobility) of the starting population of kidney cell strain 8514. Lower three panels: electrophoretic mobility distributions of cells that were present after two subcultivations of cells from fractions 96, 103, and 113 from the distribution in the top panel. (Redrawn with permission from *Adv. Space Res.*, Vol. 4, Morrison, D. R., Lewis, M. L., Cleveland, C., Kunze, M. E., Lanham, J. W., Sarnoff, B. E., and Todd, P., Properties of electrophoretic fractions of human embryonic kidney cells separated on Space Shuttle flight STS-8, ©1984, Pergamon Journals, Ltd.)

potassium penicillin G, 10 μg/ml streptomycin, and 0.25 μg/ml amphotericin B sulfate. Cells were resuspended from monolayer cultures by brief exposure to 0.25% trypsin, 0.1% Na_2EDTA, in calcium- and magnesium-free balanced salts solution. Tryspinized suspensions were washed twice in electrophoresis buffer prior to experiments.

C. BUFFERS

Two buffers were used in this study. Buffer "R-1", used for polystyrene latex particle experiments, was 2.5 mM potassium phosphate at pH 7.2; its conductivity was usually near 0.9 mmho/cm.

Triethanolamine acetate buffer was used in erythrocyte and living cell experiments. In grams per liter, its composition was 0.1 triethanolamine, 2.07 glycine, 0.02 potassium acetate, 0.006 $MgCl_2 \cdot 6H_2O$, 0.004 $CaCl_2 \cdot 2H_2O$, 2.0 glycerol, and 1.5 sucrose. This buffer was 0.295 osmol/kg and had a conductivity of ca. 0.060 mmho/cm.

D. ANALYTICAL PARTICLE ELECTROPHORESIS

EPMs were measured using a System 3000 Automated Electrokinetic Analyser (Pen Kem Inc., Bedford Hills, NY), which includes data analysis software and an IBM® AT desktop computer. Light is generated by a 5-mW HeNe laser and focused into a curtain at the front electroosmotic stationary plane of a 1 mm × 1 cm cylindrical quartz tube immersed in a recirculating water bath. The stationary plane positions in the measuring tube are determined by an automated calibration routine. The light scattered from the cells at the front stationary plane is collected 90° from the incident beam by an objective which produces an image on a rotating grating, so that the image motion is axial to the tube. When an electric field is applied across the measuring chamber, the image scattered from the electrophoresing cells moves relative to the grating, thereby causing a variation in the intensity of transmitted light. When the cell image moves in the same direction as the grating, it crosses fewer lines in a given time; therefore, the light signal has a lower frequency than that from the reference detector, which senses only the frequency of the rotating grating. Likewise, if the cell image moves opposite to the direction of rotation of the grating, the signal acquires a higher frequency than the reference. These signals are then analyzed by a Fast Fourier transform in order to produce a frequency spectrum (Fourier coefficient vs. frequency) that corresponds to the EPM distribution of the cells. The Pen Kem System 3000 makes a number of measurements of cell motion in alternating directions (16 in all cases presented here) in order to average the Fourier transforms to obtain a reproducible histogram of intensity vs. EPM.

The computer calculates the EPM from the equation EPM = fL/ME, where f = frequency shift of light passing through the grating, L = periodicity of the grating, M = optical magnification, and E = field strength. The variable f is given as the abscissa of the Fourier transform, and L and M are instrument constants, so the absolute velocity is fL/M. The instrument measures and reports conductivity and automatically adjusts the applied current, so that the absolute electric field strength E is determined automatically. The mean EPM, the EPM histogram, and its full width at half maximum are reported, along with physical constants of the determination on two screens simultaneously (graphic and alphanumeric).[25] Since the signal intensity depends on the intensity of light received by the photomultiplier tube, a few large, very bright cells would register as strongly as many smaller ones. When the particle concentration is low, homodyne and random signals from the solvent are present in the power spectrum, and graphic background subtractions are performed manually.

E. PREPARATIVE ELECTROPHORESIS

The McDonnell Douglas Continuous Flow Electrophoresis System was used in all preparative electrophoresis experiments. It consists of a rectangular separation chamber sandwiched between two electrode chambers which are also the cooling panels.[12,13,26,27] These are separated from the chamber by semipermeable ion-exchange membranes. A diagram is given elsewhere in this volume.[17] A stable electric field is maintained by ion flow through the membranes. Carrier buffer flows upward through the separation chamber at a rate of 20 ml/min. The electrodes and separation chambers are pumped differentially, and flow balancing tends to eliminate transverse flow and pressure gradients in the sample chamber, thereby enhancing cell stream stability and reproducibility of results. Dimensions are 110 cm high × 6 cm wide × 1.5 mm thick. The lateral migration distance of the cells depends on the residence time (t) spent by the cells in the chamber, the applied field E, and the cell surface charge density. The product, Et (in V-min/cm), when multiplied by the EPM of the particle at the ionic strength and viscosity of the separation, predicts the migration distance.

Cells (about 3 million per milliliter) previously washed and suspended in carrier buffer were injected into the bottom of the chamber using a vertically positioned infusion pump at 4 ml/h. Cells were exposed to the electric field for 4 min, and they were collected as fractions through 197 ports at the top of the flow chamber. In the absence of applied field, the sample stream typically exited through outlet #51. Fractions of kidney cells were collected under sterile conditions, counted by hemacytometer, and pooled into three suspensions of living cells (greater than 85% viability). An aliquot of each pool was used for further cultivation, and the majority of each suspension was used for EPM determination using the System 3000 Analyser.

IV. RESULTS

A. LATEX TEST PARTICLES
Latex microspheres were used as test particles in experiments designed to characterize the CFE system with respect to sample band spreading and separand band spreading. The data of Figure 3 indicate that monodisperse submicron particles with identical EPMs (4.40 ± 0.07 μm-cm/V-sec), but different sizes, migrated identically (to outlet 113 vs. 51 with zero field) and were spread over about 6.5 (full width of peak at half maximum) outlets (each outlet is 0.8 mm wide). As expected, the migration distance did not depend upon particle size, as submicron particles sediment very little distance during the 4-min residence time. The shapes of the distributions were as predicted for conditions in which sample buffer and carrier buffer have the same conductivity.[18,26] When the mean EPM of each fraction was measured (upper curves in Figure 3), all fractions were found to be nearly identical. This finding confirmed the monodisperse nature of the latex test particles and demonstrated that the 4.6% relative standard deviation of the migration distance (standard deviation/mean migration distance) is a property of the CFE separation system.

B. FIXED ERYTHROCYTES
Fixed rabbit and goose erythrocytes were found to have EPMs of 2.26 ± 0.06 and 3.52 ± 0.32 μm-cm/V-sec, respectively, in CFE triethanolamine buffer. When these were mixed in roughly equal proportions and reseparated by CFE, postseparation evaluation of purity by microscopic cell counts showed fraction 104 (the low-mobility peak fraction) to be 98.9% rabbit erythrocytes and fraction 120 (the high-mobility peak fraction) to be 96.8% goose erythrocytes. The migration profiles of the two cell types are shown in the top panel of Figure 4. The lower three panels consist of EPM distributions determined by the System 3000 Analyser, and these indicate that reseparated pooled fractions of rabbit and goose erythrocytes had the expected EPM distribution of each pure cell type. Pooled fractions (110 to 116, in this case) known to contain both cell types had a bimodal EPM distribution in which both cell types were represented. The distance migrated was directly proportional to the EPM, and the positions of the peaks could be predicted on the basis of 21.0 ± 1.4 fractions migrated per mobility unit in the CFE triethanolamine buffer.

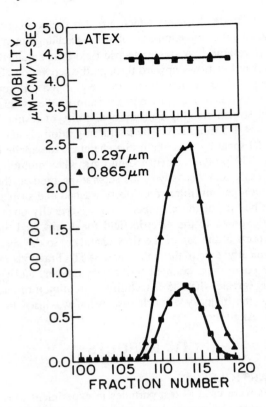

FIGURE 3.   Mobilities of polystyrene test particles separated by continuous-flow electrophoresis. Lower pair of curves: electrophoretic migration pattern of polystyrene latex microspheres in phosphate buffer in continuous-flow electrophoresis; optical density of particle suspension vs. fraction number. Migration was from left to right. Two batches of particles with identical EPM, 4.4 μ-cm/V-sec, but different diameters, 0.297 (plotted squares) and 0.865 μm (plotted triangles), were used. Upper pair of curves: mean EPM of particles collected from each fraction vs. fraction number, as determined by Pen Kem System 3000.

C. CULTURED HUMAN EMBRYONIC KIDNEY CELLS

EPM distributions of living cells immediately after separation by CFE have not been available previously. The data of Figure 5 were derived from an experiment in which CFE fractions were pooled to provide enough cells for EPM determination by the System 3000 Analyser. The top panel of Figure 5 is the CFE migration profile of HEK 1593 cells at passage five in culture. This particular population is low in high-mobility cells (thought to be high in urokinase production) when compared with other populations.[28,29] The numbers of the fractions pooled are indicated in the top panel and caption of Figure 5. The mobility histogram and mean EPM (in micron-centimeters per volt-seconds) of each pool is shown in the lower three panels of Figure 5. The ratio of distance migrated (in fractions) to mean EPM is roughly constant at 20.9 ± 1.8 fractions per EPM unit.

V. DISCUSSION

In the five cases investigated and presented here, the migration of cells in CFE was found to be consistent with the EPM distributions of the starting mixtures and the separands.

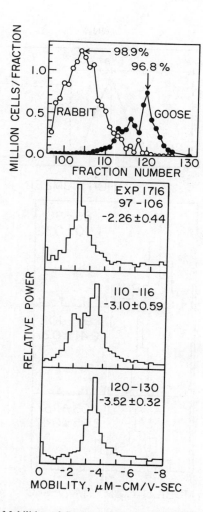

FIGURE 4. Mobilities of fixed erythrocytes separated by continuous-flow electrophoresis. Top panel: electrophoretic migration pattern of a mixture of fixed rabbit and goose red blood cells (test particles) separated by continuous-flow electrophoresis in triethanolamine buffer; number of each cell type vs. fraction number. Migration was from left to right. Lower three panels: electrophoretic mobility distributions (cell light-scattering intensity vs. mobility) of cells that were pooled from fractions 97 to 106 (rabbit cells only), 110 to 116 (roughly equal mixture), and 120 to 130 (goose cells only). Range of fraction numbers and mean mobilities (with standard deviations) are given on each panel.

Electrophoretically homogeneous particle suspensions migrated with a single peak, and heterogeneous particle suspensions were separated into fractions of which EPM distributions were, in turn, proportional to the distance they migrated in CFE. In experiments that were performed in the presence of gravity, care was taken to maintain similar conductivities between sample and carrier buffers, and the particle concentrations were held below those that would create anomalies due to excess density[12] or conductivity.[22] Within these constraints, it is concluded that CFE separates cells primarily on the basis of electrophoretic mobility.

Future biotechnology applications of cell purification by electrophoresis will depend, in each case, upon the establishment of conditions that allow separation of cells according to

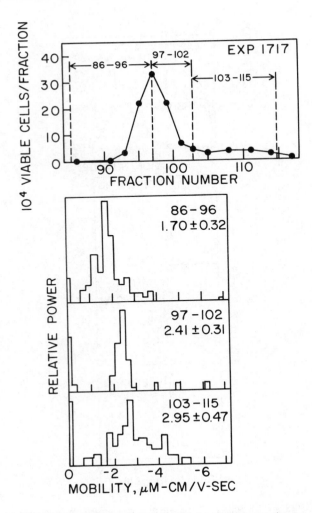

FIGURE 5. Mobilities of kidney cells separated by continuous-flow electrophoresis. Top panel: electrophoretic migration pattern of cultured human embryonic kidney cells (strain 1593) separated by continuous-flow electrophoresis in triethanolamine buffer; number of viable cells vs. fraction number. Migration was from left to right. Lower three panels: electrophoretic mobility distributions (cell light-scattering intensity vs. mobility) of cells that were pooled from fractions 86 to 96, 97 to 102, and 103 to 115 from the distribution in the top panel.

function[28,29] and that allow purification of adequate quantities of cells without compromising the physical constraints that allow separation purely according to EPM.

ACKNOWLEDGMENTS

This work was supported by National Aeronautics and Space Administration (NASA) Grant NAGW-694 and Contracts NAS9-17431 and NAS9-15584. The loan of the CFE System by McDonnell Douglas Astronautics Company and gifts of monodisperse latex particles from Interfacial Dynamics Corporation are gratefully acknowledged.

REFERENCES

1. **Heard, D. H. and Seaman, G. V. F.,** The influence of pH and ionic strength on the electrokinetic stability of the human erythrocyte membrane, *J. Gen. Physiol.,* 43, 635, 1960.
2. **Mehrishi, J. N.,** Molecular aspects of the mammalian cell surface, *Prog. Biophys. Mol. Biol.,* 25, 1, 1972.
3. **Schütt, W. and Klinkmann, H., Eds.,** *Cell Electrophoresis,* Walter de Gruyter, Berlin, 1985.
4. **Uzgiris, E. E. and Kaplan, J. H.,** Study of lymphocyte and erythrocyte electrophoretic mobility by laser Doppler spectroscopy, *Anal. Biochem.,* 60, 455, 1974.
5. **Pritchard, J. A. V., Sutherland, W. H., Moore, J. L., and Joslin, C. A. F.,** Macrophage Electrophoretic Mobility (M.E.M.) test for malignant disease, *Lancet,* 1, 672, 1972.
6. **Preece, A. W. and Light, P. A., Eds.,** *Cell Electrophoresis in Cancer and Other Clinical Research,* Elsevier/North-Holland, Amsterdam, 1981.
7. **Nordling, S., Anderson, L. C., and Hayry, P.,** Separation of T and B lymphocytes by preparative cell electrophoresis, *Eur. J. Immunol.,* 2, 405, 1972.
8. **Zeiller, K., Hannig, K., and Pascher, G.,** Free-flow electrophoretic separation of lymphocytes: separation of graft vs. host reactive lymphocytes of rat spleens, *Hoppe-Seylers Z. Physiol. Chem.,* 352, 1168, 1971.
9. **Heidrich, G.-G. and Dew, M. E.,** Homogeneous cell populations from rabbit kidney cortex, *J. Cell Biol.,* 74, 780, 1983.
10. **Kreisberg, J. I., Sachs, G., Pretlow, T. G., II, and McGuire, R. A.,** Separation of proximal tubule cells from suspensions of rat kidney cells by free-flow electrophoresis, *J. Cell. Physiol.,* 93, 169, 1977.
11. **Trump, B. F., Sato, T., Trifillis, A., Hall-Craggs, M., Kahng, M. W., and Smith, M. W.,** in *Methods in Cell Biology,* Vol. 21B, Harris, C. C., Trump, B. F., and Stoner, G. D., Eds., Academic Press, New York, 1980, 309.
12. **Hymer, W. C., Barlow, G. H., Cleveland, C., Farrington, M., Grindeland, R., Hatfield, J. M., Kunze, M. E., Lanham, J. W., Lewis, M. L., Morrison, D. R., Olack, N., Richman, D., Rose, J., Scharp, D., Snyder, R. S., Todd, P., and Wilfinger, W.,** Continuous flow electrophoretic separation of proteins and cells from mammalian tissues, *Cell Biophys.,* 10, 65, 1987.
13. **Plank, L. D., Hymer, W. C., Kunze, M. E., and Todd, P.,** Studies on preparative cell electrophoresis as a means of purifying growth-hormone producing cells of rat pituitary, *J. Biochem. Biophys. Methods,* 8, 273, 1983.
14. **Platsoucas, C. D.,** Separation of cells by preparative density gradient electrophoresis, in *Cell Separation Methods and Selected Applications,* Vol. 2, Pretlow, T. G., II and Pretlow, T. P., Eds., Academic Press, New York, 1983, 145.
15. **Boltz, R. C., Jr. and Todd, P.,** Density gradient electrophoresis of cells in a vertical column, in *Electrokinetic Separation Methods,* Righetti, P. G., Van Oss, C. J., and Vanderhoff, J. W., Eds., Elsevier/North-Holland, Amsterdam, 1979, 229.
16. **Giddings, J. C.,** Field-flow fractionation. Extending the molecular weight range of liquid chromatography to one trillion, *J. Chromatogr.,* 125, 3, 1976.
17. **Snyder, R. S. and Rhodes, P. H.,** Electrophoresis experiments in space, in *Frontiers in Bioprocessing,* Sikdar, S., Todd, P., and Bier, M., Eds., CRC Press, Boca Raton, FL, chap. 21, 1989.
18. **Strickler, A. and Sacks, T.,** Continuous free-film electrophoresis: the crescent phenomenon, *Prep. Biochem.,* 3, 269, 1973.
19. **Snyder, R. S., Bier, M., Griffin, R. N., Johnson, A. J., Leidheiser, H., Micale, F. J., Ross, S., and van Oss, C. J.,** Free fluid particle electrophoresis on Apollo 16, *Sep. Purif. Methods,* 2, 258, 1973.
20. **Barlow, G. H., Lazer, S. L., Rueter, A., and Allen, R.,** Electrophoretic separation of human kidney cells at zero gravity, in Bioprocessing in Space, NASA TM X-58191, Morrison, D. R., Ed., Lyndon B. Johnson Space Center, Houston, TX, January 1977, 125.
21. **Allen, R. E., Rhodes, P. H., Snyder, R. S., Barlow, G. H., Bier, M., Bigazzi, P. E., van Oss, C. J., Knox, R. J., Seaman, G. V. F., Micale, F. J., and Vanderhoff, J. W.,** Column electrophoresis on the Apollo-Soyuz Test Project, *Sep. Purif. Methods,* 6, 1, 1977.
22. **Morrison, D. R., Barlow, G. H., Cleveland, C., Grindeland, R., Hymer, W. C., Kunze, M. E., Lanham, J. W., Lewis, M. L., Sarnoff, B. E., Todd, P., and Wilfinger, W.,** Electrophoretic separation of kidney and pituitary cells on STS-8, *Adv. Space Res.,* 4(5), 67, 1984.
23. **Morrison, D. R., Lewis, M. L., Cleveland, C., Kunze, M. E., Lanham, J. W., Sarnoff, B. E., and Todd, P.,** Properties of electrophoretic fractions of human embryonic kidney cells separated on Space Shuttle flight STS-8, *Adv. Space Res.,* 4(5), 77, 1984.
24. **Heard, D. H. and Seaman, G. V. F.,** The action of lower aldehydes on the human erythrocyte, *Biochim. Biophys. Acta,* 53, 366, 1961.
25. **Goetz, P.,** System 3000 Automated Electrokinetic Analyser for biomedical applications, in *Cell Electrophoresis,* Schütt, W. and Klinkmann, H., Eds., Walter de Gruyter, Berlin, 1985, 261.

26. **Snyder, R. S., Rhodes, P. H., Miller, T. Y., Micale, F. J., Mann, R. V., and Seaman, G. V. F.,** Polystyrene latex separations by continuous flow electrophoresis on the space shuttle, *Sep. Sci. Technol.,* 21, 157, 1986.

27. **Rose, A. L. and Richman, D. W.,** U.S. Patent 4,310,408, 1979.

28. **Lewis, M. L., Morrison, D. R., Todd, P., and Barlow, G. H.,** Characterization of plasminogen activators produced by human kidney cells separated by electrophoresis in space, *Int. J. Haemostasis Thromb. Res.,* 14, 142, 1984.

29. **Todd, P., Plank, L. D., Kunze, M. E., Lewis, M. L., Morrison, D. R., Barlow, G. H., Lanham, J. W., and Cleveland, C.,** Electrophoretic separation and analysis of living cells from solid tissues by several methods. Human embryonic kidney cell cultures as a model, *J. Chromatogr.,* 64, 11, 1986.

Chapter 20

EFFECTIVE PRINCIPLES FOR SCALEUP OF ELECTROPHORESIS

M. Bier

TABLE OF CONTENTS

I. INTRODUCTION

Most researchers identify electrophoresis with the polyacrylamide or agarose gels so often used for analytical purposes. Such separations, whether by isoelectric focusing, disk, SDS, or two-dimensional electrophoresis, offer excellent resolutions and are widely used in biomedical research and in quality control. It is quite clear that one of the main reasons for the effectiveness of such gels or of other supporting media is their suppression of fluid convection, which otherwise may cause remixing of the separated fractions.

Scaleup of such techniques is difficult. The Joule heat generated by the electric field has to be dissipated externally. The heat is proportional to the volume of the device, while cooling is proportional to its surface. Thus, in a cylindrical device, for example, heat is generated as a function of the square of the radius, while cooling is only proportional to the radius. This imposes a definite limitation on the maximum dimension of any such instrument. Scaling up of devices with porous support media is further complicated by the difficulty of uniform packing.

Thus, for scaling beyond that obtainable in flat slabs of gels or granular media, separations are carried out in free fluids. It is of historic interest that the first demonstration of the resolving power of electrophoresis for proteins was demonstrated in free solution, in the now nearly forgotten Tiselius apparatus.[1] It is also not generally appreciated that Pauli[2a,b] developed an effective approach to preparative electrophoresis, also in free solution, nearly two decades before the Tiselius Nobel-awarded discovery.

Fractionation in free solution with continuous flow permits the use of relatively small instruments, in effect trading processing time for instrument size. In many instruments of this type, laminar flow is essential, which requires consideration of the hydrodynamics of fluid flow. In addition, distortion of the fractionation may result from gravity and/or electrically driven convections, as well as from inequalities of temperature, electrical field, and/or residence time. In other instruments, laminar flow is maintained in only one direction — perpendicular to that of migration — while mixing in other directions may actually be advantageous. Such arrangements are obtained by channeling of flows within the chamber by means of membrane or screen partitions.

In free fluids, density gradients arise from the very presence of the solute to be separated, as well as from temperature gradients. Both the Tiselius and the Pauli instruments utilize gravity in a constructive manner. In the Tiselius apparatus, the protein-free buffer is carefully layered on top of the protein solution, creating a density gradient that stabilizes the migrating protein boundaries. In the Pauli apparatus, gravity is utilized as a driving force, the colloidal solutes being driven electrophoretically against a membrane, where they accumulate in sufficient concentration to cause rapid decantation.

In other instruments, the presence of even minor density gradients causes destructive gravity-driven convection. Realizing the importance of gravity, the National Aeronautics and Space Administration (NASA) initiated about 15 years ago an ambitious program of exploring the potential usefulness of the reduced gravity prevailing in orbiting spacecraft for the advancement of electrophoretic technology.[3] Though the progress has been limited by the few opportunities for flight experimentation, it has brought significant advances in electrophoresis through ground-based as well as space research. In particular, the McDonnell Douglas Corporation has developed and tested its well-publicized continuous-flow electrophoresis apparatus.[4]

There are also significant differences in resolution obtainable by zone electrophoresis, isoelectric focusing, and isotachophoresis. Zone electrophoresis is carried out in the presence of a uniform background of buffer, proteins separating by virtue of differences in their characteristic mobilities. As a rule, isoelectric focusing offers much higher resolution, as it separates components according to their isoelectric points in a pH gradient. Isotachophoresis requires discontinuous buffer systems. It has the potential of high resolution because of its self-stabilizing boundaries which are immune to diffusion, but its application to proteins has been lagging. Mathematical modeling and computer simulation of these electrophoretic models have been utilized to point out their characteristic features.[5] Of course, the increased resolution due to molecular sieving of high-density gels cannot be achieved in free solution.

The challenge of large-scale preparative electrophoresis has brought numerous researchers to develop instruments based on diverse principles for fluid stabilization. A consistent classification of these instruments is difficult. The present paper will report on some of the more successful approaches to the solution of a complex problem.

II. GRAVITY-STABILIZED METHODS

The stabilizing effect of gravity on separations in the Tiselius cell has already been mentioned. By analogy, vertical liquid columns can be stabilized by imposition of a density gradient through layering of increased concentrations of sucrose or other nonionic solutes. This method received brief attention in the early days of isoelectric focusing but has limited capacity, as "droplet sedimentation" is obtained as soon as the density gradient (due to the focused solute) exceeds the sucrose gradient. On the positive side, even minor density gradients seem to eliminate electroosmosis, though the mechanism of this effect is not well understood.

A continuous-flow apparatus stabilized by solute density gradients has been thoroughly investigated by Mel.[6] Despite good results, his apparatus has not been commercialized, and there have been few recent applications of this principle.

III. CONTINUOUS-FLOW PARALLEL-PLATE DESIGNS

This type of continuous-flow instrument was popularized by Hannig[7] and his collaborators. The operating principle appears disarmingly simple, but is in fact quite complex. The flow of carrier buffer is channeled through a narrow gap between two plates, at least one of which is refrigerated. The electric field is applied perpendicularly to the direction of buffer flow. In most applications, zone electrophoresis is utilized. The components are separated as a result of differences in their characteristic electrophoretic mobilities within a homogeneous buffer. In this mode, the sample to be fractionated is introduced as a continuous stream at a given location within the flowing curtain of buffer, the components migrating diagonally toward the collection ports.

To successfully implement this principle, some classic hydrodynamic factors need to be considered. The gap between the plates has to be quite narrow, on the order of 1 mm or less, to minimize temperature gradients as well as the natural convection due to density differences. Thus, a Poiseuille parabolic flow prevails, with much faster flow through the center of the gap than adjacent to the walls. The differences in residence time cause comparable differences in migration distances of individual sample components.

In addition, electroosmosis at the walls is strong and engenders a secondary flow within the chamber in a direction perpendicular to the primary flow of the buffer. This distorts the pathway of the migrating sample. Fortunately, the distortion due to electroosmosis is in a direction opposite to that caused by the parabolic flow. Each partially compensates for the other, at least in the central region of the gap. The combination of these effects gives rise to the well-known crescent-shaped distortion of sample streams.

Strickler and Sacks[8] have analyzed these effects and have shown that if the electroosmotic fluid velocity at the wall is equal to the electrophoretic mobility of the component to be purified, distortion is avoided. Unfortunately, modulation of electroosmosis is difficult; one usually must rely on coating of the wall. The electrophoretic mobility can be adjusted, within limits, by an appropriate selection of pH and buffer.

In the hands of Hannig and others, this type of instrument has provided the bulk of the information on properties of electrophoretic subpopulations of a number of systems of living cells, including lymphocytes. The apparatus has also proven most valuable for the isolation of certain subcellular organelles and cellular membrane proteins not obtainable by other separation methods.[9] A preparative apparatus is currently commercially available from Bender & Hobein in Munich and an analytical apparatus from Hirschmann (Unterhaching, West Germany).

Field step electrophoresis is a modification of the operating usage of the apparatus developed by Wagner and Kessler.[10] Throughput is increased, rendering the apparatus suitable for protein purification. A recent application has been the purification of human tissue plasminogen activator, a recombinant DNA protein.[11] The field step method can also be combined with pH and/or multiple concentration steps. Further modifications of the processing modes incorporate isoelectric focusing and isotachophoresis. Thus, this type of apparatus is substantially more versatile than originally assumed.

IV. OPERATION IN MICROGRAVITY

A limiting factor in the conventional usage of the above apparatus for zone electrophoresis is gravity-caused convection. This can be minimized in the reduced gravity of NASA's orbiting spacecraft.[3,4] These spacecrafts are in a state of orbiting free-fall, and the gravity force is reduced by a factor of about 10^4. It can be readily shown that operation in microgravity permits an increase in the thickness of the flowing curtain of buffer, thereby permitting an increase in throughput and/or resolution. Some of the results of the space experiments are reported by Todd et al. in Chapter 19 of this text and by Snyder and Rhodes in Chapter 21.

We wish only to mention that the reduction of gravity does not eliminate the problems of electroosmosis and parabolic flow. Rhodes and Snyder[12] have therefore proposed a most ingenious continuous-flow apparatus with diagonally moving walls. This wall movement corrects for the effect of electroosmosis and is synchronized with the rate of buffer flow, thus eliminating both distortions. This apparatus has not yet been tested due to lack of space flight opportunities.

V. FLOW-CHANNELING METHODS AND RECYCLING ISOELECTRIC FOCUSING

The effective control of electrophoretic separations by means of membrane or filter elements for the channeling of flow patterns has been demonstrated in the older preparative processes of electrodecantation, electrophoresis-convection, and forced-flow electrophoresis. These three processes have in common very large throughputs. Only two fractions are obtained, a concentrated fraction of mobile components and an isoelectric fraction at its original concentration. Space does not permit an analysis of these approaches, which have been reviewed previously.[13,14]

It may suffice to mention a few of the more esoteric applications of these processes. Electrodecantation has been used for the creaming of rubber latex in the largest electrophoretic installation to date. Constructed as a replacement for centrifugation in prewar Malaya,[13] its fate is not known to the author. Besides conventional protein separations, forced-flow electrophoresis has also been utilized to demonstrate first electrofiltration, i.e., electrically assisted continuous filtration without filter-cake buildup.[14,15] It has also been adopted for *in vivo* extracorporeal processing of the blood of experimental animals, with simultaneous filtration of blood elements and fractionation of plasma components. This process, termed selective plasmapheresis, resulted in the selective removal of immunoglobulins from circulating blood.[16] The immunoglobulins are the isoelectric component of plasma under phys-

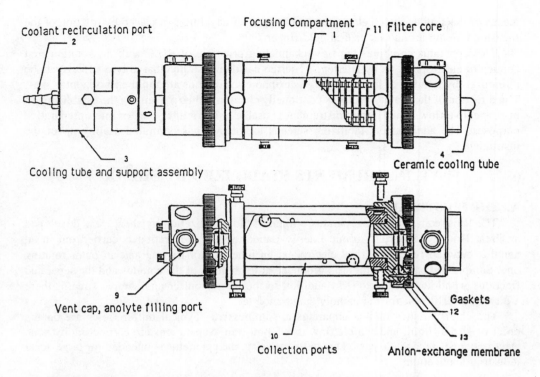

Coolant recirculation port — 2

Focusing Compartment 1 — 11 Filter core

Cooling tube and support assembly — 3

Ceramic cooling tube — 4

Vent cap, anolyte filling port — 9

Collection ports — 10

Gaskets — 12

Anion-exchange membrane — 13

FIGURE 1. Schematic presentation of the rotationally stabilized focusing apparatus. The main body of the apparatus is subdivided into 20 compartments by means of screen elements. Total volume capacity is about 40 ml. (From Bier, M., Egen, N. B., Allgyer, T. T., Twitty, G. E., and Mosher, R. A., in *Peptides,* Gross, E. and Meienhofer, J., Eds., Pierce Chemical Co., Rockford, IL, 1979, 79. With permission.)

iologic conditions. More recently, electrodecantation has also been used for this purpose in human blood donors by Sanchez.[17]

High resolution through channeling is obtainable in the recycling isoelectric focusing (RIEF) apparatus.[18] The focusing cell is subdivided into the desired number of compartments by means of monofilament nylon screen elements. These assure laminarity of liquid flow in the direction perpendicular to the electric field. The content of each subcompartment is recycled through an external heat-exchange reservoir, which provides the volume capacity of the apparatus and dissipates the Joule heat. Optionally, the instrument can be complemented by pH and ultraviolet absorption sensor arrays, under computer control.[18,19] These provide documentation of the separation process and can also be utilized for feedback optimization of the process.

Ampholine® or a similar carrier electrolyte is utilized for the control of the pH gradient, generated by the passage of the electric current itself. In favorable cases, suitable pH gradients can also be generated by mixtures of chemically well-defined ampholytes, such as amino acids.[20] Recycling is continued until each sample component has focused to its isoelectric pH value. A stationary steady state is finally obtained, wherein all components cease to migrate. All that is left at this point is the simultaneous collection of all fractions.

The apparatus is modular — the number of subcompartments, the effective cross section of the cell, and the volume of the heat-exchange reservoirs can be varied at will. Volumes ranging from 200 to 10,000 ml have been processed in a few hours. Instruments with a larger capacity can be designed if the need arises.

For the processing of smaller volumes, an alternate apparatus has been constructed, as shown in Figure 1. The horizontal focusing column is subdivided into narrow segments by means of the similar screen elements used in RIEF. Recycling is replaced by rotation of the focusing column around its central cold finger.[21a,b]

This apparatus combines the flow-channeling principle of RIEF with the rotation first applied to electrophoresis by Hjerten.[22] Hjerten has demonstrated that gravity effects can be minimized by rotation of narrow-bore electrophoretic columns around their horizontal axes. The direction of the gravity vector is continually changing, thus preventing the establishment of convective flow. The Hjerten instrument is usable only for analytical or micropreparative purposes. The addition of stabilizing screens to rotation has permitted scaling up of the instrument.

VI. INSTRUMENTS STABILIZED BY SHEAR

A. THE BIOSTREAM

The Biostream, the first apparatus utilizing fluid stabilization by shear, was developed at Great Britain's Harwell Atomic Energy Laboratories.[23] Separation is carried out in an annulus between two cylindrical electrodes, an inner stationary one and an outer rotating one. Sample and carrier buffer are introduced at the bottom of the annulus and the separated fractions withdrawn at the top. In addition to the carrier buffer and sample stream, there are two buffer flows, for the catholyte and anolyte.

The characteristics of this apparatus are impressive. Typical buffer flows are on the order of 1 to 2 l/min, and sample flow is ≤50 ml/min. Separations are carried out by zone electrophoresis in a single pass. Due to its capacity, the apparatus is intended for large-scale industrial applications.

B. RECYCLING FREE-FLOW FOCUSING (RF3)

We have recently designed another apparatus which utilizes shear for fluid stabilization in recycling isoelectric focusing.[24] Stabilizing shear is obtained by rapid recirculation of the solution through a narrow channel between two parallel surfaces of the focusing chamber. A batch process is utilized, and in every pass only a small shift toward the final steady state is obtained. The residence times are on the order of only seconds. The matched in- and outflow ports are connected to a multichannel heat exchanger which provides the volume capacity of the apparatus. A schematic of the apparatus is presented in Figure 2.

Due to the final steady state achieved in isoelectric focusing and the short residence time, this method avoids the problems of crescent phenomena, gravity-caused convection, and temperature gradients. The Joule heat is absorbed by the latent heat capacity of the process fluid during transit through the chamber and released to the heat exchanger. The allowable heating within the chamber is limited only by the heat sensitivity of the sample, rather than the stability of liquid flow.

Recently, we were able to demonstrate that the RF3 apparatus can also be utilized for recycling isotachophoresis.[25] In isotachophoresis the sample material to be separated is inserted between a leading and a terminating electrolyte, according to well-known principles. Counterflow is necessary to immobilize the migrating boundaries, in effect providing a virtually unlimited migration distance. A UV sensor locates the boundary, the counterflow being under automatic computer control.

It may be of some relevance to point out that in the Biostream the rotation-induced shear is in the direction of the electric field, while in the RF3 it is perpendicular to it. Moreover, in the Biostream, the shear is constant, the fluid velocity varying linearly between the stationary center electrode and the rotating outer one. In the RF3, shear is maximal at the two chamber walls and decreases to zero in the center of the chamber, where most of the fluid flows.

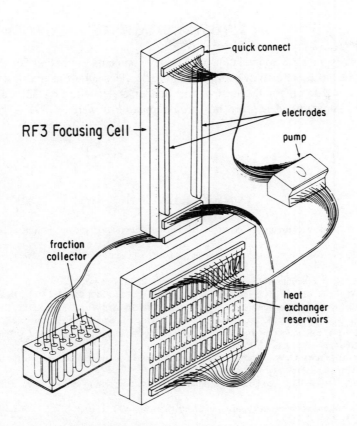

FIGURE 2. Schematic presentation of the RF3 apparatus with 48 recycling channels. Separation occurs within a narrow gap chamber, fluid stabilization being achieved through rapid recirculation. Assembly is facilitated by the multiple tubing quick-connectors. (From Egen, N. B., Thormann, W., Twitty, G. E., and Bier, M., in *Electrophoresis '83*, Hirai, H., Ed., Walter de Gruyter, Berlin, 1984, 547. With permission.)

VII. CONCLUSION

There are no obvious reasons why electrophoresis could not be carried out on as large a scale as other electrochemical processes, such as electrodialysis, electrolysis, electrocoating, etc. In fact, the huge installation in Malaya for the creaming of rubber latex by electrodecantation, mentioned above, is an example of the potential of electrophoresis.

While preparative electrophoresis has not yet gained widespread acceptance in the biotechnology industry, this chapter describes a variety of instruments, most of which are now available at the Center for Separation Science at the University of Arizona, Tucson. This is the first time that such a range of instruments, encompassing a wide range of throughputs, is available in a single laboratory.

Much work remains to be done to assess the potential advantages of electrophoresis in comparison to the well-established chromatographic processes. An important difference may be that electrophoresis is carried out in free solution, without supporting media. Thus, adsorptive losses are minimized, and even living cells or subcellular organelles can be processed. In the long run, electrophoresis may also be shown to be cheaper and to permit cleaner operations, also because of the lack of supporting media.

The main obstacle to wider use of electrophoresis may well be the lack of information and experience. The Center is dedicated to the advancement of this technology.

ACKNOWLEDGMENTS

We wish to express our appreciation to CJB Developments Limited of Portsmouth, U.K. for their loan of the Biostream system for large-scale electrophoresis and to Protein Technologies, Inc. of Tucson for the loan of the Bender & Hobein Vap 22 continuous-flow electrophoresis apparatus. The work has been supported in part by NASA grant NAGW 693.

REFERENCES

1. **Tiselius, A.,** A new apparatus for electrophoretic analysis of colloidal mixtures, *Trans. Faraday Soc.,* 33, 524, 1937.
2a. **Pauli, W.,** Untersuchungen an elektrolytfreien, wasserloeslichen Proteinkoerper, *Biochem. Z.,* 152, 355, 1924.
2b. **Pauli, W.,** Hochgereinigte Kolloide, *Helv. Chim. Acta,* 25, 137, 1942.
3. **Snyder, R. S., Griffin, R. N., Johnson, A. J., Leidheiser, H., Jr., Micale, F. J., Ross, S., van Oss, C. J., and Bier, M.,** Free fluid electrophoresis on Apollo 16, *Sep. Purif. Methods,* 2, 259, 1973.
4. **Morrison, D. R., Lewis, M. L., Barlow, G. H., Todd, P., Kunze, M. E., Sarnoff, B. E., and Li, Z.,** Properties of electrophoretic fractions of human embryonic kidney cells separated on space shuttle flight STS-8, *Adv. Space Res.,* 4, 77, 1984.
5. **Bier, M., Palusinski, O. A., Mosher, R. A., and Saville, D. A.,** Electrophoresis: mathematical modeling and computer simulation, *Science,* 219, 1281, 1983.
6. **Mel, H. C.,** Stablie-flow free boundary migration and fractionation of cell mixtures, *J. Theor. Biol.,* 6, 307, 1964.
7. **Hannig, K.,** Preparative electrophoresis, in *Electrophoresis,* Vol. 2, Bier, M., Ed., Academic Press, New York, 1967, 423.
8. **Strickler, A. and Sacks, T.,** Focusing in continuous-flow electrophoresis systems by electrical control of effective cell wall zeta potential, *Ann. N. Y. Acad. Sci.,* 209, 497, 1973.
9. **Morré, D. J., Morré, D. M., and Heidrich, H. G.,** Subfractionation of rat liver Golgi apparatus by free-flow electrophoresis, *Eur. J. Cell Biol.,* 31, 263, 1983.
10. **Wagner, H. and Kessler, R.,** *GIT Lab. Med.,* 7, 30, 1984.
11. **Barth, F., Gruetter, M. K., Kessler, R., and Manz, H. -J.,** The use of free flow electrophoresis in the purification of recombinant human tissue plasminogen activator expressed in yeast, *Electrophoresis,* 7, 372, 1986.
12. **Rhodes, P. H. and Snyder, R. S.,** U.S. Patent 4,349,429, 1982.
13. **Bier, M.,** Preparative electrophoresis without supporting media, in *Electrophoresis,* Vol. 1, Bier, M., Ed., Academic Press, New York, 1967, 263.
14. **Bier, M.,** Electrokinetic membrane processes, in *Membrane Processes in Industry and Biomedicine,* Bier, M., Ed., Plenum Press, New York, 1971, 233.
15. **Moulik, S. P., Cooper, F. C., and Bier, M.,** Forced-flow electrophoretic filtration of clay suspensions — filtration in an electric field, *J. Colloid Interface Sci.,* 24, 427, 1967.
16. **Bier, M., Beavers, C. D., Merriman, W. G., Merkel, F. K., Eiseman, B., and Starzl, T. Z.,** Selective plasmapheresis in dogs for delay of heterograft response, *Trans. Am. Soc. Artif. Intern. Organs,* 16, 325, 1970.
17. **Sanchez, V.,** personal communication.
18. **Bier, M., Egen, N. B., Allgyer, T. T., Twitty, G. E., and Mosher, R. A.,** New developments in isoelectic focusing, in *Peptides,* Gross, E. and Meienhofer, J., Eds., Pierce Chemical Co., Rockford, IL, 1979, 79.
19. **Nagabhushan, T. L., Sharma, B., and Trotta, P. P.,** Application of recycling isoelectric focusing for purification of recombinant human leukocyte interferons, *Electrophoresis,* 7, 552, 1986.
20. **Palusinski, O. A., Allgyer, T. T., Mosher, R. A., and Bier, M.,** Mathematical modeling and computer simulation of isoelectric focusing with electrochemically defined ampholytes, *Biophys. Chem.,* 13, 193, 1981.
21a. **Egen, N. B., Thormann, W., Twitty, G. E., and Bier, M.,** A new preparative isoelectric focusing apparatus, in *Electrophoresis '83,* Hirai, H., Ed., Walter de Gruyter, Berlin, 1984, 547.
21b. **Bier, M.,** U.S. Patent 4,588,492, 1986.

22. **Hjerten, S.,** *Free Zone Electrophoresis,* Almqvist & Wiksell, Stockholm, 1967.
23. **Mattock, P., Aitchison, G. F., and Thomson, A. R.,** Velocity gradient stabilised, continuous, free flow electrophoresis, *Sep. Purif. Methods,* 9, 1, 1980.
24. **Bier, M. and Twitty, G. E.,** Patent pending.
25. **Sloan, J. E., Thormann, W., Bier, M., Twitty, G. E., and Mosher, R.,** Recycling isotachophoresis: a novel approach to preparative protein fractionation, in *Electrophoresis '86,* Dunn, M. J., Ed., VCH Press, Deerfield Beach, FL, 1986, 696.

Chapter 21

ELECTROPHORESIS EXPERIMENTS IN SPACE

Robert S. Snyder and Percy H. Rhodes

TABLE OF CONTENTS

I. INTRODUCTION

Electrophoresis experiments on Apollo,[1] Apollo Soyuz Test Project (ASTP),[2] and the Space Shuttle[3] have been carried out in space to show that disturbances due to buoyancy-induced thermal convection and sample sedimentation during the separation process are negligible in reduced gravity. The experiments to date have been small-scale demonstrations of specific principles that have increased our knowledge of electrokinetic and fluid dynamic phenomena and have supported our long-range electrophoresis goals.[4] Simultaneously, the limitations of ground-based electrophoretic separators have been documented,[5] and new concepts have been proposed for future experiments.[6]

Several laboratory instruments have been constructed utilizing past developments in the design and operation of continuous-flow electrophoretic separators by Strickler[7] and Hannig[8] combined with innovative developments in the field of fluids analysis by Saville[9] and Ostrach.[10] Building on these developments, the McDonnell Douglas Astronautics Company (MDAC) designed a continuous-flow electrophoresis system (CFES) that evolved from a detailed survey of the requirements for fractionation of biological materials.

In 1978, MDAC began discussions with the National Aeronautics and Space Administration (NASA) on opportunities to develop a space CFES that would incorporate specific modifications to their laboratory instruments to take advantage of weightlessness. The first MDAC flight experiments with the CFES on STS-4 in June 1982 fractionated a proprietary tissue culture medium and evaluated the effect of sample concentration using mixtures of rat serum albumin and ovalbumin. MDAC concluded that there was no loss of resolution at the higher concentrations processed in space. In addition, they reported that the quantity of albumin that could be fractionated in the CFES in space was significantly higher (over 400 times) than the quantity that could be processed in their ground laboratory instrument during the same time interval. They proposed that these improvements originated from both instrument modifications and increased sample concentrations permitted by weightlessness.

Under the terms of the Joint Endeavor Agreement (JEA), NASA was provided an opportunity to process two samples on Space Shuttle flight STS-6 in April 1983.[11] All experiment objectives and operational parameters (such as applied field, sample residence time in the field, and buffer composition) had to fit MDAC capabilities and NASA flight constraints. The NASA objectives were formulated so as to include validation of the sample concentration effects reported by MDAC on Space Shuttle flight STS-4. The specific objectives were (1) to use a model sample material at a high concentration to evaluate the continuous-flow electrophoresis (CFE) process in the MDAC CFES instrument and compare its separation and sample throughput resolution with related devices on earth and (2) to expand our basic knowledge of the limitations imposed by fluid flows and particle concentration effects on the electrophoresis process by careful design and evaluation of the space experiment. Hemoglobin and a polysaccharide were selected as primary samples for the STS-6 flight experiment by the Marshall Space Flight Center (MSFC, Huntsville, AL) and a review group organized by the Universities Space Research Association.

The NASA experiments on STS-7 in June 1983 were intended to build upon the results obtained on STS-6.[12] Because MDAC selected a propionate buffer (pH 5.2) for use on STS-7 rather than barbital buffer (pH 8.3) as on STS-6, it was not possible to perform a verification experiment using hemoglobin and polysaccharide as processed on STS-6. The change in pH would have resulted in both the hemoglobin and polysaccharide becoming positively charged and migrating toward the cathode. Polystyrene latex (PSL) particles were therefore chosen for separation on STS-7, since they are known to be negatively charged at pH 5.2 and are produced in a range of sizes with different surface charge groups and surface charge densities.

II. MATERIALS AND METHODS

A. CONTINUOUS-FLOW ELECTROPHORESIS

CFE takes place within a flowing curtain of aqueous electrolyte contained in a long rectangular chamber of high aspect ratio. The narrow gap between the broad faces of the chamber confines the flow to a thin curtain. The sample is continuously injected into this curtain as a finely drawn stream (filament) and fractionated under the influence of a lateral electric field produced by flanking electrodes. The fractionated sample bands are subsequently collected in a uniform array of collection ports situated along the exit end of the chamber. The no-slip condition at the chamber wall causes a parabolic profile to develop in the direction of buffer flow. In addition, a lateral flow across the width of the chamber (perpendicular to buffer flow), called electroosmosis, exists when charged walls are present. These independent phenomena are inherent in the CFE process, and they combine to produce crescent-shaped distortions, the curvature of which is determined by the flow that predominates (either laminar flow or electroosmosis). This mechanism of sample stream distortion has always been thought to be a major cause of resolution degradation in CFE devices.[13] The term "wall effects" has frequently been used to describe the phenomenon.

The space MDAC CFES (Figure 1) is essentially a rectangular polycarbonate separation (flow) chamber with internal dimensions 16 cm wide, 120 cm long, and 0.3 cm thick. A cooling jacket covers each broad chamber face, with one electrode in each cooling chamber positioned diagonally across from the other to provide an electric field across the length of the chamber. The platinum electrodes are in contact with the separation chamber via slots cut into the chamber plates and are covered with a proprietary, porous membrane. The circulation of cooled electrolyte, from the bottom (flow entrance) of the chamber to the top through a serpentine passageway, establishes a uniform lateral temperature profile and removes the bubbles formed at each electrode. Rectangular struts used to form the coolant passage also provide some support to the thin separation chamber plates. The sample enters the chamber through a thin-wall glass tube (0.1-cm I.D.) located 11 cm from the curtain buffer entrance of the chamber. The buffer and separated sample fractions exit at the top of the chamber through a collection array of 197 Tygon® tubes (0.068-cm I.D., 0.078-cm O.D.) which span the width of the chamber.

B. SAMPLES AND BUFFER FOR STS-6 EXPERIMENTS

Hemoglobin was selected as the primary sample candidate based on the following characteristics: (1) availability in large quantities as a single molecular species, (2) visibility for easy analysis, (3) utility as an electrophoresis standard in laboratories, (4) availability as variants with different electrophoretic mobilities, and (5) stability in the cyanmethemoglobin forms. Human hemoglobin A (HbA) provided for ground and space use was prepared at the Centers for Disease Control, Atlanta, GA. Selected from a single donor, the HbA was purified on an ion-exchange column, converted to cyanmethemoglobin, and concentrated to nearly 11%. The hemoglobin was sent to MSFC, where it was then dialyzed against the flight buffer and stored for use in the planned experiments.

The second sample was one type of pneumococcal capsular polysaccharide (PCP) obtained from Lederle Laboratories, Pearl River, NY. PCP type 6 had a distinctly higher mobility than HbA with minimal variation and, although not colored, could be detected immunologically in low concentrations. The purity of the PCP was determined by specific antibody tests. Since it is known that the PCPs have multiple repeating units of a simple saccharide chain, an attempt was made to obtain PCP with constant molecular weight using chromatography and laboratory CFE.

The buffer selected by MDAC for laboratory and flight experiments was 2 mM barbital consisting of 0.386 mg/ml sodium barbiturate, 0.070 mg/ml barbituric acid, and 50 μg/ml gentamycin sulfate, pH 8.3, with an electrical conductivity of 160 μmho/cm at 25°C.

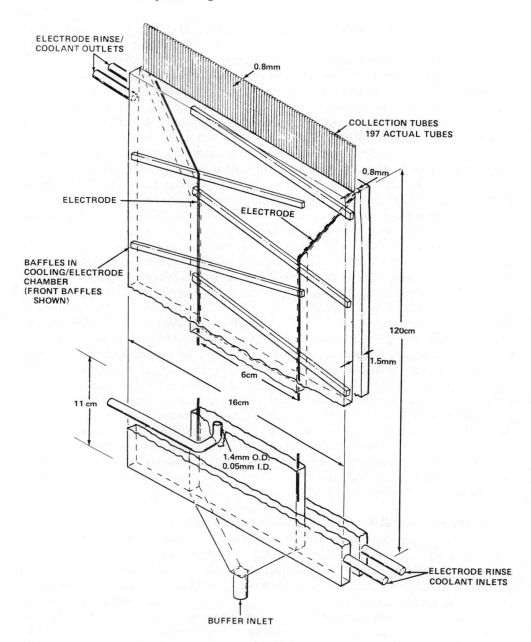

FIGURE 1.   Continuous-flow electrophoresis system (CFES).

C. SAMPLES AND BUFFER FOR STS-7 EXPERIMENTS

Monodisperse PSL particles less than 1.0 μm in diameter were chosen to minimize sedimentation and eliminate any requirement for resuspending the sample during flight. Three particle sizes were ultimately chosen, and two of the particle populations were dyed to enhance photographic detail and aid in experimental analysis.

The latex particles with the highest mobility (nominal diameter, 0.56 μm) were dyed red; the particles with the lowest mobility (nominal diameter, 0.80 μm) were dyed blue. The particles with intermediate mobility (nominal diameter, 0.30 μm) were not dyed. The latexes were then suspended in the curtain buffer containing Brij® 35 (polyoxyethylene lauryl ether 35) using a procedure of buffer exchange with filtration. Appropriate volumes of each latex were combined to yield equal concentrations for the final flight sample with the following properties: total latex concentration, 5.0%; pH, 5.6; and conductivity, 155 ± 5 μmho/cm. The conductivity of a second sample of PSL was increased approximately three times that of the initial sample, to 455 ± 5 μmho/cm, by adding 0.10 M NaCl while maintaining a total latex concentration of 5.0% by weight.

The curtain buffer was prepared from a 100× stock solution of 225 mM sodium propionate, pH 5.2, by 100-fold dilution with distilled water to give a 2.25 mM solution with a pH of 5.0 and conductivity of 140 ± 5 μmho/cm. Laboratory tests were initially performed with a nonionic surfactant, 0.05% w/v Brij® 35, added to the sample and curtain buffer. Shortly before flight, however, it became necessary for MDAC to omit the Brij® from the curtain buffer. Since the sample latex had been selected according to separation in buffer with Brij®, the flight samples included the surfactant in the suspension medium, although it was excluded from the curtain buffer.

III. RESULTS

The two samples processed in the first NASA space experiment on CFES were (1) a high concentration of hemoglobin alone and (2) a mixture of polysaccharide and hemoglobin at a lower concentration. The band behavior obtained from processing the single species was used to define the performance of the space instrument. The low-concentration mixture, 1.9% HbA and 0.5% PCP-6, although still higher than could be processed at high resolution on earth, would permit a comparison of the separations achieved before flight in the various laboratory units.

In orbit, the NASA hemoglobin sample of 8.7% concentration was processed first to establish an initial CFES performance. The hemoglobin experiments each started with an initial passage of the sample at zero voltage. The sample band spread only slightly, as expected on the basis of diffusion, during its 8-min passage through the chamber. After the zero run, an electric field of 25 V/cm was applied for an interval of about 32 min in order to establish steady-state conditions. Photographs of the electrophoresis runs were taken at the point of sample insertion, midway up the chamber, and at the collection end. A portion of each fraction was also collected for later analysis.

Photographs of the high-concentration HbA, taken at the same chamber region, are shown in Figure 2, with the time in hours and minutes shown in the lower right-hand corner of each picture. Figure 2A shows the zero voltage position of the sample stream. With the cathode on the left, the sample should proceed only toward the right to the anode when the electric field is applied. However, Figures 2B to D, taken over a period of 10 min, show that the band spreading was so extensive that some of the sample underwent a retrograde migration. This condition could only exist if the sample thickness spread beyond the electroosmotic stationary layer of the chamber. Based on comparison of the photographs, there appeared to be a buildup of sample near the walls, as the more intense color of the later photograph seems to suggest.

After half the MDAC proprietary samples were run, the second NASA preparation, the HbA and PCP mixture, was processed. The stream, although faint, was more compact, with little indication of spreading. The migration of the leading edge of the sample was roughly the same as in the previous experiment. However, there did not appear to be the large retrograde migration observed in the high-concentration hemoglobin experiment.

In addition to the photographic data, the entire fluid output of the space chamber was collected in fractions, as noted previously, during the middle of each separation run in 197 small (1.5-ml) polypropylene pockets arranged in a steel tray. The NASA samples were analyzed for the amount of hemoglobin and polysaccharide in each; the results are shown in Figures 3 (HbA alone) and 4 (HbA plus PCP).

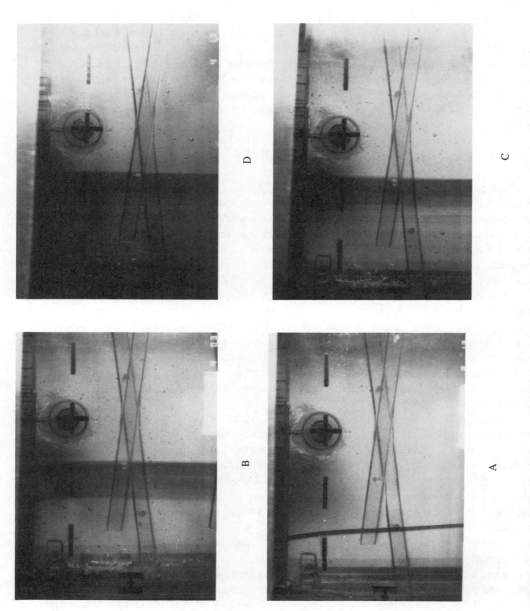

FIGURE 2.   Photographs of the deflection of high concentration sample of hemoglobin in CFES in microgravity. (A) Zero voltage; (B) deflection at 4 h, 3 min; (C) deflection at 4 h, 10 min; (D) deflection at 4 h, 13 min.

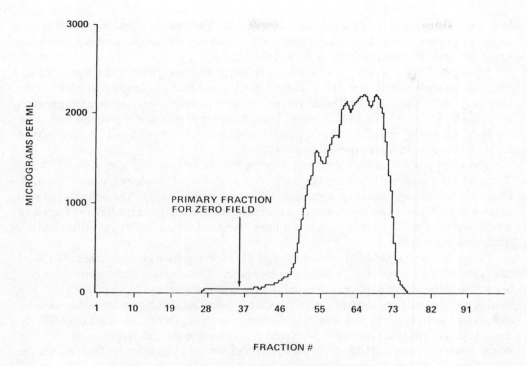

FIGURE 3. Band spread of high-concentration hemoglobin sample.

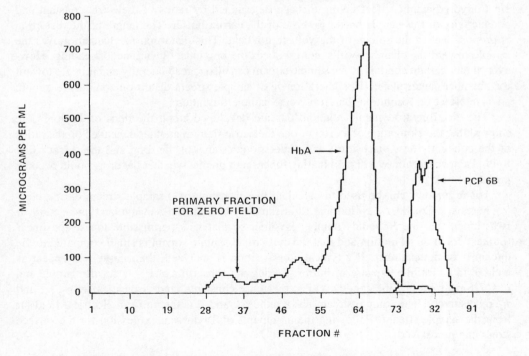

FIGURE 4. Band spread of low-concentration hemoglobin and polysaccharide sample.

Figure 3 shows a very broad single peak with some of the HbA sample deflected toward the cathode side of the zero field location. However, this does not show the total amount of retrograde sample migration, since sample near the walls remained in the chamber for a longer time because of the Posueille flow distribution. The sample retention time increased as the walls were approached. This is apparent by observing the sample intensity increasing to the left (cathode) wall in Figures 2C and D.

Figure 4 shows a separation of the low-concentration hemoglobin and polysaccharide. The separation, however, was not as distinct as obtained in the laboratory CFES or in the Desaga FF48.[11] Also, none of the earth-based laboratory experiments had sample appear on the cathode side of the zero field location. It is interesting to note that Figure 4 shows about the same amount of retarded and retrograde sample between tubes 28 and 55, as observed for the high-concentration experiment in Figure 3.

The same procedures were followed with the PSL samples on the next flight, STS-7.[12] The results of the earlier flight demonstrated considerable band broadening when the sample conductivity was approximately three times that of the curtain buffer. These results, as well as prior ground-based data, provided the justification for the design of the STS-7 experiment which evaluated the relationships between the sample and curtain buffer properties and the fractionation resolution.

Figure 5 shows a series of photographs of the experiments with unmatched (left side) and with matched (right side) sample conductivities. These photographs were taken just before the three separated latexes entered the collection tray at the top of the CFES. The band distortion and spreading of the PSL samples when the conductivities of the samples and curtain were not matched was evident from the photographs and was confirmed by analysis of the collected fractions. In Figure 6 it is seen that band spreading is far more severe when the sample conductivity is three times that of the carrier buffer, and sample migration also appears to be anomalous.

It has been determined, however, that the actual initial sample stream distortion is an electrohydrodynamic effect giving a shape determined by ratios of dielectric constant and conductivity of the sample buffer to those of the carrier buffer. The length of the distortion is proportional to the square of the voltage gradient. This distortion mechanism moves the sample toward the chamber walls and produces the previously mentioned spreading. However, under certain conditions the sample stream can also spread laterally and migrate toward the chamber center plane. This lateral mode of sample stream distortion could have a more adverse effect on resolution than transverse sample migration.

The electrohydrodynamic relationships are developed along the lines of Taylor.[14] He showed that the elongation of a drop of one dielectric fluid in another depended on the ratios of the conductivities, viscosities, and dielectric constants of the drop and the surrounding fluid. Taylor determined a discriminating function to predict whether the drop would become prolate or oblate.

His arguments can be used to calculate the distortion of a sample stream during electrophoresis.[15] Consider a cylindrical fluid sample filament in a rectangular flow section. It is assumed that the Reynolds number is small so that it is permissible to use the linear (Stokes) equation of motion and that the cylindrical sample stream is small compared to the thickness of the chamber. If a typical cross section is analyzed, the sample will appear as a circle in an infinite expanse of buffer. Taking the subscripts of $i = 1$ for the sample and $i = 2$ for the buffer, the conductivities, dielectric constants, and viscosities are σ_i, k_i, and η_i, respectively. The sample stream has a radius a, and E is the uniform electrical field far from the sample stream. Following the arguments of Taylor, an expression for the surface stress can be derived:

$$C = \frac{E^2 K_2}{4\pi(R^2 + 1)^2} [S(R^2 + R + 1) - 3]\cos 2\theta$$

where C = surface stress; E = electric field strength; R = ratio of conductivity of sample

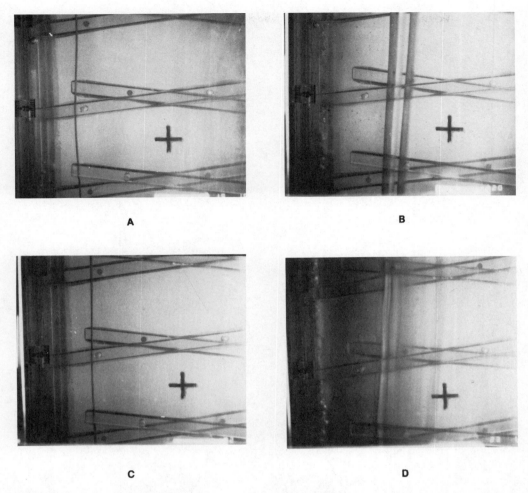

FIGURE 5. Photographs of polystyrene latex samples with unmatched conductivity (A and C) and matched conductivity (B and D).

to that of buffer, σ_1/σ_2; K_2 = dielectric constant of buffer; and S = ratio of dielectric constant of buffer to that of sample, $-K_2/K_1$.

The term inside the brackets can be considered a discriminating factor, since it determines the sign of the surface stress. The ratio of dielectric constants, S, will be near unity for experiments with dilute PSL samples in buffers made of the same constituents as the flowing curtain. When the conductivity of the sample equals that of the buffer, R = 1 and the surface stress is zero, signifying an undistorted sample stream. When R is greater than 1.0 (sample stream conductivity is higher), the discriminating factor and surface stress are positive and the surface stress varies with θ, as shown in Figure 7. When a relatively low-conductivity sample buffer is used, R is less than 1.0 and the surface stress vectors shown in Figure 7 are reversed.

The major difference between Taylor's case of the immiscible fluid drop and this case of the sample stream in a buffer solution is that no restoring force such as surface tension is present to limit the distortion of the sample stream. Therefore, it follows that the deformation will continue until a ribbon is formed. This formation was confirmed by the following experiments. Figure 8 is a view of the buffer-collection end of a thick horizontal electrophoresis chamber designed to permit observation and photographs of the collection cross

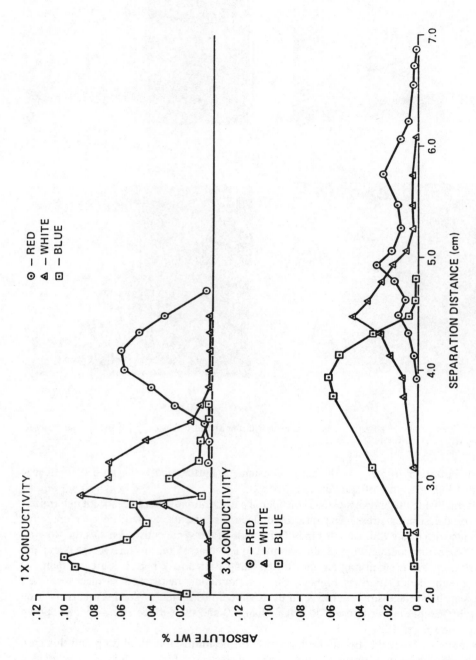

FIGURE 6.   Band spread of polystyrene latex samples.

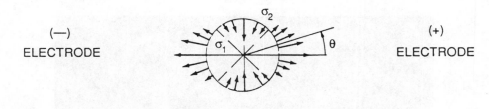

(—)
ELECTRODE

(+)
ELECTRODE

HIGH CONDUCTIVITY SAMPLE

FIGURE 7. Cross-sectional diagram of high conductivity sample showing surface stress vectors acting on a cylindrical sample stream with long axis perpendicular to the electric field.

section. A monodisperse suspension of white PSL was inserted into the input end of the chamber (not in view), and the thin cylindrical sample stream could be seen (Figure 8A). The electric field was produced by electrodes that extend the length of the 15.24-cm-long chamber. Figure 8B shows an example of the low-conductivity latex sample distorted into a vertical (transverse) ribbon in the high-conductivity buffer. Figure 8C shows a horizontal (lateral) ribbon obtained when a high-conductivity sample stream was inserted into a low-conductivity curtain buffer.

IV. DISCUSSION AND CONCLUSION

These experiments in the MDAC CFES were intended to confirm two previous observations on space electrophoresis: that only electrophoresis and electroosmosis determined particle migration in an electric field during zone electrophoresis on Apollo 16, and that electrophoresis was independent of sample concentration with up to 20% albumin on STS-4. Instead, these experiments showed that the previous conclusions were too simplistic.

First, the sample insertion disk used for the Apollo zone electrophoresis experiments is not a valid model for the sample filament configuration of CFE. Second, increasing the quantity of protein added to the electrophoresis buffer to constitute the sample also proportionally changes the electrical properties of the sample, and this has a significant impact on the subsequent electrophoresis of that sample by CFE.

Hemoglobin electrophoresis was the first of these experiments in space, and the pattern of the collected sample did not fit our model. Although hemoglobin had a single electrophoretic mobility, this mobility could not be measured with precision in the barbital buffer, and any estimate of electroosmosis was subject to error. Thus, the band structure of the high concentration of hemoglobin could not be readily understood.

During the past year, a more detailed analysis of the hemoglobin experiment has led to the following probable sequence of events. The 9% hemoglobin sample added to the 2 mM barbital buffer (pH 8.2) reduced the overall sample pH to 7.5. Since the pK of the barbituric ion is 7.8, less than half of the sodium barbiturate molecules in the sample stream were ionized. As soon as the sample stream entered the flowing chamber buffer and the region of the electric field, the positive sodium ions carried most of the current within the sample stream, so Na$^+$ rapidly built up at the trailing edge of the sample stream. This led to a hemoglobin sample with reduced conductivity in the front of the band and higher conductivity in the rear. Electrohydrodynamic forces then acted on the different cross-sectional regions of the hemoglobin band to induce the flows and resultant band structure. The sequence of events is shown in Figure 9. The significance of this conductivity distribution in causing the dispersion of the hemoglobin in space was not determined until after the PSL particle electrophoretic distributions in space were analyzed. Thus, the analysis of the PSL flight results in the article by Snyder et al.[12] is incomplete.

A

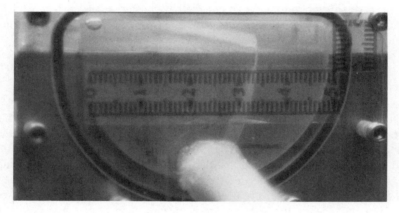

B

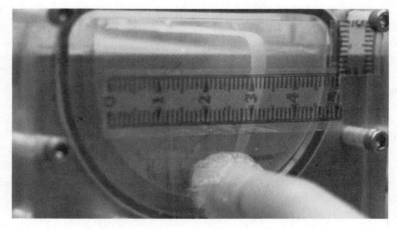

C

FIGURE 8.   End-on view of polystyrene latex sample stream at zero voltage in thick electrophoresis chamber. A, with no stream distortion; B, with vertical ribbon formation when
field is applied to low-conductivity sample streams; and C, with horizontal ribbon formation
when field is applied to high-conductivity sample stream.

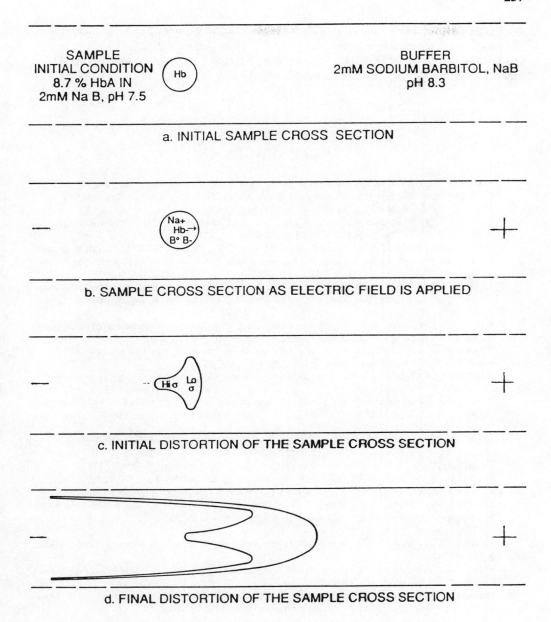

FIGURE 9. Electrohydrodynamic distortion of high-concentration hemoglobin sample.

Although the removal of thermal convection and sedimentation does not result in optimized separations, the electrohydrodynamic distortions would not have been observed and measured without access to space with the McDonnell Douglas CFES. These distortions are independent of gravity and must be controlled in any future space electrophoresis system as well as in separations at earth's gravity.

ACKNOWLEDGMENTS

The authors appreciate discussions with Drs. Geoffrey V. F. Seaman, Oregon Health Sciences University and Glyn O. Roberts, Roberts Associates. Dr. John L. Sloyer conducted the ground-based tests with the PCP, with the assistance and support of Lederle Laboratories. The hemoglobin samples were provided by Mr. Danny Ju of the Centers for Disease Control. The support of McDonnell Douglas Astronautics Company personnel, especially Mr. David Richman and Dr. Wayne Lanham, was also appreciated.

REFERENCES

1. **Snyder, R. S., Griffin, R. N., Johnson, A. J., Leidheiser, H., Jr., Micale, F. J., Ross, S., van Oss, C. J., and Bier, M.,** Free fluid electrophoresis on Apollo 16, *Sep. Purif. Methods,* 2, 259, 1973.
2. **Allen, R. E., Rhodes, P. H., Snyder, R. S., Barlow, G. H., Bier, M., Bigazzi, P. E., van Oss, C. J., Knox, R. J., Seaman, G. V. F., Micale, F. J., and Vanderhoff, J. M.,** Column electrophoresis on the Apollo-Soyuz Test Project, *Sep. Purif. Methods,* 6, 1, 1977.
3. **Snyder, R. S., Rhodes, P. H., Herren, B. J., Miller, T. Y., Seaman, G. V. F., Todd, P., Kunze, M. E., and Sarnoff, B. E.,** Analysis of free zone electrophoresis of fixed erythrocytes performed in microgravity, *Electrophoresis,* 6, 3, 1985.
4. **Snyder, R. S.,** Separation techniques, in *Materials Sciences in Space — A Contribution to the Scientific Basis of Space Processing,* Feuerbacher, B., Hamacher, H., and Naumann, R. J., Eds., Springer-Verlag, Berlin, 1986, 465.
5. **Miller, T. Y., Williams, G. P., and Snyder, R. S.,** Effect of conductivity and concentration on the sample stream in the transverse axis of a continuous flow electrophoresis chamber, *Electrophoresis,* 6, 377, 1985.
6. **Rhodes, P. H. and Snyder, R. S.,** Sample band spreading phenomena in ground and space-based electrophoretic separators, *Electrophoresis,* 7, 113, 1986.
7. **Strickler, A.,** Continuous particle electrophoresis: a new analytical and preparative capability, *Sep. Sci.,* 2, 335, 1967.
8. **Hannig, K.,** Separation of cells and particles by continuous free flow electrophoresis, in *Techniques of Biochemical and Biophysical Morphology,* Vol. 1, Glick, D. and Rosenbaum, R. M., Eds., John Wiley & Sons, New York, 1972, 191.
9. **Saville, D. A.,** The fluid mechanics of continuous flow electrophoresis, *Physicochem. Hydrodyn.,* 2, 832, 1978.
10. **Ostrach, S.,** Convection in continuous flow electrophoresis, *J. Chromatogr.,* 140, 187, 1977.
11. **Snyder, R. S., Rhodes, P. H., Miller, T. Y., Sloyer, J. A., and Seaman, G. V. F.,** Unexpected electrophoretic behavior of macromolecular sample on Space Shuttle Flight STS-6, in preparation.
12. **Snyder, R. S., Rhodes, P. H., Miller, T. Y., Micale, F. J., Mann, R. V., and Seaman, G. V. F.,** Polystyrene latex separations by continuous flow electrophoresis on the Space Shuttle, *Sep. Sci. Technol.,* 22, 157, 1986.
13. **Strickler, A. and Sacks, T.,** Continuous free-film electrophoresis: the crescent phenomenon, *Prep. Biochem.,* 3, 269, 1973.
14. **Taylor, G. I.,** Studies in electrohydrodynamics in the circulation produced in a drop by an electric field, *Proc. R. Soc. London Ser. A,* 291, 159, 1966.
15. **Rhodes, P. H., Snyder, R. S., and Roberts, G. O.,** Electrohydrodynamic distortion of sample streams in continuous flow electrophoresis, *J. Colloid Interface Sci.,* 129, 78, 1989.

Chapter 22

CELL PARTITIONING IN TWO-POLYMER-PHASE SYSTEMS: TOWARD HIGHER RESOLUTION SEPARATIONS

Donald Elliott Brooks

TABLE OF CONTENTS

I. INTRODUCTION

Aqueous mixtures of dextran and polyethylene glycol (PEG) form two phases when mixed at concentrations above a few percent each, the top phase rich in PEG and the bottom containing most of the dextran. These solutions can be buffered and used as separation media, via preferential partitioning, for biological macromolecules, organelles, and cells. Because of the biocompatibility of the polymers, their modest concentrations, and the low interfacial tension between the phases, such systems provide an extremely benign environment for living cells which supports retention of viability over extended periods.

Particles with diameters greater than a few hundred angstroms generally distribute themselves between one phase and the interface between the two liquids, rather than between the two bulk phases, because the loss of interface area which results from such adsorption is sufficient to significantly reduce the free energy of the system. The partition coefficient for a population of cells or particles, therefore, ought to depend on the magnitude of the free energy of adsorption of cells in the interface, i.e., the decrease in free energy which occurs when a cell adsorbs in the phase boundary. This free energy difference depends on the interfacial tension, the cell surface area, and the differential compatibility of the cell surface with the components in the two phases.[1]

If cells distributed themselves according to thermodynamic equilibrium, as molecules do, the partition coefficient, K (defined as the number of cells free in the bulk phase divided by the number adsorbed in the interface), would be determined predominantly by the cell adsorption free energy, ΔG^0, and the absolute temperature, T, according to the Boltzmann distribution:

$$K = \exp (\Delta G^0/kT) \tag{1}$$

where k is the Boltzmann constant. It is possible to test this relationship,[1-3] since ΔG^0 can be calculated if the interfacial tension and the angle subtended by the tangent to the cell surface and the phase boundary at the line of contact, the contact angle, are measured for representative cells in a population. By altering the composition of the phase systems, ΔG^0 can be varied over a wide range. Parallel measurement of K under each set of conditions allows Equation 1 to be tested. The results indicate that, while cell partitioning is a stochastic process and K depends roughly exponentially on ΔG^0, the apparent value for kT is several orders of magnitude too large. The systems behave as if the characteristic random energy responsible for distributing cells between the interface and a bulk phase was much larger than thermal energies. Stated another way, the energy required to remove a cell from the interface is much larger than kT; yet, during the partitioning process, a significant fraction of cells are removed into one of the bulk phases by unidentified forces. Cell partitioning, therefore, is not an equilibrium process.

If the nonequilibrium contributions to the mechanism of cell partitioning can be determined, it should be possible to design processes which minimize these kinetic effects and provide partition constants which more closely follow thermodynamic behavior. This would be a significant advance, since it would mean that K would more closely obey Equation 1. The value of the contact angle is under experimental control[1] and can be manipulated in a variety of ways to maximize the differences in ΔG^0 between, for instance, two types of cells in a mixture. Since we know[2] that the magnitude of ΔG^0 is much greater than kT, Equation 1 states that moderate differences in ΔG^0 should produce large differences in K, hence greatly increasing the resolution of separation. In the sections that follow, three approaches that we are taking, aimed at enhancing the resolution of cell partition, will be discussed.

II. DEMIXING OF PHASE-SEPARATED SYSTEMS IN A REDUCED-GRAVITY ENVIRONMENT

A cell partition step is carried out by introducing cells to the two-phase system, mixing it to expose the cells to both phases, then permitting the system to demix to allow isolation of the phases. Because the interfacial tension between the phases is very low (typically ≤10 μN/m), emulsification of the system occurs readily, and manual mixing reduces the characteristic dimensions of the domains to the order of cellular dimensions. Hence, immediately upon cessation of mixing, all cells will be in contact with the interface between the phases. If the equilibrium contact angle was formed at this stage and maintained throughout demixing, presumably the resulting partition coefficient would obey Equation 1.

However, the experimental results alluded to above imply that this contact is not maintained, suggesting that during the demixing process forces act at the interface to remove adsorbed cells which at equilibrium would remain there. These forces have not been identified, but we hypothesize that they result from the complex hydrodynamic environment to which the cells are exposed during the demixing process. In the laboratory, following a brief lag period during which the local phase domains grow by coalescence or diffusion processes, the system demixes chiefly via strong convective streams as the dextran-rich phase sinks and the PEG-rich phase rises due to density differences. The convection sets up secondary flows and creates a chaotic environment which produces relatively strong, unsteady shear forces throughout the system. Following the convective streaming, sedimentation and floatation of phase drops to which cells are adsorbed also occurs, a process which likewise produces shear forces at the drop surface. At many stages in the demixing process, then, shear forces are present which could remove adsorbed cells from the interface.

In order to test our hypothesis and carry out cell partitioning experiments in the absence of convection and droplet sedimentation, we have begun a series of experiments aimed at measuring cell partition coefficients in the reduced gravity environment of the Space Shuttle.[4] The first requirement for such an experiment, of course, is that the phases demix in space in a reasonable length of time. We examined this point in April 1985 with an experiment in which the time course of demixing was followed by photographing at intervals a small unit containing a number of different aqueous two-phase systems in transparent chambers. Each chamber was equipped with a mixing ball so that when the unit was shaken manually the systems were emulsified. The interesting result was that in all cases the systems eventually demixed, although considerably more slowly than on the ground. The disposition of the demixed phases was such that the PEG-rich phase, which wet the polymethyl methacrylate container walls, was in contact with the wall, with the dextran-rich phase floating like a spherical egg yolk in the center of the chamber (Figure 1). The mechanism by which the demixing took place is not known, but the kinetics suggest that coalescence of phase drops was involved.[4] Demixing occurred over periods of 10 to 60 min in systems useful for cell partitioning on the ground. Therefore, it appears that it will be possible to carry out a cell partitioning experiment in space in due course to provide a test of the above ideas.

III. CELL PARTITION WITH THE LOWER PHASE IMMOBILIZED ON CHROMATOGRAPHY BEADS

If the hypothesis proposed above is correct, the resolution of cell partitioning might be expected to be different if the partitioning step was to be carried out in a different mechanical environment from that present when the system is simply mixed and allowed to demix by settling. One method by which such a change in the process might be realized would be to immobilize one of the phases on a chromatography bead and perform the separation in a column. Since Müller has developed a method for immobilizing a dextran-rich phase on

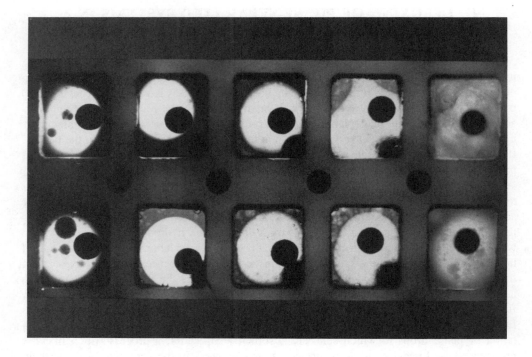

FIGURE 1. The appearance of ten dextran/PEG two-phase systems after several days exposure to the reduced-gravity environment of space. The dark phases are PEG rich, dyed with trypan blue. The dark sphere in each is a stainless steel mixing ball. Each chamber is 1.5 cm wide. For a description of the properties of each system, see the original paper.[4] The compositions of the ten systems described below are designated (X/Y/Z)S, T/B. X = concentration of dextran (% w/w); Y = concentration of PEG, 8,000 mol wt (% w/w); Z = concentration of Ficoll®, 400,000 mol wt (% w/w); S indicates the buffer present: I = 109 mM NaH$_2$PO$_4$, 35 mM Na$_2$HPO$_4$, pH 7.2; II = 150 mM NaCl, 7.3 mM Na$_2$HPO$_4$, 2.3 mM NaH$_2$PO$_4$, pH 7.2; III = 84.6 mM Na$_2$HPO$_4$, 25.4 mM NaH$_2$PO$_4$, pH 7.5; and T/B indicates the ratio of PEG-rich to dextran-rich phase volumes present. All systems contained dextran of ~500,000 mol wt except the (7/5/0) system, which contained dextran of ~40,000 mol wt. The compositions of the ten systems illustrated are: first row, left to right — (8/4/0)I, 2/3; (8/4/0)I, 3/2; (7/0.29/12)III, 1/1; (7/5/0)II, 1/1; and (7/5/0)II, 1/1; second row — (8/4/0)I, 1/1; (6/4/0)I, 1/1; (5/4/0)I, 1/1; (5/3.5/0)I, 1/1; and (5/4/0)I, 1/1.

chromatography beads derivatized with polyacrylamide,[5] we attempted to utilize this approach for cell separations.[6,7]

In the studies to be outlined, a model separation of dog and human erythrocytes from mixtures of the two cell types was attempted. These two cells are similar in their surface properties, but show a differential ability to adsorb fatty acid-PEG conjugates which results in large differences in partition constants in systems in which appropriate concentrations of such conjugates are included.[7] This effect is illustrated in Figure 2 for erythrocytes fixed with glutaraldehyde, where cell partitioning measured in a series of conventional experiments is shown as a function of concentration of the PEG ester of oleic acid. The conjugate was included in a phase system composed of 3% dextran T40 (40,000 mol wt), 7.7% PEG 8000 (8000 mol wt), and 0.079 M dibasic and 0.025 M monobasic sodium phosphate, pH 7.2. The results suggest that concentrations near 1 μM ester should be most effective in separating the two cell types.

In utilizing Müller's technique for immobilizing the dextran-rich phase, two types of beads were derivatized with polyacrylamide: agarose (200 to 300 μm diam; Pharmacia, Piscataway, NJ) and Fractogel®, a silica-based material (TSK HW-65F, 30 to 65 μm diam, BDH). The original papers should be consulted for experimental details.[5-7] The technique

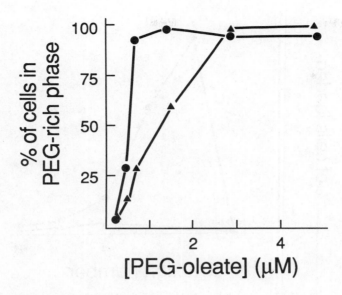

FIGURE 2. The partition behavior of glutaraldehyde-fixed dog and human erythrocytes in a 3.0% dextran 40, 7.7% PEG 8000, isotonic sodium phosphate buffer, pH 7.2 phase system as a function of the concentration of PEG-oleate present, measured in a conventional single-tube experiment; (●) dog cells, (▲) human cells.

essentially is to incubate the derivatized beads in the bottom, dextran-rich phase, which is thus absorbed into the gel matrix of the bead. The polyacrylamide derivatization makes the beads incompatible with PEG-containing phases,[8] effectively immobilizing the dextran-rich phase when the PEG-rich phase is used as the eluent. PEG-oleate can be incorporated into the elution medium either at constant concentration or in a gradient. Both approaches were attempted.

Separations were achieved with both the agarose and Fractogel® systems. In the former case, the best results were obtained when the cells were incubated on the column in the PEG-rich top phase in the absence of PEG-oleate, the column then washed with ester-free top phase to remove unbound cells (which comprised about 25% of the number loaded), and an optimal concentration of PEG-oleate (4.6 μm for unfixed cells) then applied isocratically. This protocol released approximately 50% of the dog cells loaded, at a purity of >75% (not shown). Phosphate-buffered saline (PBS) was then added to produce a single phase which eluted 30% of the human cells at a purity of >70%. This protocol was found to be more effective than any combination of gradients of PEG-oleate or other esters that interact even more strongly with the cells.[6]

The Fractogel®-based preparations proved to be somewhat more effective than the agarose beads, largely because there was essentially complete adsorption of the cells loaded on the column if no PEG-ester was present. It is possible that the smaller size of the Fractogel® was responsible for this behavior, producing smaller interbead distances and higher frequencies of cell-bead contact than in the agarose system. However, increasing the time over which the cells were incubated on the agarose columns before eluting with ester did not improve the situation.

The easiest and most efficient procedure for the Fractogel® columns proved to be simply to load the cell mixture on the column at the desired PEG-oleate concentration and elute isocratically with no preincubation. Figure 3 shows the rest of loading 0.1 ml containing 5 × 10⁷ cells/ml of ^{125}I-labeled fixed human erythrocytes and an equal concentration of ^{51}Cr-labeled fixed dog red cells on a 0.7 cm² × 1-cm column in a PEG-rich phase containing

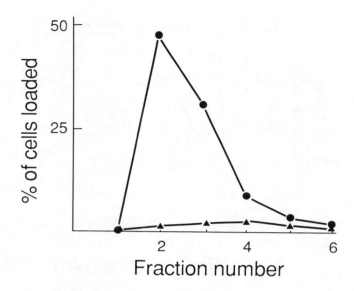

FIGURE 3.   The elution profile resulting from the elution of a mixture of human and dog erythrocytes by PEG-rich phase containing 1.25 × 10^{-6} *M* PEG-oleate from the column described in the text; (●) dog cells, (▲) human cells.

1.25×10^{-6} *M* PEG-oleate and eluting with the same medium, with no incubation period. Of the dog cells, 92% were eluted with only 7% of the human cells; the remainder were released with addition of pure buffer. The purest fraction contained 49% of the dog cells contaminated by only 0.4% of the human cells.

Two interesting features of this type of separation emerged from a detailed analysis of a series of experiments like those described above. First, the elution volume of the peak for either cell type, measured in experiments on single cell types, did not decrease with increasing PEG-oleate concentration except under conditions that led to poor recoveries. Hence, multistep partitioning does not appear to be occurring with cells. The system behaves more like a single-step affinity or ion-exchange column than a true chromatography column. Second, and of particular interest in the context of the present paper, the resolution of separation was considerably higher than would be obtained from a single partitioning step of the conventional type. This can be seen by comparing Figures 2 and 4. The latter shows the results of studies in which the total recovery measured in a series of isocratic elution experiments at different PEG-oleate concentrations is plotted for the two cell types. It is clear that the results obtained in these experiments indicate much higher separation on the columns.

Since multiple partitioning steps evidently do not occur, some other feature must be responsible for the enhancement in performance found with the columns. It is quite possible that the fluid mechanical environment in the columns is responsible. Certainly with a constant elution rate the local shear stresses, acting at the surface of the beads holding the dextran-rich phase, would be steadier and probably weaker than those present during demixing of phase emulsions. During demixing via droplet coalescence, for instance, relatively high shear might be expected to occur near the region in which the two involved interfaces merge and retract. Since the separations obtained with columns more closely approach thermodynamic behavior than do conventional single-step partitions, the above results support the idea that the hydrodynamic environment in which particle partitioning takes place plays an important role in determining K.

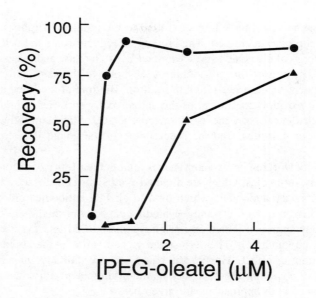

FIGURE 4. The total recovery of cells eluted in separate runs from a series of columns to which were applied either dog (●) or human (▲) erythrocytes suspended in PEG-rich phase containing the concentration of PEG-oleate indicated and eluted with the suspending medium.

IV. IMMUNOAFFINITY PARTITIONING OF CELLS

The third approach we are taking to improve the resolution of cell partitioning is to develop very specific methods of altering the surface free energy of the subpopulation of cells which is to be isolated from the total population in which it resides; that is (referring to Equation 1), the aim is to make the value of ΔG^0 much larger in magnitude for the subpopulation of interest than for the remainder of the cells present.

The principle being followed here is known as affinity partitioning,[9] in analogy to affinity chromatography. The idea is to use the specificity of some biorecognition reaction appropriate to the subpopulation to be isolated as a basis of the separation. The most generally applicable reaction is antigen-antibody binding, since surface antigenicity is the characteristic most frequently used to define cell subpopulations. In the context of partitioning, the challenge is to develop methods by which the binding of antibodies to specific surface antigens will cause those cells to partition into the phase from which the rest of the population is excluded. To accomplish this, conditions must be found under which the antibody of interest strongly partitions to one phase while the cell mixture partitions to the opposite phase or to the interface between them. In fortunate cases such a disposition may be found without modifying the ligand. For instance, in some viral systems binding of antibody significantly modifies partitioning of the cells.[10] However, more frequently some chemical modification of the ligand is necessary to concentrate it into one phase. The limited work on immunoaffinity partitioning that has been completed to date[11,12] has utilized dextran/PEG systems. Hence, derivatizing antibodies with PEG is a natural approach, particularly since it is almost always possible, by manipulating the buffer composition, polymer molecular weight, or interfacial tension between the phases, to direct cells into the interface or dextran-rich bottom phase in the absence of ligand. Lower molecular weight materials may also be used, however, as is shown below. In fact, there may be some advantage to modifying antibodies with small molecules, as in some cases they interfere less with antigen binding.[11]

PEG-modified antibodies accumulate predominantly in the top, PEG-rich phase. When the phase system is mixed, the cells are exposed to the modified antibodies which bind to cells bearing the appropriate antigen. The binding effectively coats the cell with PEG. This lowers the free energy of the coated surface in the PEG-rich phase and raises it in the dextran-rich phase; i.e., it increases the magnitude of ΔG^0 due to the same polymer-polymer incompatibility that drives phase separation in solution. Binding of modified antibodies, therefore, increases the partition coefficient of the antigenically specified subpopulation of cells to a degree which depends on the partition coefficient of the antibody, the number of molecules bound per unit area, and the accessibility of the bound PEG to the phase polymers.[1,11]

We are currently working on two approaches to immunoaffinity partitioning, using either derivatized primary antibodies, which are directed against the cell surface antigen of interest, or derivatized secondary antibodies which bind to all IgG antibodies produced by a given species without interfering with the antigen-binding capacity of the latter. The advantage of the second antibody approach is that any primary, unmodified IgG may be used to bind to the specified cell subpopulation. The derivatized second antibody then binds to it to produce the affinity partitioning effect. Hence, the same second antibody may be used with all primary antibodies, obviating the necessity of modifying each individual immunoglobulin species. Examples of both approaches are given below.

A. STUDIES USING PEG-MODIFIED PRIMARY ANTIBODIES

Derivitization of rabbit anti-(human red cell) IgG (rαhRBC) with monomethyl PEG 1900 (Aldrich, Milwaukee) was carried out as follows. The cyanuric chloride derivative of each was synthesized[13] and incubated at mole ratios from 0.2 to 5 (with respect to lysine) with approximately 5 mg protein. The reaction was allowed to proceed for 40 min at room temperature in 0.1 M borate buffer, pH 9. The degree of substitution was estimated by measuring the number of free amines able to react with fluorescamine before and after the reaction.[16] Details may be obtained from the original publication.[11] The phase system utilized consisted of 5% dextran T500 (Pharmacia), 3.4% PEG 8000 (sentry grade, Union Carbide, New York), 130 mM NaCl, and 10 mM phosphate buffer, pH 7.2, prepared as described.[14] Cells were obtained by venipuncture, collected into 0.38% trisodium citrate, and washed three times in 130 mM NaCl, 10 mM phosphate buffer, pH 7.2 (PBS). In some cases washed cells were fixed in 1% glutaraldehyde in PBS. Control experiments were run with nonspecific rabbit IgG or rabbit erythrocytes which did not react with the anti-(human red cell) IgG. Cell partition coefficients were determined by resistive pulse cell counting, typically from 2-ml phase systems of equal phase volumes containing 10^7 cells. All partitioning experiments were carried out at room temperature.

In approaching the problem of developing effective immunoaffinity partitioning techniques for cells, three types of reactions must be kept in mind. First, coupling modifying agents to the IgG must not eliminate antibody binding to the antigen. Reducing the binding constant somewhat is acceptable, perhaps even desirable, if the affinity ligand is to be removed from the cell surface following partitioning. The second reaction of concern is agglutination caused by simultaneous binding of the divalent IgG to two cells. Agglutination leads to sedimentation of aggregates in the phase systems into or through the interface, reducing the resolution of the separation. Hence, it is to be avoided if possible. Depending on the location of the antigen in the membrane, glycocalyx IgG may or may not cause agglutination. However, again, reduction of the association constant can be beneficial from this perspective. The final relevant reaction is, of course, the partitioning of the cell-affinity ligand complex. The partition coefficient of the complex strongly depends on the partition coefficient of the affinity ligand and on the density of antigens present on the cell surface, although the latter dependence is weaker than would be expected. Further discussion of these issues is provided elsewhere.[11]

TABLE 1
Effect of PEG-IgG on the Partition of Human Erythrocytes

IgG Concentration (mg/ml)				
rαhRBC	Nonspec.[a]	PEG-rαhRBC	K_{cell}[b]	$K_{cell}/K_{control}$
—	—	—	15	—
—	—	0.15	35 ± 5	2.3
0.3	—	—	18.5	1.23
0.3	—	0.15	19	1.27
—	0.5	—	15	1.0
—	0.5	0.15	24	1.60
—	—	—	24	—
—	—	0.15	25.5[c]	1.06

[a] Nonspecific IgG.
[b] Percentage of cells in system partitioning into the top phase.
[c] Fixed rabbit red cells were used instead of human.

The rαhRBC derivatized with PEG 1900 to a level of 31 mol of PEG per mole of IgG was tested for its ability to affect the partitioning behavior of human red blood cells. The results are given in Table 1. It is seen that PEG-rαhRBC more than doubled the human, but not the rabbit, cell partition in a manner that was inhibited by rαhRBC, but not by nonspecific antibody. This partition increase was sufficient to allow easy separation of human from rabbit red cells by countercurrent distribution, but did not produce sufficient resolution to give a single-step separation.[11] Although the experiments carried out with rαhRBC were successful, sufficient experimentation was required to determine acceptable coupling and partitioning conditions so that a more general approach was clearly desirable. In addition, since a polyclonal antibody preparation was utilized, there was some uncertainty regarding the details of the results because of the mixture of molecular species present, not all of which would bind to the human cells. For these reasons, an approach utilizing monoclonal antibodies as a second reagent was investigated.

B. STUDIES USING PEG-DERIVATIZED SECONDARY ANTIBODIES

In attempts to develop a more general immunoaffinity reagent, we have been working with antibodies directed against the Fc fragment of IgG, a region of the protein which is not involved in antigen recognition. By derivatizing such a secondary antibody it was thought possible to increase the partition of any cell to which a nonderivatized IgG was specifically bound. A monoclonal antibody against the Fc fragment of rabbit IgG (mαrFc) was raised by standard techniques[15] and derivatized with monomethyl PEG 1900 using cyanuric chloride as above. Polyclonal rabbit anti-(human NN glycophorin) IgG (rαglyco) was obtained by raising an antiserum against human glycophorin, an erythrocyte membrane glycoprotein isolated from human NN red cells,[17] and isolating the IgG by fast protein liquid chromatography (FPLC) on a Mono-Q column (Pharmacia).[18] Cells and phase systems were obtained and used as described above.

The results of a series of experiments testing this approach are summarized in Figure 5.[18] It is seen that there is almost a fourfold increase in the immunospecific partitioning of the human red cells only in the presence of both the underivatized rabbit anti-glycophorin antibody and the PEG-mαrFc. Hence, this approach was also effective and certainly produced a sufficient change in the partition to allow separation of the two cell types in a few extractions.

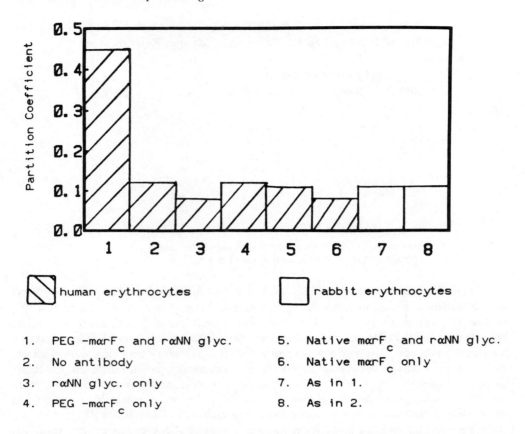

1. PEG $-m\alpha rF_c$ and $r\alpha NN$ glyc.
2. No antibody
3. $r\alpha NN$ glyc. only
4. PEG $-m\alpha rF_c$ only
5. Native $m\alpha rF_c$ and $r\alpha NN$ glyc.
6. Native $m\alpha rF_c$ only
7. As in 1.
8. As in 2.

FIGURE 5. The effect of PEG-modified $m\alpha rF_c$ and $r\alpha$glyco on the partition coefficient of human and rabbit erythrocytes. (From Brooks, D. E., Sharp, K. A., and Stocks, S. J., *Makromol. Chem. Macromol. Symp.*, 17, 387, 1988. With permission.)

C. STUDIES WITH TRYPAN BLUE-DERIVATIZED ANTIBODIES

Finally, we have combined the use of a small, nonpolymeric modifying agent with a secondary antibody. In this case we used trypan blue (TB) as the agent to be coupled to the IgG to increase its partition. TB has a partition coefficient of ten in the phase system used, much higher than PEG.

TB (10 mg/ml) was coupled to sheep anti-(mouse Fc) ($s\alpha mFc$, 5 mg/ml) by stirring with 1-ethyl-3(3-diaminopropyl)carbodiimide HCl (40 mg/ml) in sodium acetate buffer, pH 4.8, for 4 h at room temperature, followed by dialysis against 1 M NaCl. The extent of modification was estimated using ^{125}I-IgG and dye absorption.

The results given in Table 2 were obtained when 28 mol/mol of TB were coupled to $s\alpha mFc$ and the conjugate used as a secondary antibody in a system in which an unmodified monoclonal mouse anti-(human NN glycophorin) ($m\alpha$glyco) was included as a primary antibody. Although there was some nonspecific interaction between the double antibody complex and rabbit erythrocytes, in this system there was still a much larger increase in the partitioning of the cells toward which the primary antibody was directed, the ratio of the two partition coefficients being greater than seven. It may be that the dye has a tendency to adsorb to cells nonspecifically, since it is highly charged in solution. In any event, the results in Table 2 clearly indicate that the approach taken would again be ample to allow separation of the two illustrated cell types with a few sequential extractions.

TABLE 2
Partitioning Behavior of Systems Containing TB-sαmFc and mαglyco

Cell type	1° Ligand	2° Ligand	K^a	$K_{control}{}^b$	$K/K_{control}$
—	—	TB-sαmFc	5.0	0.3	17
—	mαglyco	TB-sαmFc	3.4	0.6	7
Human RBC	mαglyco	TB-sαmFc	2.6	0.1	26
Rabbit RBC	mαglyco	TB-sαmFc	0.9	0.25	3.6

[a] Partition coefficient of protein(s), equal to concentration ratio in phases, or cells, equal to ratio of number of cells in top phase to remainder.

[b] Partition coefficient of underivatized sαmFc, mαglyco in the absence of TB-sαmFc, or cells in the absence of antibody, as appropriate.

V. CONCLUSIONS

Both the use of immobilized dextran-rich phases on beads and the development of immunoaffinity partitioning have significantly increased the ability of dextran/PEG systems to separate closely related cell types. Studies in the reduced gravity, convection-free environment of space hold the promise of further advances in this direction, as well as providing unique information on the nonthermodynamic determinants of cell partitioning. We are continuing efforts in all three areas and anticipate that further improvements will allow virtually any cell separation problem to be solved by appropriate application of the partitioning technique.

ACKNOWLEDGMENTS

This work was supported by grant MT-5759 from the Medical Research Council of Canada and contract NAS8-35333 from the Microgravity Science and Applications Division of the National Aeronautics and Space Administration (NASA).

REFERENCES

1. **Brooks, D. E., Sharp, K. A., and Fisher, D.,** Theoretical aspects of partitioning, in *Partitioning in Aqueous Two-Phase Systems. Theory, Methods, Uses and Applications in Biotechnology,* Walter, H., Brooks, D. E., and Fisher, D., Eds., Academic Press, Orlando, FL, 1986, chap. 2.
2. **Sharp, K. A.,** Theoretical and Experimental Studies on Erythrocyte Partition in Aqueous Polymer Two Phase Systems, Ph.D. thesis, University of British Columbia, Vancouver, 1985.
3. **Brooks, D. E., Bamberger, S. B., Harris, J. M., and Van Alstine, J.,** Rationale for two phase polymer system microgravity separation experiments, in Proc. 5th Eur. Symp. Material Sciences under Microgravity, ESA SP-222, Schloss Elmau, Federal Republic of Germany, November 5 to 7, 1984, 315.
4. **Brooks, D. E., Bamberger, S. B., Harris, J. M., Van Alstine, J., and Snyder, R.,** Demixing kinetics of phase separated polymer solutions in microgravity, in Proc. 6th Eur. Symp. Material Sciences under Microgravity, ESA SP-256, Bordeaux, France, December 2 to 5, 1986, 131.
5. **Müller, W.,** New phase supports for liquid-liquid partition chromatography of bio-polymers in aqueous poly(ethyleneglycol)-dextran systems: synthesis and application for the fractionation of DNA restriction fragments, *Eur. J. Biochem.,* 155, 213, 1986.
6. **Skuse, D., Müller, W., and Brooks, D. E.,** Column chromatographic separation of cells using aqueous polymeric two-phase systems, *Anal. Biochem.,* 174, 628, 1988.
7. **Skuse, D. and Brooks, D. E.,** Column based liquid/liquid separation of cells using aqueous polymeric two-phase systems, in *Advances in Separations Using Aqueous Phase Systems in Biology and Biotechnology,* Fisher, D. and Sutherland, I. A., Eds., Plenum Press, London, in press.

8. **Müller, W.,** personal communication.
9. **Flanagan, S. D. and Barondes, S. H.,** Affinity partitioning method for purification of proteins using specific polymer-ligands in aqueous polymer 2-phase systems, *J. Biol. Chem.,* 250, 1484, 1975.
10. **Phillipson, L., Killander, J., and Albertsson, P.-A.,** Interaction between poliovirus and immunoglobulins. I. Detection of virus antibodies by partition in aqueous polymer phase systems, *Virology,* 28, 22, 1966.
11. **Sharp, K. A., Yalpani, M., Howard, S. J., and Brooks, D. E.,** Synthesis and application of a poly(ethylene glycol)-antibody affinity ligand for cell separations in two polymer aqueous phase systems, *Anal. Biochem.,* 154, 110, 1986.
12. **Karr, L. J., Shafer, S. G., Harris, J. M., Van Alstine, J., and Snyder, R. S.,** Immuno-affinity partition of cells in aqueous polymer 2-phase systems, *J. Chromatogr.,* 354, 269, 1986.
13. **Abuchowski, A., van Es, T., Palczuk, N. C., and Davis, F. F.,** Alteration of immunological properties of bovine serum albumin by covalent attachment of polyethylene-glycol, *J. Biol. Chem.,* 252, 3578, 1977.
14. **Bamberger, S., Brooks, D. E., Sharp, K. A., Van Alstine, J. M., and Webber, T. J.,** Preparation of phase systems and measurement of their physicochemical properties, in *Partitioning in Aqueous Two-Phase Systems. Theory, Methods, Uses and Applications in Biotechnology,* Walter, H., Brooks, D. E., and Fisher, D., Eds., Academic Press, Orlando, FL, 1986, chap. 3.
15. **Gooding, J. W.,** *Monoclonal Antibodies: Principles and Practice,* Academic Press, London, 1983.
16. **Stocks, S. J., Jones, A. J. M., Ramey, C. W., and Brooks, D. E.,** A fluorimetric assay of the degree of modification of protein primary amines with polyethylene glycol, *Anal. Biochem.,* 154, 232, 1986.
17. **Marchesi, V. T. and Andrews, E. P.,** Glycoproteins: isolation from cell membranes with lithium diiodogalicylate, *Science,* 174, 1247, 1971.
18. **Stocks, S. J.,** Development of a General Ligand for Immunoaffinity Partitioning, M.Sc. thesis, University of British Columbia, Vancouver, 1986.

Chapter 23

LARGE-SCALE AFFINITY PARTITIONING OF ENZYMES IN AQUEOUS TWO-PHASE SYSTEMS

Göte Johansson

TABLE OF CONTENTS

I. INTRODUCTION

A. TWO-PHASE SYSTEMS

Liquid-liquid two-phase systems for enzyme purification are mainly composed either of a salt and a polymer (e.g., polyethylene glycol [PEG] and potassium phosphate) dissolved in water or of a mixture of two polymers in water (e.g., dextran and PEG). The salt-polymer systems, because of their low price and rapid phase settling, have been applied for large-scale enzyme extractions.[1,2] The systems composed of two polymers have other advantages; for instance, the partition can be adjusted between the phases by directing the separation properties toward charge, hydrophobic pockets, or specific ligand binding of the target enzyme.[3-6] By anchoring an affinity ligand to one of the phase-forming polymers, an enzyme (with affinity for this ligand) can be extracted into the corresponding phase.

The compositions of the two phases as well as their volume ratio depend on the amounts and molecular weights of the two polymers. A number of polymer pairs have been studied and the composition of the phases described in the form of phase diagrams.[7] Under certain concentrations of polymers only a single phase is obtained. With increasing concentrations of the two polymers, the compositions of the upper and lower phases will be different, as shown in Table 1.

B. POLYMER-BOUND TRIAZINE DYES

A number of triazine dyes, e.g., Cibacron blue F3G-A and Procion red HE-3B, have been found to act as pseudoaffinity ligands for a number of proteins, especially for enzymes in the kinase and dehydrogenase groups. The low cost and the ease of attachment to a matrix have made these dyes popular as ligands for affinity chromatography.[8,9] These reactive dyes have mainly been attached to the phase-forming polymers (PEG, dextran, Aquaphase® PPT) in two ways: (1) reaction of the dye in alkaline water solution with the polymer, followed by removal of free dye and eventually unsubstituted polymer;[10-12] or (2) by introducing amino groups on the polymer molecules which, in turn, react swiftly with the reactive dyes.[13,14] To obtain the amino derivatives of the polymers they are activated by bromination ($SOBr_2$), tosylation, or tresylation.[15,16] The use of diaminoalkanes makes it possible to introduce a spacer between the dye ligand and the polymer. When added to an aqueous (two-polymer) two-phase system, the dye-polymer is concentrated in the phase rich in the parent polymer if the degree of substitution is low (1:1 molar ratio). More heavily substituted polymers show a change in partition behavior due to the introduction of larger amounts of hydrophobic as well as charged groups. An increasing number of dye groups per polymer molecule makes the partition of dye-polymer more sensitive to electrolyte conditions of the system (i.e., the kind of salts and the pH).[17]

C. PARTITIONING OF PROTEINS

Proteins as well as other cell components (nucleic acids, membranes, organelles) can be included in the system and partitioned between the two phases. The two-polymer-phase systems have a high capacity for proteins, and up to 150 g protein per liter can be incorporated, though loadings of 5 to 50 g/l are more normal. Due to the low interfacial tension and small difference in the densities of the phases, mixing and equilibration are achieved extremely rapidly (within a few seconds). The separation (settling) of the phases is slow for the same reasons, but is speeded up by the presence of cell debris (which, in general, partition toward the lower phase). For large-scale extractions, phase separation can be shortened by centrifugation.

TABLE 1
Compositions of the Phases of Some Two-Phase Systems Composed of Water, Dextran T-500, and PEG 8000 at 20°C[a]

Values Are Expressed as % (w/w)

Total composition		Upper phase		Lower phase	
Dextran	PEG	Dextran	PEG	Dextran	PEG
4.0	4.0	1.8	4.9	7.3	2.6
5.0	5.0	0.3	7.2	13.3	1.1
7.0	7.0	0.1	10.5	20.3	0.4
9.0	9.0	0.0	13.6	25.5	0.2

[a] Calculated from Albertsson.[7]

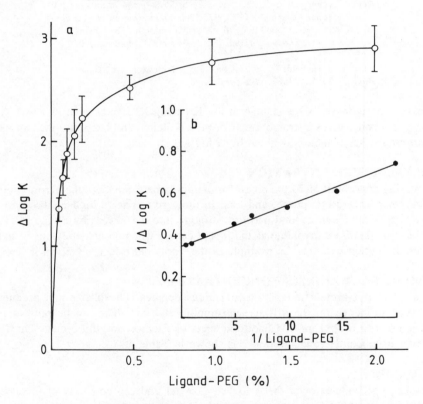

FIGURE 1.   (a) Increase in the logarithmic partition coefficient of phosphofructokinase from baker's yeast, Δ log K, as a function of concentration of Cibacron blue F3G-A PEG (given in percentage of total PEG). System as in Table 3, but with 50 mM sodium phosphate buffer. (b) Reciprocal plot. (From Johansson, G., Kopperschläger, G., and Albertsson, P.-Å., *Eur. J. Biochem.*, 131, 589, 1983. With permission.)

II. PARAMETERS DETERMINING AFFINITY PARTITIONING

A. LIGAND CONCENTRATION

The partition coefficient, K, increases (if the ligand is in the upper phase) or decreases (ligand in the lower phase) with increasing concentration of polymer-bound ligand and asymptotically approaches a ''saturation'' value, as shown in Figure 1. The effectiveness

TABLE 2
Affinity Partitioning of Lactate Dehydrogenase from an Extract of Swine Muscle at Various Polymer Concentrations

Aquaphase® PPT (% [w/w])	PEG 8000 (% [w/w])	Percentage in upper phase		Purification factor
		LDH	Protein	
5.0	13.0	86	48	1.8
5.7	15.0	91	35	2.6
6.5	17.0	95	24	3.9
7.2	19.0	84	15	5.7
8.0	21.0	68	8.2	8.3
8.8	23.0	48	5.2	9.2

Note: Procion yellow HE-3G was used as PEG-bound affinity ligand. System: Aquaphase® PPT, PEG 8000, Procion yellow HE-3G PEG 8000 (1% of total PEG), 40 mM sodium phosphate buffer (pH 7.9), and 3.0 g protein per liter (muscle extract). Temperature, 22°C.

Data from Tjerneld, F., Johansson, G., and Joelsson, M., *Biotechnol. Bioeng.*, 30, 809, 1987. With permission.

of the extraction is taken as the change in log K (Δ log K). An inverse plot of $1/\Delta$ log K vs. $1/c_{\text{ligand}}$ generally gives a straight line (for pure proteins) and facilitates the extrapolation to the limiting value, which is denoted by Δ log K_{max}.

B. POLYMER CONCENTRATION

Increasing concentration of the phase-forming polymers enhances the partitioning of the ligand-polymer toward one phase, and this, in turn, gives rise to higher (absolute) values of Δ log K_{max}. Proteins in general partition strongly into the lower phase, and PEG (in the upper phase) is therefore favorable as the ligand carrier when an enzyme is to be extracted from a crude protein mixture. An example of the "polymer effect" is given in Table 2.

C. KIND OF SALT AND ITS CONCENTRATION

Salts affect dye-protein interaction to various degrees. Phosphates and acetates have only a weak influence on the affinity partitioning and can therefore be present in high concentrations (up to 250 mM). Chaotropic salts like iodide and thiocyanate show strong interference at moderate concentrations, as shown in Table 3.

D. pH VALUE

Increasing pH values lower the Δ log K_{max} values and can be a way of increasing the specificity in extraction. Examples are found in References 11, 18, and 20.

E. TEMPERATURE

The Δ log K_{max} values generally increase with decreasing temperature.[11,18]

F. FREE LIGANDS

Addition of natural ligands, e.g., ATP or NAD (especially in combination with Mg^{2+} or sulfite), has a strong effect on affinity partitioning.[11,18] It can be used for group-specific "stripping" while more "nonspecific" dye-protein interactions are not affected. In the case of glucose-6-phosphate dehydrogenase, a 2.7-fold purification has been obtained by the use of NADP.[21]

TABLE 3
Effects of Various Salts on the Affinity Partitioning Effect: Partition of Phosphofructokinase from Baker's Yeast in Systems with and without a Saturating Amount of Cibacron Blue F3G-A PEG

Salt	$\Delta \log K_{max}$		
	25 mM	63 mM	250 mM
Sodium phosphate, pH 7.0 additional	2.8	2.9	n.d.
Sodium acetate	2.9	2.7	2.5
Sodium sulfate	2.7	2.4	0.9
Potassium chloride	2.9	2.2	0.6
Potassium iodide	2.7	1.9	<0.1
Sodium thiocyanate	n.d.	1.9	n.d.
Potassium phthalate, pH 7.0	n.d.	1.3	n.d.
Sodium perchlorate	n.d.	1.2	n.d.

Note: System: 7% dextran 500, 5% PEG 8000, 10 mM sodium phosphate buffer, pH 7.0, 0.5 mM EDTA, 5 mM 2-mercaptoethanol, and 4 μkat enzyme/l (1 μkat = 60 U). Temperature, 0°C; n.d. = not determined.

Adapted from Johansson, G., Kopperschläger, G., and Albertsson, P.-Å., *Eur. J. Biochem.*, 131, 589, 1983. With permission.

G. LIGAND/POLYMER MOLAR RATIO

Increasing the number of ligands per polymer molecule has interesting effects on the partitioning of the dye-polymer complex; this also influences the effect of the complex on the partitioning of proteins. This has been studied on dextran.[17] Highly substituted dye-dextran can easily be partitioned to either the upper or lower phase, or it can be equally distributed between the phases by choice of salt in the system. This is due to the interfacial potential[7] caused by differences in the ions of the salt for the two phases. Since the dyes are charged molecules, the polymer distribution will be affected due to the high number of charges.

III. COUNTERCURRENT DISTRIBUTION OF CRUDE PROTEIN EXTRACTS

The heterogeneity of a crude protein extract can be analyzed by repeated partitioning steps carried out in a manner similar to chromatographic processes. Most popular is countercurrent distribution (CCD) according to Craig.[22] The principle of CCD is illustrated in Figure 2. The fractions obtained are analyzed for protein content as well as enzymes of interest. The obtained CCD diagram shows whether an enzyme partitions as a single component. If so, its CCD curve should be given by Equation 1,[22]

$$T_i = \frac{n!}{i!(n-i)!} \cdot \frac{G^i}{(1+G)^n}$$ (1)

where T_i is the fraction of a single component with partition ratio G in tube number i (i varies between 0 and n) after n transfers. G is equal to $K \cdot V_U/V_L$, where V_U and V_L are the volumes of the upper (mobile) and lower (stationary) phases, respectively. The G value,

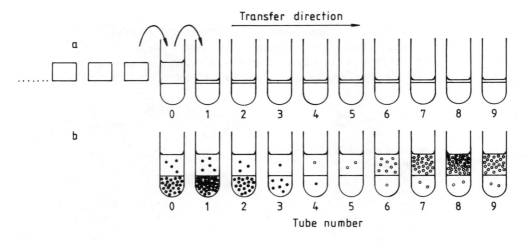

FIGURE 2.   Principle of countercurrent distribution. (a) Arrangement for a nine-transfer countercurrent distribution. Sample is included in tube 0. Curved arrows indicate the first transfer. (b) The ten systems obtained after
nine transfers. The dots show the distribution of homogeneous proteins with K = 0.1 (●) and K = 10 (○),
respectively. The volume ratio (upper to lower phase) is 1.5, and 9% of the upper phase is stationary. (From
Johansson, G. and Andersson, M., *J. Chromatogr.*, 291, 175, 1984. With permission.)

and therefore also the K value, can be estimated from the position of the peak i_{max} (tube
with maximal amount) in the CCD diagram, and Equation 2:

$$G = \frac{i_{max}}{n - i_{max}} \tag{2}$$

A CCD resolution of a protein extract from baker's yeast is shown in Figure 3.

IV. PREPARATIVE EXTRACTION

A major advantage of the aqueous two-phase systems is that their properties can be
determined on a small scale (using systems of 2- to 5-ml volume) and that the results can
then be directly applied on a large scale. A preliminary test, on a small scale, for the affinity
extraction (with Procion yellow HE-3G PEG) of lactate dehydrogenase (LDH) is shown in
Table 2. When protein extract from swine muscle was partitioned in a system containing
Aquaphase® PPT (a hydroxypropyl starch) and PEG plus ligand-PEG (ratio 99:1), the
purification factor increased with increasing concentration of the two polymers. The highest
concentrations tested, however, reduced the amount of enzyme extracted into the upper
phase.

The presence of debris and of proteins influences the effect of the affinity extraction,
as shown in Table 4.

Three methods can be applied for the extraction. First is the affinity partitioning of the
crude homogenate, including debris. Despite the need for a higher concentration of the
affinity ligand and the relatively poor recovery, this method allows extraction in a single
step. Eventually, the homogenization of the tissue can be carried out directly in the two-
phase system. Second, the tissue is homogenized in a buffer and the debris is removed by
centrifugation before the affinity partitioning. This reduces the required amount of ligand-
PEG. The third method, involving a fractional precipitation of proteins, including LDH, by
treating the supernatant liquid with PEG, introduces more steps before the extraction, but
offers a concentration of the protein and therefore reduces the necessary volume of the two-

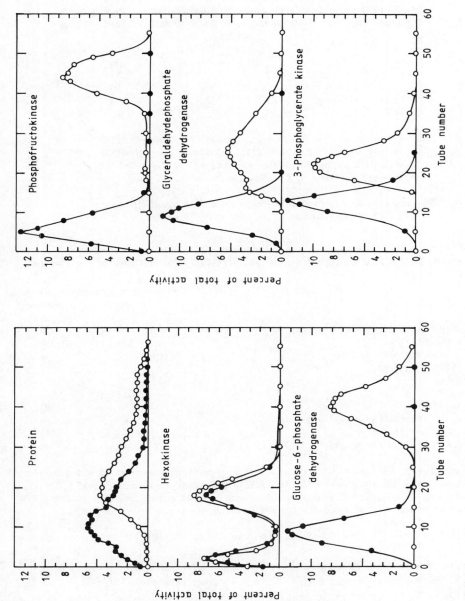

FIGURE 3. Countercurrent distribution of extract of baker's yeast in systems without ligand-PEG (●) and with Procion olive MX-3G PEG (○). System composition: 7% dextran 500, 5% PEG 8000 (with or without ligand-PEG, 1% of total PEG), 50 mM sodium phosphate buffer (pH 7.0), 5 mM 2-mercaptoethanol, and 0.2 mM EDTA. Temperature, 3°C; number of transfers, 55. (From Johansson, G., Andersson, M., and Åkerlund, H.-E., *J. Chromatogr.*, 298, 483, 1984. With permission.)

TABLE 4
Effect of the Concentration of PEG-Bound Procion Yellow on the Partitioning of Lactate Dehydrogenase (LDH)

Procion yellow HE-3G PEG (% of total PEG)	Percentage of LDH in upper phase		
	Muscle homogenate (with debris)	Muscle extract (supernatant)	Pure LDH
0	0	0	0
0.2	2	14	34
0.5	6	33	71
1.0	18	69	93
2.0	30	85	n.d.
6.0	50	94	n.d.

Note: System composition: 20% (w/w) Aquaphase® PPT, 7.6% (w/w) PEG including Procion yellow HE-3G PEG and 40 mM sodium phosphate buffer, pH 7.9. In all cases the concentration of LDH was 100 U/g system. Temperature, 22°C; n.d. = not determined.

Data from Tjerneld, F., Johansson, G., and Joelsson, M., *Biotechnol. Bioeng.*, 30, 809, 1987. With permission.

TABLE 5
Large-Scale Extraction Using a Centrifugal Separator for Settling of the Phases

	Muscle homogenized in system	Muscle extract
Concentration of Procion yellow PEG (% of total PEG)	2	1.5
Muscle (g) used per 1 system	100	160
Initial LDH concentration (kU/l system)	66	70
Recovery of LDH in top phase:		
(kU/l system)	32	58
(kU/kg muscle)	312	364
(kU/g ligand-PEG)	21	51
Total recovery of LDH in both phases (kU/l system)	63	68
Specific activity (U/mg protein)	114	126
Degree of purification (relative to protein)	10×	11×

Note: The extraction of LDH from muscle and muscle extract, respectively, was carried out with a two-phase system containing 20% Aquaphase® PPT, 7.6% PEG, 40 mM sodium phosphate buffer, pH 7.9, and either muscle (10%) that was homogenized directly in the system, or muscle extract (20%, corresponding to 16% of meat). Temperature, 25°C; speed of separation, 0.5 l/min; volume ratio (top/bottom), 0.93.

From Tjerneld, F., Johansson, G., and Joelsson, M., *Biotechnol. Bioeng.*, 30, 809, 1987. With permission.

phase system. The first two alternatives have been compared[19] on a large scale by using an Alfa-Laval® centrifugal separator, LAPX 202; results are presented in Table 5.

The enzyme in the upper phase can be further purified by washing the phase with a fresh lower phase. A simple way of removing the PEG and ligand-PEG is addition of a phosphate mixture to the recovered upper phase. A two-phase PEG-salt system is obtained where the enzyme (95% or more) is recovered in the salt phase. The ligand-PEG (together with PEG) is recovered with 96% efficiency and can be reused without purification.[19] A schematic of a large-scale process for extracting the enzyme, "stripping" the enzyme with salt, and recycling the ligand-PEG is suggested in Figure 4. A washing step is included,

279

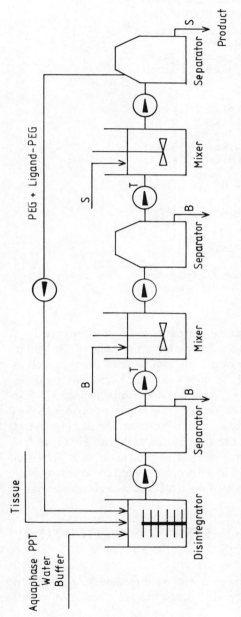

FIGURE 4. Scheme for continuous extraction of enzyme by affinity partitioning. The tissue is disintegrated and mixed with the phase-forming components. After separation the top phase (T) is washed once with pure bottom phase (B). Enzyme and PEG are separated by addition of salt (S) to the top phase. PEG and ligand-PEG are recycled. (From Tjerneld, F., Johansson, G., and Joelsson, M., *Biotechnol. Bioeng.*, 30, 809, 1987. With permission.)

TABLE 6
Process Data for Economical Evaluation
of the Large-Scale Enzyme Extraction
Described in Figure 4

	Consumed and produced amounts of material per hour	
	Muscle homogenized in system	Muscle extract
Material consumption		
Aquaphase® PPT	6.64 kg	6.64 kg
PEG	0.11 kg	0.11 kg
Ligand-PEG	10.7 g	1.5 g
Salt (phosphate)	3.99 kg	3.99 kg
Tissue	3.0 kg	4.8 kg
Enzyme recovery		
In first top phase	960 kU	1740 kU
After wash step	867 kU	1572 kU
In final bottom phase	824 kU	1493 kU

Note: The data are based on the values in Table 5.

Data from Tjerneld, F., Johansson, G., and Joelsson, M., *Biotechnol. Bioeng.,* 30, 809, 1987. With permission.

and the only slightly contaminated lower phase can be reused for the first extraction. Data for the material consumption and the recovery of enzymes for a flow rate of 30 l/h are presented in Table 6. The extraction can be made from both homogenated tissue as well as the supernatant liquid obtained by centrifugation. In both cases the extractions yield the same cost, but the latter one is more effective (i.e., more enzyme can be recovered). The contributions to the total cost in both cases are as follows: the lower phase polymer (Aquaphase, not recovered), 75%; other chemicals and tissue, 13%; labor costs, 7%; and investment and energy costs, 5%. It is worthwhile to observe that the cost for the ligand-PEG is as low as 0.5% (of total costs) for affinity extraction from debris-free protein solution. The low cost is due to the effective recovery of this polymer in the process. When crude homogenate is used, the consumption of ligand-PEG rises to 4% of the total cost because of adsorption of the polymer to the debris.

V. AFFINITY PARTITIONING IN COMBINATION WITH OTHER METHODS

In most cases affinity partitioning must be combined with other methods of purification in order to obtain enzyme preparations of required purity. A few examples are listed below.

A. PURIFICATION OF GLUCOSE-6-PHOSPHATE DEHYDROGENASE

This enzyme has been purified from baker's yeast in three main steps, including fractional precipitation with PEG, affinity partitioning with Procion yellow HE-3G, and batchwise sorption/desorption on DEAE cellulose. The results are shown in Table 7. The purification can also be followed, as shown in Figure 5, from the sodium dodecyl sulfate (SDS) electrophoresis pattern of the original extract and the product obtained from each step.

TABLE 7
Purification Scheme for Isolation of Glucose-6-Phosphate Dehydrogenase from 1 kg of Baker's Yeast by Using Procion Yellow HE-3G as a PEG-Bound Affinity Ligand

Purification step	Volume (ml)	Protein (g)	Specific activity (mkat/kg protein)	Recovery (%)	Purification factor
Homogenate	3540	44.5	2.8	100	1
Fractional precipitation with PEG	201	16.2	5.8	76	2.1
Affinity partitioning					
1. With Procion yellow HE-3G PEG					
Upper phase	645	4.7	18.5	71	6.6
2. Upper phase from 1 + fresh lower phase					
Upper phase	645	3.7	24.5	74	8.8
3. Upper phase from 2 + fresh lower phase containing Cibacron blue F3G-A dextran					
Upper phase	615	2.7	34.1	74	12.3
4. Upper phase from 3 + fresh lower phase + NADP and sulfite					
Lower phase	510	0.75	121	74	43.6
DEAE cellulose treatment, 75—170 mM KCl	500	0.065	915	48	329

Note: System composition: 10% (w/w) dextran 500, 7% (w/w) PEG 8000, and 25 mM sodium phosphate buffer, pH 7.0. Temperature, 3°C. 1 mkat = 60 kU.

From Johansson, G. and Joelsson, M., *Enzyme Microb. Technol.*, 7, 629, 1985. With permission.

B. PURIFICATION OF PHOSPHOFRUCTOKINASE FROM BAKER'S YEAST

Phosphofructokinase was extracted in a similar way from baker's yeast. After a final gel filtration step, pure enzyme was obtained. The rapid removal of 99% of (noninhibitable) proteases by PEG precipitation and affinity partition is of certain interest. This yeast enzyme is very sensitive, but showed very good stability as long as the phase polymers were present.

C. PURIFICATION OF FORMATE DEHYDROGENASE

Formate dehydrogenase has been isolated from *Candida boidinii* by Cordes and Kula[14] using Procion red HE-3G. They treated 10 kg of cells in a 50-l system.

VI. DISCUSSION

The experiments described above show that

1.   Affinity partitioning can be used for the extraction of enzymes from crude mixtures. Removal of debris and/or concentration of the proteins by fractional precipitation (with PEG) may enhance the effectiveness as well as the capacity of the procedure.

2.   Scaling up the extraction is possible by using commercial centrifugal separators for the separation of the phases. Low interfacial tension allows a very rapid equilibration of the partition. Usually 1 to 2 min of careful mixing is enough. In some cases, e.g., when a large amount of debris is present, the phases settle rapidly in normal gravity.

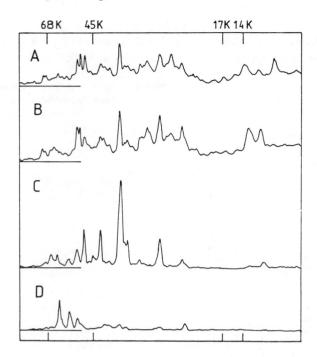

FIGURE 5. Polypeptide pattern (obtained by SDS gel electrophoresis and analyzed by photometric scanning) of the glucose-6-
phosphate dehydrogenase-containing fractions in Table 7. (A) Extract of yeast; (B) after fractionated precipitation with PEG; (C)
after affinity partitioning, system no. 4, lower phase; (D) after
treatment with DEAE cellulose (concentrated 50×). Molecular
weights are indicated. (From Johansson, G. and Joelsson, M.,
Enzyme Microb. Technol., 7, 629, 1985. With permission.)

The procedure is therefore rapid, and extractions can be carried out within a couple
of minutes. This eliminates the need for cold temperatures, which are required for
procedures with long processing times.

3. The recoveries of both ligand-PEG and PEG give good economy to the process. This
is of great importance if more specific (and more costly) affinity ligands are used.

4. In the experiment described above (with the exception of the extraction of formate
dehydrogenase), only a small percentage of the possible extracting capacity has been
utilized. The capacity for protein is at least 150 g/l.[11] A preconcentration of protein
and the use of *all* PEG as ligand carrier would give an excellent volume capacity. It
has been verified experimentally[11,14] that the systems work perfectly at high ligand
concentrations (10 to 30 mM) and that no irreversible binding occurs.

5. The effective and rapid removal of proteases, shown in the case of the preparation of
phosphofructokinase (Table 8), should be of great advantage in many enzyme extraction
procedures.

6. By using a series of PEG-bound ligands in sequence, several enzymes may be extracted
stepwise from the same lower phase. Experimental data in this direction have been
presented by Joelsson and Johansson.[26]

7. The partitioning can be adjusted by variation of a number of parameters. Systems with
suitable partition properties can be formulated by the laboratory-scale tests and then
used directly in full-scale applications.

TABLE 8
Purification of Phosphofructokinase from 1 kg of Baker's Yeast

Purification step	Volume (ml)	Total activity (U)	Total protein (mg)	Specific activity (U/mg)	Purification factor	Total proteolytic activity (U)
Homogenate	1,370	5,400	13,170	0.41	1	3,690[a]
Fractionated precipitation with PEG 8000	120	4,810	1,836	2.62	6.4	670
Affinity partitioning, including prewashing of lower phase, extraction with Cibacron blue F3G-A PEG, and washing with lower phase	120	3,610	153	23.6	58	34
DEAE-cellulose batch treatment	40	2,520	63	40	98	15
Gel filtration	4	1,625	28	58	142	1.7

Note: Composition of two-phase system: 7.5% (w/w) dextran 500, 5% (w/w) PEG 8000, and 25 m*M* sodium phosphate buffer, pH 7.1. The protease inhibitor phenylmethylsulfonyl fluoride was included in all steps.

[a] Proteolytic activity in absence of phenylmethylsulfonyl fluoride was 13,370 U.

From Kopperschläger, G. and Johansson, G., *Anal. Biochem.*, 124, 117, 1982. With permission.

ACKNOWLEDGMENTS

Work in the author's laboratory was supported by the National Swedish Board for Technical Development (STU). Alfa-Laval Separation AB is thanked for supporting the experiments by providing an LAPX 202 separator.

REFERENCES

1. **Kula, M.-R., Kroner, K. H., and Hustedt, H.,** Purification of enzymes by liquid-liquid extraction, in *Advances in Biochemical Engineering,* Vol. 24, Fiechter, A., Ed., Springer-Verlag, Berlin, 1982, 73.
2. **Hustedt, H., Kroner, K. H., and Kula, M.-R.,** Applications of phase partitioning in biotechnology, in *Partitioning in Aqueous Two-Phase Systems: Theory, Methods, Uses, and Applications to Biotechnology,* Walter, H., Brooks, D. E., and Fisher, D., Eds., Academic Press, Orlando, FL, 1985, 529.
3. **Johansson, G., Hartman, A., and Albertsson, P.-Å.,** Partition of proteins in two-phase systems containing charged poly(ethylene glycol), *Eur. J. Biochem.*, 33, 379, 1973.
4. **Shanbhag, V. P. and Axelsson, C.-G.,** Hydrophobic interactions determined by partition in aqueous two-phase systems. Partition of proteins in systems containing fatty-acid esters of poly(ethylene glycol), *Eur. J. Biochem.*, 60, 17, 1975.
5. **Flanagan, S. D. and Barondes, S. H.,** Affinity partitioning — a method for purification of proteins using specific polymer-ligands in aqueous polymer two-phase systems, *J. Biol. Chem.*, 250, 1484, 1975.
6. **Johansson, G.,** Aqueous two-phase systems in protein purification, *J. Biotechnol.*, 3, 11, 1985.
7. **Albertsson, P.-Å.,** *Partition of Cell Particles and Macromolecules,* 3rd ed., John Wiley & Sons, New York, 1986.
8. **Dean, P. D. G. and Watson, D. H.,** Protein purification using immobilised triazine dyes, *J. Chromatogr.*, 165, 301, 1979.
9. **Lowe, C. R. and Pearson, J. C.,** Affinity chromatography on immobilised dyes, *Methods Enzymol.*, 104, 97, 1984.
10. **Johansson, G. and Joelsson, M.,** Preparation of Cibacron blue F3G-A (polyethylene glycol) in large scale for use in affinity partitioning, *Biotechnol. Bioeng.*, 27, 621, 1985.
11. **Johansson, G. and Andersson, M.,** Parameters determining affinity partitioning of yeast enzymes using polymer-bound triazine dye ligands, *J. Chromatogr.*, 303, 39, 1984.

12. **Gemeiner, P., Mislovicova, D., Zemek, J., and Kuniak, L.,** Anthraquinone-triazine derivatives of polysaccharides. Relation between structure and affinity to lactate dehydrogenase, *Collect. Czech. Chem. Commun.*, 46, 419, 1981.

13. **Bückmann, A. F., Morr, M., and Johansson, G.,** Functionalization of poly(ethylene glycol) and mono-methoxypoly(ethylene glycol), *Makromol. Chem.*, 182, 1379, 1981.

14. **Cordes, A. and Kula, M.-R.,** Process design for large-scale purification of formate dehydrogenase from *Candida boidinii* by affinity partition, *J. Chromatogr.*, 376, 375, 1986.

15. **Harris, J. M.,** Laboratory synthesis of polyethylene glycol derivatives, *J. Macromol. Sci.*, C-25, 325, 1985.

16. **Olde, B. and Johansson, G.,** Affinity partitioning and centrifugal counter-current distribution of membrane-bound opiate receptors using naloxone-poly(ethylene glycol), *Neuroscience*, 15, 1247, 1985.

17. **Johansson, G. and Joelsson, M.,** Affinity partitioning of enzymes using dextran-bound Procion yellow HE-3G: influence of dye-ligand density, *J. Chromatogr.*, 393, 195, 1987.

18. **Johansson, G., Kopperschläger, G., and Albertsson, P.-Å.,** Affinity partitioning of phosphofructokinase from baker's yeast using polymer-bound Cibacron blue F3G-A, *Eur. J. Biochem.*, 131, 589, 1983.

19. **Tjerneld, F., Johansson, G., and Joelsson, M.,** Affinity liquid-liquid extraction of lactate dehydrogenase in large scale, *Biotechnol. Bioeng.*, 30, 809, 1987.

20. **Johansson, G.,** Affinity partitioning of enzymes, *Methods Enzymol.*, 104, 356, 1984.

21. **Johansson, G. and Joelsson, M.,** Partial purification of glucose 6-phosphate dehydrogenase from baker's yeast by affinity partitioning using polymer-bound triazine dyes, *Enzyme Microb. Technol.*, 7, 629, 1985.

22. **Craig, L. C.,** Countercurrent distribution, in *Comprehensive Biochemistry*, Vol. 4, Florkin, M. and Stotz, E. H., Eds., Elsevier, Amsterdam, 1962, 1.

23. **Johansson, G. and Andersson, M.,** Liquid-liquid extraction of glycolytic enzymes from baker's yeast using triazine dye ligands, *J. Chromatogr.*, 291, 175, 1984.

24. **Johansson, G., Andersson, M., and Åkerlund, H.-E.,** Counter-current distribution of yeast enzymes with polymer-bound triazine dye affinity ligands, *J. Chromatogr.*, 298, 483, 1984.

25. **Kopperschläger, G. and Johansson, G.,** Affinity partitioning with polymer-bound Cibacron blue F3G-A for rapid large-scale purification of phosphofructokinase from baker's yeast, *Anal. Biochem.*, 124, 117, 1982.

26. **Joelsson, M. and Johansson, G.,** Sequential liquid-liquid extraction of some enzymes from porcine muscle using polymer-bound triazine dyes, *Enzyme Microb. Technol.*, 9, 233, 1987.

Chapter 24

A COMPUTERIZED STRATEGY FOR ESTIMATION OF PROTEIN PURITY

D. W. Sammons, R. C. Humphreys, N. H. Lin, K. H. Kwan, W. J. Ko, N. Egen, F. S. Markland, and K. Chan

TABLE OF CONTENTS

I. INTRODUCTION

Evaluation of protein components within a sample derived from genetic engineering and fermentation processes, downstream processing, solid-phase peptide synthesis, and isolation of protein from tissues requires accurate quantitation of all components. Several methods have been used; however, they lack sensitivity and adequate resolution, and they yield relative concentrations of the detected components instead of absolute concentrations. In this chapter, we present a strategy which overcomes the difficulties of the presently used methods for determining the purity of protein samples.

The approach presented is exemplified using an enzyme isolated from a snake venom. The enzyme was purified by two independent separation methods: recycling isoelectric focusing and chromatography. Similar levels of purity were achieved, as demonstrated with two-dimensional (2-D) gel electrophoresis.[1] The purified enzyme isolated from the chromatographic methods was selected for demonstrating the strategy.

II. METHODS AND MATERIALS

A. MATERIALS

Snake venom of *Agkistrodon contortrix contortrix* was purchased from Biotoxins, Inc., St. Cloud, FL. Carrier ampholytes for the preparative gel electrophoresis were purchased from LKB Products, Pleasant Hill, CA. Acrylamide, bis-acrylamide, ammonium persulfate, and urea were purchased from Bio-Rad, Richmond, CA. Ampholytes for the 2-D electrophoresis, TEMED, and NP-40 were purchased from Sigma Chemical Co., St. Louis, MO. Sephadex® G-100, CM® cellulose, *p*-aminobenzamidine agarose, and the carbonic anhydrase pI markers came from Pharmacia, Piscataway, NJ. Glass pipettes (0.2 ml) were purchased from American Scientific Products, Phoenix, AZ. BCA protein reagent was from Pierce Chemical Company, Rockford, IL. All other chemicals were reagent grade.

B. PURIFICATION

The enzyme was isolated from the crude snake venom using preparative-scale isoelectric focusing in the recycling isoelectric focusing (RIEF) apparatus[1] and three different, but sequential, chromatographic methods.[2] With the RIEF method, a few high molecular weight contaminants were present with isoelectric points close to that of fibrolase. These were subsequently removed by gel filtration on Sephadex® G-100. In the chromatographic method, a three-step scheme was followed utilizing CM® cellulose cation exchange, gel filtration on Sephadex® G-100, and then affinity on *p*-aminobenzamidine agarose to remove serine proteinase contaminants. The RIEF method and subsequent chromatography were performed within 1 d, whereas the chromatographic steps required 2 weeks to process 1 g of venom. The first RIEF was performed in a pH 4 to 6 gradient; the second RIEF was performed in a 0.7 pH range gradient.

C. TWO-DIMENSIONAL GEL ELECTROPHORESIS

Gel electrophoresis was performed as generally described by Anderson and Anderson.[3,4] The equipment is of the design distributed by Health Products, Inc., Rockford, IL. The 2-D procedures followed a modified version of the user's manual for 2-D gel electrophoresis furnished by Health Products, Inc. Briefly, the samples were mixed with the standard Iso sample buffer and electrophoresis was performed at 700 V for 17 h. The Iso gels were then equilibrated for 15 min and applied to 10 to 20% gradient acrylamide slab gels. The sodium dodecyl sulfate (SDS) gel samples were dissolved in the standard SDS buffer, mixed 1:1 with a melted agarose solution, and pipetted into an inverted 0.2-ml glass pipette to solidify. A 2-mm portion was sliced from the solidified gel and applied to the top of the SDS in a manner similar to that used for tube Iso gels. Electrophoresis was performed at 250 V for 5 h. The staining was done by the color silver stain GELCODE® method.[5]

D. DATA ACQUISITION

The scanning of the 2-D gels was performed with the BioImage Corporation Visage® system using the single channel without the color filter wheel. The spot location and segmentation were automatically performed as directed by the BioImage EQ software. The spot list was generated with the global program, thereby giving the maximum accuracy possible for the integrated intensity.

E. DATA TRANSFORMATION

The EQ-generated spot lists are filtered into a program called GETPIMW™ (copyright, University of Arizona) and the x,y coordinates are converted to pI and molecular weight. The files of GETPIMW™ are loaded into a SPOTMATCH™ program (copyright, University of Arizona), and all spots of the gels in the study are matched. The matched spots are listed for the replicate gels of each dilution, and the integrated intensities are correspondingly attached in the data base program. The data base analysis is performed on a VAX® computer under the data base management system "Accent R".

F. DATA ANALYSIS

The standard reference protein is glyceraldehyde-3-phosphate dehydrogenase (G3PDH). It is a brown protein and stains the same color as the venom enzyme and the unknown contaminants in the 2-D gel after silver staining. The protein is serially diluted, with a range of protein concentration from 160 to 10 ng per spot. The integrated intensities are likewise determined in the EQ global software and then listed. The integrated intensity is plotted vs. the protein concentration and the least squares line drawn through the data collected from three separate gels. The line is linear and is used to derive the concentration of the protein of each spot located in the gels of each dilution. Percentage purity is calculated after normalizing the extrapolated concentration with the appropriate dilution factor.

G. PROTOCOL

The following instructions were followed:

1. Perform replicate 2-D gel electrophoresis and stain with GELCODE® silver stain on any fraction of a purification scheme; dilutions (9) of the fractions of interest should span at least a 500-fold concentration range.
2. Include a reference standard which stains the same color with GELCODE® silver stain as the proteins of interest and spans the linear range of the unknown proteins in the gels.
3. Perform data acquisition by scanning the gels on the Visage system, and create spot lists containing x,y coordinates and integrated intensities of all spots.
4. Perform data transformation via algorithms to determine relative isoelectric point, apparent molecular weight, and a listing of all known spots; create a composite map of all unknown spots from the range of dilutions.
5. Perform data analysis by utilizing the color standard reference protein to quantitate the unknown protein of each spot.
6. Calculate the percent purity of each of the proteins present in the final fraction and express the values on the composite map.

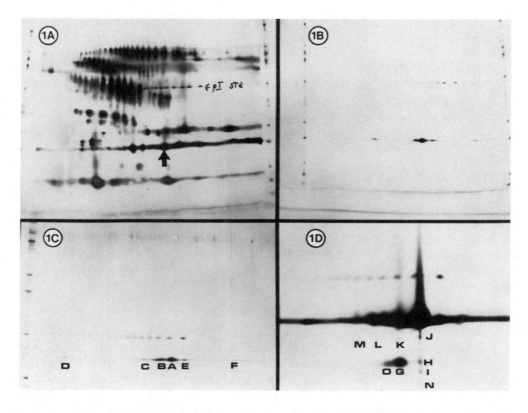

FIGURE 1. (A) GELCODE®-stained 2-D gels of whole venom; (B) RIEF purified and G-100 treated; (C) chromatographically purified; (D) 50-fold more concentrated than C. The same activity was added to B and C. The letters represent all spots located over a 500-fold concentration range.

III. RESULTS

A. PURIFICATION AND SEPARATION BY TWO-DIMENSIONAL GEL ELECTROPHORESIS

The enzyme was isolated from the venom of the snake *Agkistrodon contortrix contortrix* using preparative-scale isoelectric focusing in the RIEF apparatus[1] and three sequential, but different, chromatographic methods.[2] In both purification approaches described above, the protein was isolated to near homogeneity.

Figure 1A, a 2-D gel protein pattern, illustrates the complexity of the starting material. Figure 1B is the purified protein after two sequential RIEF fractionations followed by size exclusion chromatography. For comparison, Figure 1C is the same protein after extensive chromatography. The qualitative 2-D gels show comparable degrees of purity between the electrophoretically and chromatographically isolated protein samples. Without independent confirmation, it is impossible to document which of the three major bands is indeed the venom enzyme under study. It is assumed in this paper that the most intense band is the enzyme and the other spots are contaminants. Clearly, it is just as plausible that the other spots are posttranslational forms with similar enzymatic activity. The spots in Figures 1C and D are lettered to identify them throughout the analysis. The sample in Figure 1D is the same as 1C, except it has a 50-fold higher protein concentration. With the higher concentration applied to the 2-D gel in Figure 1D, additional contaminants are revealed.

TABLE 1
Computerized Strategy: Algorithms for Three Phases of Data Processing

Data acquisition (BioImage)	Data transformation (University of Arizona)	Data base analysis (University of Arizona)
EQ list	GETPIMW™	System R-SPOTPLOT™
x,y Coordinates	SPOTMATCH™	Normalized protein concentration
Integrated intensities	GETPURITY	Percentage of each component
	Relative isoelectric point	
	Apparent molecular weight	
	Protein concentration	

Note: The spot lists created by data acquisition programs are utilized for transformation and analysis. Input data result from scans of four different dilutions — $1/2$, $1/64$, $1/512$, and $1/1024$ — of the purified sample and five replicate 2-D gels from each dilution. Only integrated intensity values of spots identified as matched spots in the SPOTMATCH program are included in the GETPURITY algorithm.

B. COMPUTERIZED STRATEGY AND ANALYSIS OF DATA FOR PURITY

Table 1 lists the major steps required for the computerized strategy. This strategy can be applied to a purified sample or to any step during purification. The 2-D gels are central to the strategy, since the high resolution is paramount to separating each component into a single spot in order to utilize the full power of the computerized algorithms. Dilution of the sample across a 500-fold range allows the intensities of major and trace components to fall in the linear range of the standard reference curve.

Selection of a reference protein of the same color as the unknown component is an important step for transformation of integrated intensity data into nanograms per spot. This ensures accurate quantitation (nanograms per spot) from integrated intensities.[6]

The next step is data acquisition with the BioImage Visage system. The image is captured with a solid-state photodiode scanning camera, and the EQ software is utilized for the creation of a spot list which contains x,y coordinates and the integrated intensity of each spot.

True automation is approached with data transformation. Conversion of the x,y coordinates is achieved by the GETPIMW™ algorithm, which utilizes the internal pI and molecular weight standards within the 2-D gel pattern. Execution of this algorithm normalizes gel patterns, and by running the gel SPOTMATCH™ algorithm, each spot in the transformed spot list from all gels is matched to the identical standard spot in a reference gel. From the EQ-generated spot list stored in the RVAX data base, matched spots are used to extract the corresponding integrated intensities of the matched spots. The GETPURITY™ algorithm is used to generate the quantity of protein per spot (Table 2). The transformed data are expressed by the SPOTPLOT™ algorithm, and the nanograms per spot of each component are presented in Table 2. The spots are assigned component classes relating to the principal component of interest, the isolated snake venom protein. Spots having similar molecular weights, but different pIs, are designated ISO-SIZE (for molecular weight isoforms). Those having similar isoelectric points, but different molecular weights, are designated ISO-PI (for pI isoforms). Spots which have pI and molecular weight dissimilar to the principal component are termed PI-SIZE. ISO-PI spots J, H, I, and O of the composite map of Figure 2 are probably degradation artifacts.

IV. DISCUSSION

Analysis of trace components in purified protein samples has always been of utmost importance to the researcher studying protein structure and function. The evaluation of all components within a protein sample has recently assumed more stringent requirements, since

TABLE 2
Data Transformation and Analysis

Spot	pI	Molecular weight (kDa)	Dilution factor	Integrated intensity	ng/Spot (diluted)	ng/Spot (normalized)
A	−0.64	25.77	512	1.73	66.77	34,418
B	−1.57	26.14	256	0.36	13.51	5,041
C	−3.65	25.91	32	1.06	43.75	1,400
D	—[a]	—	32	0.86	36.87	1,180
E	—[a]	—	32	1.36	54.01	1,728
F	—[a]	—	32	0.54	25.88	828
G	−1.60	19.40	32	0.23	15.22	487
H	−0.79	18.87	1	1.27	50.96	51
I	−0.73	17.23	1	0.77	33.78	34
J	−0.77	23.03	1	0.57	27.01	27
K	−1.87	22.12	1	1.43	56.46	56
L	−3.41	22.47	1	0.55	26.22	26
M	−4.86	22.41	1	0.37	20.03	20
N	—	—[b]	1	0.38	20.37	20
O[c]	−2.65	18.70	2	1.50	58.87	118

Note: The purified protein sample was diluted serially from 1 to 1/512. The pI and molecular weight of each spot are listed; they represent the spots which were matched within a dilution gel set. The integrated intensities were obtained from the spot lists and were transformed to ng/spot by extrapolating from a dilution series of the reference protein, G3PDH. Initial protein concentration of the reference was determined by BCA assay. A linear regression analysis was performed on the concentration and integrated intensity values of the dilution series on three replicate gels. The correlation coefficient was 0.99. The normalized concentration is derived from the following expression: ng/spot × dilution factor. The total protein concentration in all spots of the normalized gel is 45,199 ng.

[a] Falls outside pI range.
[b] Falls outside MW range.
[c] Integrated intensity fell outside linear range and was divided by 2 to use the reference line.

many of the genetically engineered and natural compounds are injected into human subjects and small amounts of potent inhibitors may be present. These events place an even greater importance on the evaluation of purity.

As a result of interest in protein manufacturing by the pharmaceutical and food industries, there is an increasing interest in the scaleup of purification methods. Many possible approaches exist to the purification of a given protein. The selection of scaleup equipment and procedures in bioprocessing is often based on analytical-scale experiments. During the optimization phase of a purification procedure, the fractionated samples must be monitored by measuring the relative concentrations of components in each fraction. Even the best purification scheme has contaminating proteins, and optimization of the purification is dependent upon estimation of purity at each step, especially in the final fraction. Presently, methods for estimation of protein purity are inadequate, as they lack the capability of detecting trace amounts of contaminants.

One method for estimating protein purity is to determine the increase in specific activity. By comparing the specific activities at various steps and among different methods, it is possible to evaluate the various purification procedures. The success of this method depends on protein estimation and activity measurements, both of which are affected by the analytical method chosen and the presence of interfering contaminants.

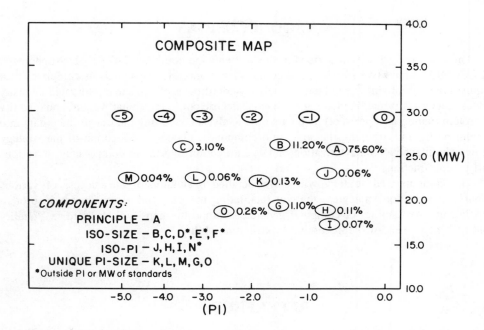

FIGURE 2.   The composite map of all components. Spots 0, −1, −2, −3, −4, and −5 represent the internal isoelectric markers (carbonic anhydrase). The spots with letters represent the unknown spots which were located by pI and molecular weight. Principle component, venom enzyme A; ISO-SIZE B, C, D, E, F; ISO-PI H, I, J, N; unique PI-SIZE G, K, L, M, O. The percentage of the total is indicated to the right of each spot. Venom enzyme A represents 75.6%, ISO-SIZE represents 22.50%, ISO-PI represents 0.27%, and PI-SIZE represents 1.59%. The % purity was calculated from data in Table 1 with the following expression:

$$\frac{\text{Normalized spot concentration}}{\text{Total normalized spot concentration}} \times 100$$

Another standard method utilizes one-dimensional gel electrophoresis by comparing the contaminant bands to the band of interest. This approach only discriminates on the basis of molecular weight or isoelectric point and is unable to detect all contaminants. In addition, it is not possible to estimate the absolute concentration of the contaminants. Frequently, the estimated purity is higher than the actual value due to different stoichiometric relationships of dye or stain binding between the standard and the unknown protein. Often the concentration of contaminants is below the level of detection of the particular method.

In this chapter we have demonstrated accurate quantitation of unknown proteins using a standard protein having the same color (after GELCODE® staining) as the unknown proteins. In the present case, the venom enzyme stained brown, as did glyceraldehyde-3-phosphate dehydrogenase, the standard reference protein. Utilizing gel-matching algorithms, the protein of interest and the contaminating proteins were located. The matched spots were mapped according to relative isoelectric point and apparent molecular weight, and their integrated intensities were transformed to nanograms of protein. Using the standard color protein reference, the protein quantity vs. integrated intensity line was constructed, and from these measurements the purity was accurately determined.

The increased appearance with time of the minor closely related spots has been observed, and the present strategy will be applied to elucidate the nature of their formation. Whether the contaminating spots have similar enzymatic activity will be the subject of a future study.

V. CONCLUSIONS

The analytical strategy described in this presentation combines 2-D gel electrophoresis, GELCODE® color silver staining, and computerized analysis. This analysis comprises data acquisition by digitizing the gel pattern into x,y coordinates of spots and integrated intensity values; data transformation yielding relative isoelectric point, molecular weight, and quantity of protein (nanograms per spot); and data base analysis converting the transformed data into percent purity and grouping the spots into component classes. Application of the strategy allows simultaneous characterization and quantification of sample components in various purification fractions.

The computerized strategy will allow detailed evaluation of the success of various possible purification strategies, thereby facilitating the trial and error process in protein purification. Availability of computer-generated composite maps will standardize "quality control" monitoring of bioprocessing lot-to-lot variability.

REFERENCES

1. **Egen, N. B., Russell, F. E., Sammons, D. W., Humphreys, R. C., Guan, A. L., and Markland, F. S.,** Isolation by preparative isoelectric focusing of a direct fibrinolytic enzyme from the venom of *Agkistrodon contortrix contortrix, Toxicon,* 25(11), 1189, 1987.
2. **Markland, F. S., Reddy, K. N. N., and Guan, A. L.,** Purification and characterization of a direct-acting fibrinolytic enzyme from southern cooperhead venom, in *Animal Venoms and Hemostasis,* Pirkle, H. and Markland, F. S., Eds., Marcel Dekker, New York, 1988, 173.
3. **Anderson, N. G. and Anderson, N. L.,** Analytical techniques for cell fractionation. XXI. Two-dimensional analysis of serum and tissue proteins; multiple IEF, *Anal. Biochem.,* 85, 331, 1978.
4. **Anderson, N. L. and Anderson, N. G.,** Analytical techniques for cell fractionation. XXII. Two-dimensional analysis of serum and tissue proteins; multiple gradient, *Anal. Biochem.,* 85, 341, 1978.
5. **Sammons, D. W., Adams, L. D., and Nishizawa, E. E.,** Ultrasensitive silverbased color staining of polypeptides in polyacrylamide gels, *Electrophoresis,* 2, 141, 1981.
6. **Sammons, D. W., Vidmar, T. J., Adams, L. D., Jones, D. H., Hatfield, C. A., Chuba, P. J., and Crooks, S. W.,** Applicability of color silver stain GELCODE® system to protein mapping with 2-D gel electrophoresis, in *Methods and Applications of Two-Dimensional Gel Electrophoresis of Proteins,* Celis, J. E. and Brave, R., Eds., Academic Press, New York, 1984, 111.

Part 5. Bioseparations II — Chromatographic Methods

Chromatography is still the "workhorse" of downstream processing, and the field continues to grow. While various chromatographic techniques are extensively applied to high-resolution analyses, more progress is critically needed in process-scale applications of chromatography for purifying high-value proteins. In this section, representatives from industry and academia discuss the recent and promising developments in the key areas of adsorption, ion exchange, and affinity chromatography. Also, the chapter by Hedman et al. deals with the need to understand the processing and scaleup issues through modeling and optimization.

Chapter 25

PURIFICATION OF PROTEINS OF LOW VALUE IN HIGH VOLUME

John G. Watt

TABLE OF CONTENTS

I. INTRODUCTION

Setting aside the production of bread, cheese, or beer, biotechnology is, or seems to be, preoccupied with the production of proteins and similar molecular species of very high value combined with a relatively low production volume. In many cases, this means that the development through entry to the market for a particular product is slower than the time required for dissemination of knowledge regarding that same material. For the fortunate few this means that there may be high rewards to be gained, but on balance there may well be massive losses. This is not to say that it is not important that the high-value materials be produced, but simply that this is almost certainly not where the real future of biotechnology lies. Somewhat like the chemical industry around 1935, biotechnology has to scale up to survive.

A. BACKGROUND

It is almost exactly 150 years since Mulder[1] first suggested the title "protein" for those substances characterized by possession of what we now recognize as a peptide linkage. Interest in separation and purification of such substances from biological precursors preceded this event. It was, however, nearly 50 years before interest was established in the large-scale purification of proteins in significant quantities, i.e., when techniques such as salting in and salting out came to be used in purification of antisera or for recovery of killed or attenuated bacteria for the early vaccines developed through the successes of Pasteur and others.

Actually, these developments arrived a little late in the U.K. for a number of interesting reasons, which include the existence of the Cruelty to Animals Act of 1876 and a peculiar sort of chauvinism which has persisted to the present. Such chauvinism is not peculiar to the U.K.; there are plenty of countries and institutions for which "Not Made Here" is considered a good reason for not accepting a particular technical system.

My own entry into the field of protein separation and purification, 28 years ago, was related to the separation of mammalian plasma proteins, and it is from this precedent that I shall try to develop some examples of what I believe must happen in the next few years within those diverse manufacturing activities grouped under the general definition of bio-technology. Plasma fractionation as a systematic activity has a century of experience behind it and possesses most of the problems (with only some of the solutions) now facing the industrialization of modern biologics production.

II. DEVELOPMENT OF PLASMA PROCESSING

By the end of the 19th Century, the production of relatively large volumes of semipurified antisera was quite commonplace, and the techniques of precipitation using salts such as ammonium sulfate had been established as almost the standard means of achieving the levels of purity considered essential. However, other separation systems were being developed and used, e.g., ethyl ether precipitation by Sclavo or the beginnings of the complex ethyl alcohol system by Mellanby[2] at Oxford. These tended to remain unique to only one or two laboratories because already there was a tendency for individual processors to remain faithful to the systems they understood, even if these were not the best suited to the purpose.

A. ALTERNATIVE SYSTEMS

For a time, under the influence of the Swedish school originating from the work of Svedberg and continuing through Tiselius to Porath, it seemed that a more rational approach to fitting the separation system to the purpose might have appeared, but this was blocked by the formal U.S. adoption of the methods elucidated and adopted by Cohn and his co-workers[3,4,5] for the systematic recovery of moderately pure proteins from human plasma. The appearance of this system provoked a watershed in the processing of plasma and, to some degree, in all protein recovery systems. Interestingly, little credit has been given to Mellanby, who instigated this form of separation technology, and his contribution has tended to be forgotten.

It would be quite wrong to underestimate the contribution of the Boston group headed by Cohn, for this was a major development in many ways, apart from the detailed chemical systems which were defined. The most vital elements of the study were the adoption and creation of many of the tools essential to protein separation science. We forget today, with digital pH and temperature control easily available and inexpensive, that the forerunners of these systems were developed for separation of plasma proteins and that the parameters laid down for these activities were dictated in major part by the limitations of these new measuring systems.

B. CUSTOM EQUIPMENT AND SYSTEM DEVELOPMENT

While the principles of the tube centrifuge were well understood, the machines used in the Cambridge laboratory were modified from systems used for dewatering engine sump oil in ships. Rotating weirs and other devices for bringing coolant inside the walls of rotating centrifuge bowls are quite new inventions.[6] The original system called for coolant passing inside the bowl frame in a manner which meant that the cooling capacity was involved only incidentally in removal of heat from the bowl contents by removing the radiant frictional heat from the bowl surface and in cooling the huge volumes of air which passed through the bowl frame.

The fact that very similar machines remain in use today is a credit to the original scheme and the tenacity of the users. Certainly, tube centrifuges are simple to use and easily maintained, but essentially they are uncontrollable in terms of temperature regulation and demand a uniformity in use which precludes simple modification or adjustment of process parameters. Thus, they contribute to the tendency to eschew development and change, which has bedeviled the plasma processing industry for most of the past 40 years. I believe the same sort of phenomenon is beginning to be apparent in biotechnical activities, and it may be that the regulatory agencies hold a major responsibility for this. Perhaps the success of the institutionalization of the Cohn technology by the Bureau of Biologics, now part of the U.S. Food and Drug Administration (FDA), has been a precedent for this attitude. I see symptoms of the same attitude afflicting process development in biotechnology, aided perhaps by the slightly stereotyped approaches of regulatory agencies in the health care fields.

I may be accused of indulging in the fashionable tendency of blaming the conservatism of regulatory affairs bureaus for inhibition of development effort, but the speed of response to the commercial pressures created by the identification of the human immunodeficiency virus (HIV) and the consequent exaggeration of epidemic risks of the acquired immunodeficiency syndrome (AIDS) indicate that the internal lethargy of institutionalized research and development groups may be the more critical factor. However, there is some justice in the accusation, as demonstrated by the almost precipitate haste with which biosynthetic sources of human growth hormone have replaced the natural product which, implicated in Creutzfeldt-Jakob disease, might have been seen as a failure of the regulatory agencies themselves. Adoption of such rapid and careful action, if maintained, could do much to encourage and limit the cost of many modern developments.

With the arrival of chromatography[7] and electrophoresis,[8,9] it seemed certain that the older precipitation systems would be replaced by the more elegant and somewhat less damaging new technologies, but this has not happened in nearly the expected manner. Several large-scale separation systems based on chromatography have been established, but still appear to lack the knowledge and control necessary for these to become safe and reliable tools. Control of Joule heat and convection continues to prevent the adoption of electrophoresis as a dynamic preparative tool.

Discrete pore-size filtration systems, soon followed by ultrafiltration, diafiltration, and reverse osmosis, also have been slow to reach anything close to their full potential, although some of these are of increasing importance for large-scale systems. It is slightly ironic that it required public reaction to the rare, but real, dangers of asbestos to force the introduction of such technologies and that the social desire for alcohol-free motoring while continuing to drink beer is the major stimulus for adoption of the more powerful derivative systems.

C. CONTAMINATION

In the processing of therapeutic materials from human plasma, there has been a dependence on solvent precipitation systems because of their known potential for making some materials safe from viral transmission. This apparently blind conservatism has been encouraged with justification in the last 4 years, since attempts to make immunoglobulin fractions using chromatographic methods have produced materials which have been unexpectedly implicated in transmission accidents.[10,11] Whether or not this is due to an inherent danger of the method or to failure to establish adequate standard procedures of good manufacturing practice has not been established clearly. So, there remains some reasonable doubt on the safety of this type of technology for recovery of purified materials from natural mammalian sources. It is my personal view that the materials available for chromatography have been designed too slavishly toward the needs of the academic laboratory and are, in large part, less than adequately designed for the more robust and rigid environment of the factory floor. Newer materials capable of withstanding more strenuous lot separation and sterilization techniques have been slow to be adopted, possibly because of the first-class public relations efforts of the established opposition; fortunately, this position is changing.

Attempts at the harnessing of the principles of electrophoresis for large-scale separation have been much less successful than this powerful technology promises. The scale limitations imposed by the countereffects of convection have not been solved adequately, in my opinion, although the system developed with typical elegance by Thompson and colleagues of the U.K. Atomic Energy Authority at Harwell[12] from the concepts of Philpott comes close. Although realistic scale performance has not been achieved successfully, the system has considerable potential for intermediate-scale work.

D. PROCESS CONTROL

Process control technology has been applied to protein separation systems, but remains less well exploited than it should be. The first semicontinuous system was commissioned in 1974,[13,14] and eight further systems have appeared since then. However, none have developed to the extent of becoming fully continuous in the way which will be required if the fullest advantage is to be gained from fermentation and cell culture systems as feedstock sources of purified protein on a sensibly large scale.

The use of discrete microprocessor control systems, each managing a separate aspect of the full system under control from a central processor, is able to control and log data from a wide range of different process elements. Such technology is already near standard in many industries. My former colleagues in the Protein Fractionation Centre, Edinburgh have carried through a redevelopment of the original Scottish system which goes some way toward adoption of similar standards, but it still falls short of the ideal. My own attempts at further development remain untried as a fully operational system. Though early trials are encouraging, it is still too early to be certain that the full system will integrate in the expected manner. Nevertheless, it is clear that a fully automated system operating under totally aseptic conditions is achievable.[15] There should be no great pride in this when it is considered that more than 40 years have been required to reach such an elementary stage of biologics processing.

E. ENVIRONMENTAL LIMITATIONS

Perhaps one of the greatest difficulties of the large-scale preparation of protein fractions has been the need to use cold rooms as manufacturing areas. It is not easy to operate a cost-effective and acceptably safe system in such an area, although several good techniques have been developed for making such areas operate in accordance with the precepts of good manufacturing practice (GMP). Nevertheless, there has been constant pressure to devise alternative systems using jacketed kettles and tanks,[16] especially for staff comfort. These are usually cumbersome and complex structures which can be difficult to maintain in an adequate state of cleanliness. One of the advantages of continual processing has been that the need for cold areas, or for staff to spend long periods in such areas, has been reduced, and the latest attempts are to reduce the need for cooling as a part of the whole process.[15] This is achieved by using the ionic strength of the reactions in a more interactive manner than has been possible in the past, rediscovering the findings of Mellanby and Cohn with application of modern control techniques.

Many of the earlier systems were developed to operate fast, to reduce bacterial growth in cold conditions, or, as in the case of the organic solvent systems, to limit chemical damage to the desired proteins. Today, with better means for real-time control, it is possible to operate even the organic solvent systems at higher temperatures; they still require precise control but do not need to be so demanding in terms of the actual temperature of the reaction itself. Furthermore, with the advent of in-line and transverse-flow filtration systems, there can be little excuse for operating open or quasiopen systems. The regulatory ideal of a closed system accepting only sterile feedstock and discharging under closely controlled conditions is not only possible, but totally desirable, even for animal blood derived from an abattoir environment.[17]

There have been several designs for a comprehensive scheme for moderate- to large-scale purification using chromatographic methods, and these, despite their present setbacks, have to be regarded as the ideal methodology for the future. However, those plants now operating with reasonable success demand high-quality clean room environments and require skilled and dedicated attention to such matters as the microbiological quality of water and buffer feedstocks.[18] Several installations have failed to meet these demands in a convincing manner, which suggests that the requirements for GMP for such installations are still less than perfectly understood. In at least three plants, personal experience indicates that the creation and classic operation of a clean room environment is less than adequate to ensure a safe working environment; few units have so far elected to adopt real-time control of the environment, possibly because the means for measuring pressure differentials, particulates, air movements, humidities, and temperatures are either too cumbersome or too precise for the tasks involved.

III. IMPACT OF NEW DEMANDS

As attention passes more and more from natural source materials to the products of fermentors or culture vessels, these possible advantages become increasingly more important. There is a laudable and important tendency among regulatory affairs bureaus to regard the fermentation or culture stage as being separate from the strictly pharmaceutical processing stages of the overall product recovery. This allows the development of quite new design approaches to the purification process line. That such compartmentalization is seen as necessary stems in part from the recognition that use of fermentation products as source materials for pharmaceutical products is not new, but in fact predates the development of penicillin, by several thousand years in some cases. The culture of staphylococci for recovery of protein A is not particularly different from the recovery of vitamin concentrates from spent brewers yeast and should, by the very nature of the activity, be cleaner overall than the extraction of insulin from pancreatic tissue collected in an abattoir, where the adjective "clean" has an entirely different meaning.

A. TECHNICAL DESIGN

Many of the process lines which have been developed to effect recovery of the desired moieties from bacterial cultures are thinly disguised versions of laboratory processes, often with little attempt at process optimization after the culture stage. This may be due in part to the tendency toward concentration of products which have astonishingly small volumes and very high values. However, multiplication of supply sources and other factors tends to depress the sale value of many of these materials very quickly. Thus, we see the phenomenon in which years of development, almost entirely related to genetic engineering aspects of the new source, are followed by a short period in which the cost of this development has to be recouped before the value of the product falls to "ordinary" levels.

This difficulty is made no easier for the multitude of small, dedicated, and specialized organizations which have rapidly appeared in the field. A large number of these ventures are created around a single product or family of products, and they find difficulty in co-ordinating the marketing of their initial product with development of the next new material on which the organization's survival is based. Clever marketing or the use of licensing agreements may assist the spreading of the time frame, but will still leave such organizations in a worried and precarious state.

It appears inevitable that the real way forward for biotechnical development will lie in the creation of larger-volume and lower-value products which can accommodate several producers in the marketplace. Some of these will engage in production of new materials, but many will turn their attention to newer means of obtaining existing products. There is a certain qualified truth to the adage, "If it can be made in a fermentor, it's bound to be cheaper." However, there are a host of modifying factors to be taken into account, not the least being the scale at which the fermentor and its associated downstream technology will be expected to work. Not all of this will be protein oriented, since a huge market (estimated in billions of dollars) is developing for enzyme products from fermentation sources. Already we see that γ-linolinic acid from algal culture is more cost effective than collecting the seeds of the evening primrose, even if the latter may be a more attractive occupation.

B. NEW INSTRUMENTATION

The impact of new demands seems hampered by lack of access to good process control instrumentation capable of maintaining control parameters within very narrow limits and with short hysteresis factors at costs which can be borne by an industry in the making. A vast flexibility is required to meet the need for rapid change if some of the existing enterprises are to meet their own built-in demand for alteration in their product lines without a total rebuilding of the process lines. One interesting facet, difficult to understand, is the frequency with which one discovers microprocessor equipment sold in the U.S. and in the U.K. at exactly the same quoted price, without reference to the value of the currency involved. Such cynicism on the part of the instrumentation industry eventually becomes self-defeating since it encourages retention of older, less flexible technology.

Within the machine-tool engineering industry there exist many workshops which are equipped with precision machines for welding, cutting, grinding, and milling and which can be adapted to make almost anything made of metal. It seems inevitable that large-scale protein purification, perhaps including polysaccharide purification on the same process line, needs to become equally adept and adaptable. The new, biology-based industrial revolution might learn much from reexamination of the lessons of the first industrial upheaval rather than from the growth of electronics, where a new process always seems to need a new factory, usually in a new place and with new staff. The consequent social upheavals are sad and unnecessary.

C. QUALITY ASSURANCE

It may seem superfluous to mention the need for a level of quality assurance (QA) at least commensurate with the quality of the science it supports. While it is easy to reconcile the cost and need for good QA when considering the production of invasive materials such as pharmaceuticals or cosmetics, it may be less apparent when considering production of a base material for foodstuff or plantlets for crop production. However, adoption of new and strange means for obtaining the raw materials from which these are to be derived requires exercise of responsibility and public awareness at levels not usually associated with either the food or agricultural industry. The inclusion of quality and knowledge in the products would go far to allay suspicion and will usually prove to be self-financing due to the effect of such discipline on the ability to meet sales deadlines and product predictability.

IV. CONCLUSIONS

While it might be argued that comments on the organization of scientific endeavor have no place in a gathering of scientists, the lesson remains cogent. If there is an intent to produce a useful end product using any particular skill or knowledge, it is essential that the process of manufacture and recovery of the product be planned in such a manner as to realize useful (salable) material as a result.

Partly because of the attractions of small-scale production and the apparent applicability of laboratory techniques in downstream processing, there has been a rush of biotechnical talent into the "manufacturing business", and this has been accompanied by partial success which is more apparent than real. Organizations with a history of pharmaceutical production, especially of biologics, have enjoyed better success than most of those which have entered *de novo* because they understand that the necessary scientific knowledge is a fairly minor component in the problem of successfully bringing a new product to the market. Knowledge of plant design and of the more tedious elements of production housekeeping are of greater overall importance.

The long-term viability of biotechnology lies most certainly in the achievement of success in harnessing the ability to make simple and inexpensive materials in economically large volumes and in the adoption of flexible system designs to meet the challenge of changing markets. There is little evidence to suggest this is happening and considerable evidence that product designers, producers, and regulators are intent on creating the same mistakes as in earlier developments in the field of biologics production and purification.

REFERENCES

1. **Mulder,** in correspondence with Berzelius, 1838.
2. **Mellanby, J.,** Diphtheria antitoxin, *Proc. R. Soc. London Ser. B,* 80, 399, 1908.
3. **Cohn, E. J., Oncley, J. L., Strong, L. E., Hughes, W. L., and Armstrong, S. H.,** Chemical clinical and immunological studies on the products of human plasma fractionation. I. The characterisation of protein fractions of human plasma, *J. Clin. Invest.,* 23, 417, 1944.
4. **Cohn, E. J., Strong, L. E., Hughes, W. L., Mulford, D. J., Ashworth, J. N., Melin, M., and Taylor, H. L.,** Preparation and properties of serum and plasma proteins. IV. A system for the separation into fractions of the protein and lipoprotein components of biological tissues and fluids, *J. Am. Chem. Soc.,* 68, 459, 1946.
5. **Cohn, E. J.,** The separation of blood into fractions of therapeutic value, *Ann. Int. Med.,* 26, 341, 1947.
6. **Hemfort, H.,** Solid-liquid separation. The use of centrifuges in large scale human plasma fractionation, in Proc. Int. Workshop on Technology for Protein Separation and Improvement of Blood Plasma Fractionation, Sandberg, H. E., Ed., National Institutes of Health, Bethesda, MD, 1977, 81.
7. **Martin, A. J. P.,** The principles of chromatography, *Endeavour,* 11, 5, 1952.

8. **Tiselius, A.,** The Moving Boundary Method of Studying the Electrophoresis of Proteins, D.Sc. thesis, University of Uppsala, Sweden, 1930.

9. **Bier, M.,** *Electrophoresis, Theory, Methods and Applications,* Academic Press, New York, 1959.

10. **Lever, A. M. L., Webster, A. D. B., Brown, D., and Thomas, H. C.,** Non-A non-B hepatitis occurring in agammaglobulinaemic patients after intravenous gammaglobulin, *Lancet,* 2, 1062, 1984.

11. **Ochs, H. D., Fischer, S. H., Vincent, F. S., Lee, M. L., Kingdon, H. S., and Wedgewood, R. J.,** Non-A non-B hepatitis and intravenous immunoglobulin, *Lancet,* 1, 404, 1985.

12. **Thompson, A.,** personal communication, 1975.

13. **Watt, J. G.,** Continuous plasma fractionation: a preliminary report, *Vox Sang.,* 18, 42, 1970.

14. **Foster, P. R. and Watt, J. G.,** The CSVM fractionation process, in *Methods of Plasma Protein Fractionation,* Curling, J., Ed., Academic Press, London, 1980, 17.

15. **Neillie, G. K. and Watt, J. G.,** unpublished results, 1987.

16. **Kistler, P. and Nitschmann, H.,** Large scale production of human plasma fractions, *Vox Sang.,* 7, 414, 1962.

17. **Watt, J. G., Neillie, G. K., and McQuillan, T. A.,** unpublished results, 1987.

18. **Martinache, L.,** personal communication, 1987.

Chapter 26

HIGH PERFORMANCE ADSORPTION SEPARATIONS

Howard L. Levine

TABLE OF CONTENTS

I. INTRODUCTION

Traditionally, chromatographic separations aim to achieve an optimal combination of speed of elution, sample size, and separation of a mixture of components while maintaining high resolution. High resolution is obtained by controlling the differential migration rates of a group of solutes as they move down a column (column selectivity) and minimizing the extent of band broadening for each solute (column efficiency). Performance in these systems is judged by increases in the number of theoretical plates for a given separation, and chromatograms showing tens of peaks separated to baseline in short time periods demonstrate high performance.

The downstream processing of biotechnology products usually includes several column-based operations. With the exception of gel filtration, most of these separations are simple adsorption/desorption operations. In adsorption or on-off chromatography, material is fed onto the column until the desired amount of product is bound to the resin. Impurities move rapidly through the column, while the strongly adsorbed product moves slowly until it finds an unoccupied binding site, where it adsorbs and remains until the elution conditions are changed. Weakly bound impurities are displaced by the more tightly binding product. After loading, nonadsorbed or weakly bound material is washed out of the column and the product eluted by changing the elution conditions. Thus, the desired degree of resolution is relatively easily obtained; increased performance is judged in terms of practical considerations such as quantity of biologically active material purified per cycle. Furthermore, while analytical chromatography by definition must be done in a column, preparative, on-off chromatography is done in a column or fixed bed for convenience. A batch process or fluidized bed can often give results equivalent to those produced in column separations.

In developing large-scale adsorption processes, we must change our thinking about these chromatography operations. We are not concerned with separating every component in a complex mixture, but rather with isolating just one species in the highest yield and at the lowest cost. In order to maximize our yield, we must optimize the binding and elution conditions for each step in an isolation process. We must also concern ourselves with practical matters such as the cleaning and reuse of chromatography media. In this chapter we will discuss some of these parameters as they relate to the development of large-scale high performance ion exchange and affinity separations.

II. ION EXCHANGE SEPARATIONS

Commercial purification of proteins is often done by on-off ion exchange chromatography. Since proteins are zwitterionic, their net charge can be altered by changing the pH, and suitable conditions can be found to bind any protein to an ion exchange resin. Since proteins bind to ion exchange resins at multiple sites, they do not desorb until conditions are changed. The purification of monoclonal antibodies by ion exchange chromatography demonstrates the tremendous purification power of this technique. In the example shown in Figure 1, a large dilute solution containing a monoclonal antibody was applied to a cation exchange column under conditions where the antibody bound very tightly to the resin. The majority of the impurities in the sample had a pI below 7.0 and, hence, passed through the column without binding. Stepwise elution of the product resulted in a greater than 50-fold purification of the antibody. After the product was eluted, residual proteins were removed with a high-salt wash and the column reequilibrated in starting buffer for the next cycle. Numerous other examples of ion exchange separations of monoclonal antibodies are readily available.[1-5]

One advantage of using an adsorption process such as ion exchange early in a purification scheme is that the sorbent can capture solute from a dilute solution. Unlike analytical

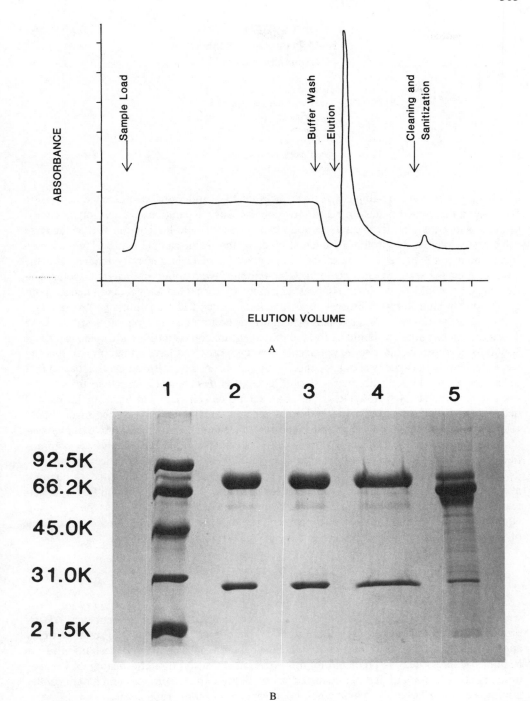

FIGURE 1.   Cation exchange purification of a monoclonal antibody. (A) *In vitro* cell culture fluid containing 0.3 g/l monoclonal antibody was loaded onto a cation exchange column equilibrated at pH 7.3 with Hepes buffer. After loading, the column was washed with buffer to remove residual contaminating proteins. The antibody was then eluted with 0.25 *M* NaCl in Hepes buffer. (B) SDS-PAGE of the separation shown in A. All samples were reduced with 5% 2-mercaptoethanol before electrophoresis. Lane 1: molecular weight markers — phosphorylase B, 92,500; bovine serum albumin (BSA), 66,200; ovalbumin, 45,000; carbonic anhydrase, 31,000; soybean trypsin inhibitor, 21,500. Lanes 2—4: purified antibody eluted from the column. Lane 5: column load.

TABLE 1
Ideal Properties of Ion
Exchange Resins

Hydrophilic support
Permeable and macroporous
High binding capacity
No nonspecific adsorption
Uniform, spherical beads
Good mechanical stability
Good chemical stability

chromatography, where sample size must be kept to a minimum to maximize separation factors and theoretical plates, in preparative ion exchange separations we are not concerned with resolution, so that large dilute samples can be loaded onto the column without adverse effects. As long as the protein of interest binds to the resin, sample can be pumped onto the column until the resin is saturated. In practice, sample is generally loaded onto the column until the resin reaches 60 to 70% saturation. Overloading an analytical column in this manner would result in gross peak asymmetry, zone broadening, and, hence, poor resolution. In an adsorption process, once a solute is bound, there is no zone broadening.

Ideal properties of an ion exchange medium for protein purification are listed in Table 1. An ion exchange resin should be hydrophilic to ensure compatibility with aqueous protein solutions. The gel should also be permeable, macroporous, and have a high protein binding capacity. This high capacity should be specific to minimize nonspecific adsorption of material to the resin. In addition, the matrix should have good mechanical strength so that it does not compress under high flow rates. The resin should be spherical and form stable, packed beds which equilibrate rapidly to the desired running conditions. Finally, the resin should tolerate the harsh conditions sometimes required to clean and sanitize chromatography systems. Currently, no ion exchange resin exhibits all of these properties. Therefore, compromises must be made in developing and optimizing a specific adsorption process.

In scaling up ion exchange separations, the ultimate goal is maximum product throughput at a minimum cost. Throughput depends on several factors, including column flow rate and resin binding capacity. In mathematical terms, flow through a packed bed obeys the following equation, derived from Darcy's law (for a more thorough discussion of the theoretical aspects of adsorption chromatography, see the text by Wankat[6]):

$$v = \frac{K\, d^2\, \Delta P}{\mu L}$$

where v = linear velocity, K = permeability constant, d = particle diameter, ΔP = pressure drop across the bed, μ = fluid viscosity, and L = bed height. The small particles and long columns typically used in analytical separations to achieve high resolution result in high backpressures, as predicted by this equation. In adsorption chromatography, resolution is not a factor and particle diameter can be large. Furthermore, column length can be minimized, since large bed heights give no advantage in on-off separations.

Recent improvements in the media for ion exchange chromatography, as well as the introduction of radial flow columns, now permit extremely high flow rates at relatively low system backpressures and, therefore, faster throughputs than were previously possible. Figure 2 shows the flow properties of several different ion exchange resins. Figures 3 and 4 show the dependence of backpressure on column geometry for two different types of ion exchange resins. These curves show that the higher the length-to-diameter (L:D) ratio, the lower the pressure at which the flow-pressure curves begin to level off. This leveling off of the flow-

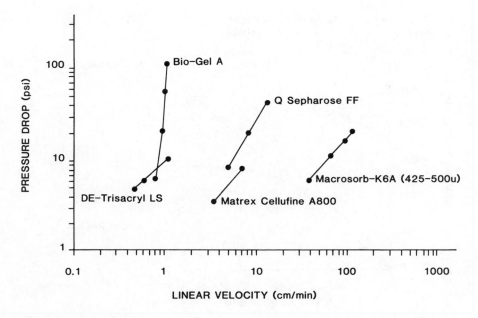

FIGURE 2. Comparative flow characteristics of some fast flow ion exchange resins. Linear flow rate vs. pressure drop in a 1-m column for five different ion exchange resins was measured in deionized water.

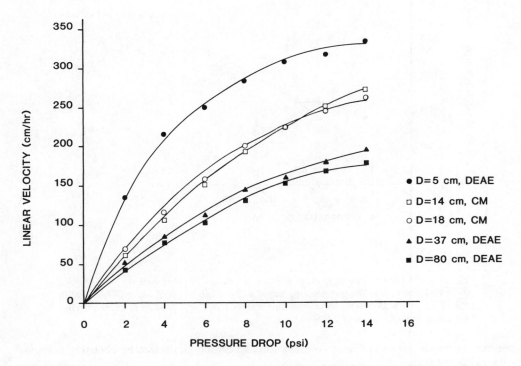

FIGURE 3. Flow rate as a function of pressure in columns of Pharmacia® CM- and DEAE-Sepharose Fast Flow. Columns of varying diameter were packed to a bed height of 15 cm with either CM- or DEAE-Sepharose Fast Flow and equilibrated in 0.15 M NaC1. The pressure drop across the bed was measured as a function of linear flow rate.

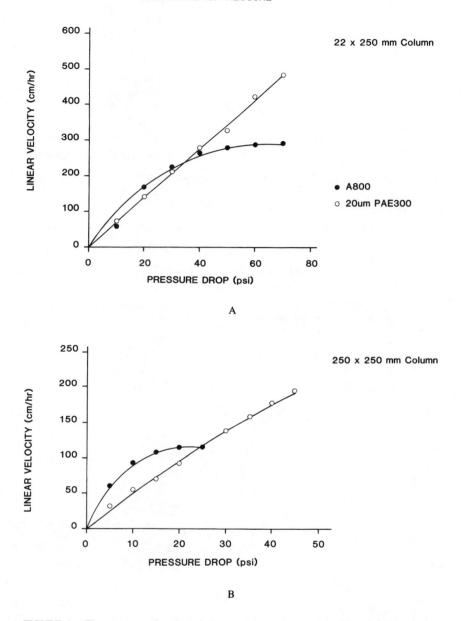

FLOW RATE vs. PRESSURE

FIGURE 4.   Flow rate as a function of pressure in columns of either Amicon Matrex® PAE 300 or Cellufine A800. The pressure drop across a 25-cm bed height of Matrex® PAE 300 in either a 2.2 (A) or 25 cm (B) diameter column was measured as a function of linear velocity. The solvent in both cases was 0.5 M NaCl.

pressure curves is a result of resin compression and can be minimized by choosing columns with small L:D ratios.

While the pressure drops observed in conventional columns are not prohibitive, radial flow columns offer some advantages over traditional axial flow columns. Since flow is distributed over the cylindrical surface of the radial flow column, the throughput at a given pressure drop is greater than that observed in an axial flow column. The AMF Zetaprep®

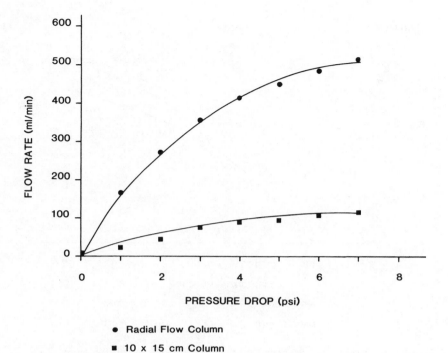

FIGURE 5. Pressure vs. flow rate in conventional and radial flow columns. Sepharose®
CL-4B (1.2 l) was packed into either a 10 × 15 cm conventional column (■) or a 35 cm
long radial flow column (●). The pressure drop as a function of absolute flow rate in both
columns was measured in 0.15 M NaCl.

system represents one approach to radial flow ion exchange separations. The system is
composed of a special ion exchange paper spirally wrapped around a core to maximize
surface area and minimize the space required to contain the matrix. The cartridge, because
of its design, has relatively low compressibility and permits extremely high flow rates. More
recently, Sepragen (San Leandro, CA) has introduced a line of radial flow columns which
can be packed with any adsorption medium. Figure 5 shows the increase in flow rates
obtainable in some of these columns packed with Sepharose® CL-4B. As discussed below,
radial flow columns can be particularly useful in affinity chromatography where small
amounts of very high-capacity resins are often used.

Because of the increased flow properties of ion exchange resins, column media compressibility no longer restricts flow rates in large-scale chromatography. With flow rate
constraints removed, column capacity becomes the limiting factor in scaling up these separations. Capacity is determined by factors at the molecular level, including matrix composition, the effectiveness of contact between product and available binding sites, and the
binding constant of the product for the adsorbent. The effective binding capacity of a column
packing material under conditions of high flow rate has been termed "capture efficiency"
by Roy et al.[7] According to Chase[8] and Arnold et al.,[9] this capture efficiency is best
determined from the breakthrough curves for a given column under different conditions. As
illustrated in Figure 6, a breakthrough curve is obtained by pumping sample solution continuously onto the column until the solute begins to appear in the effluent. The rate of
increase of solute concentration in the eluant is a measure of the adsorption efficiency of
the column. The shape of the breakthrough curve is dependent upon the rate of protein
uptake and the linear flow rate through the column. At high flow rates, solute moves through

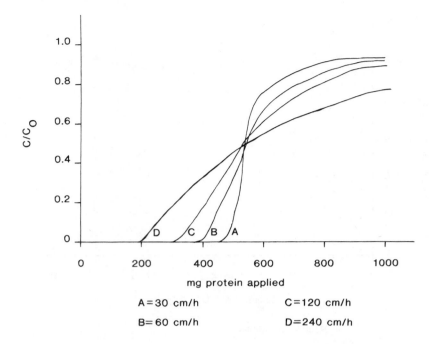

FIGURE 6.   The effect of flow rate on the capture efficiency of DEAE-Sepharose Fast
Flow. A solution of BSA (5 mg/ml) in 50 mM Tris-HCl, pH 7.2 was loaded onto a 1.6 ×
2.5 cm column at various flow rates. The amount of BSA in the column eluant as a function
of volume loaded was determined by UV absorbance. C/CO = BSA concentration in column
eluant relative to initial concentration. Flow rates: A = 30 cm/h, B = 60 cm/h, C = 120
cm/h, and D = 240 cm/h.

the bed faster than the protein adsorbs onto the resin. As a result, binding capacity is sacrificed
for high throughput and shorter cycle times.

Another approach to maximizing capture efficiency is to determine the kinetic uptake
of the protein from solution. By determining the batch binding capacity and rate of product
adsorption of the resin, the maximum flow rate that allows 100% product binding can be
calculated. The rates of protein uptake on Whatman® DE-52 cellulose and Pharmacia®
DEAE-Sepharose Fast Flow are shown in Figure 7B. These data imply that at any given
flow rate, the binding capacity (and thus the overall throughput) of a given amount of DE-
52 is higher than that of DEAE-Sepharose Fast Flow. Therefore, all other factors being
equal, to achieve the same yield of product per cycle, larger columns of DEAE-Sepharose
Fast Flow are necessary than of DE-52. A similar comparison between weak cation exchangers is shown in Figure 7A.

Scaleup of an ion exchange adsorption process is achieved by maintaining a constant
ratio of adsorbent to product. Since long column beds are unnecessary in adsorption chromatography, bed heights are usually chosen for convenience rather than increased separation
power. As seen in Figure 4, short, wide columns permit higher linear flow rates at a given
pressure drop than do long, narrow columns. In practice, bed heights of 10 to 20 cm permit
high flow rates at relatively low pressure drops. After fixing the bed height in this region,
an adsorption process can be scaled up by increasing the column diameter. Since column
volume increases with the square of the diameter, doubling the column diameter quadruples
the column volume. Columns of stainless steel, glass, or acrylic are available in a wide
range of diameters.

One of the most important features in a process column is the design of the flow
distribution system to ensure uniform flow over the entire bed surface. One method of

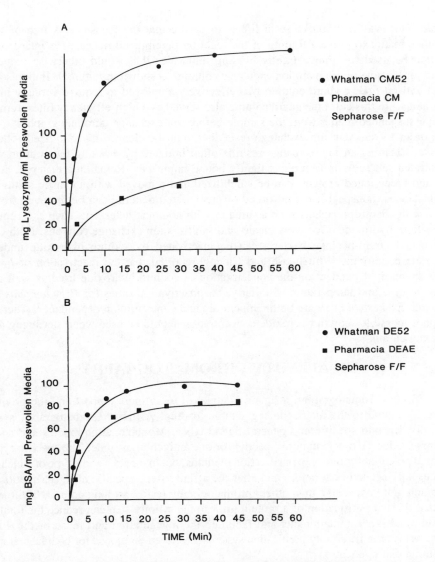

FIGURE 7. A comparison of the relative rates of protein absorption and capacities for (A) Whatman® CM-52 vs. Pharmacia® CM-Sepharose Fast Flow and (B) Whatman® DE-52 and Pharmacia® DEAE-Sepharose Fast Flow. The rate of uptake of BSA (A) or lysozyme (B) was determined from continuously stirred buffered solutions containing excess protein. Sodium acetate (0.01 M, pH 4.4) buffer was used for the cation exchange resins and 0.01 M sodium phosphate (pH 6.0) was used for the anion exchange resins. The original protein concentrations were 1 mg/ml in all cases.

maintaining a uniform flow distribution over the surface of the gel bed is illustrated in Figure 8. The entering flow is divided by a special radial end-cell design which evenly distributes the stream over the bed surface. To avoid impact of the fast-flowing mobile phase on the packing, an antijetting device disperses the entering stream. The exit flow is collected by an identical end cell. The combination of antijetting device, radial flow distribution, and sintered bed support produces an infinite multiport entry system with low dead volume, minimal mixing, and high resolution.

Since the cost of chromatographic media can be a substantial portion of the cost of manufacturing a therapeutic protein, it is important to reuse the media as many times as possible. One way to improve resin life is to ensure that the feedstock is free of visible particulate matter to prevent fouling of the resin by precipitated material. A guard column can often be used to remove tightly binding impurities that would otherwise reduce the capacity or lifetime of the main ion exchange column, as shown in Figure 9. In the isolation of goat anti-rat IgG, a guard column was effectively employed to remove strongly binding components.[2] In this case, the guard column also served as a high-efficiency filter, removing colloidal suspended matter from the sample before ion exchange chromatography.[2]

In order to reuse an ion exchange resin, it should be cleaned and sanitized following each use. Cleaning an ion exchange resin is often initiated by washing the column with a concentrated salt solution to remove tightly bound impurities. Remaining material, such as lipids and precipitated protein, can be solubilized and removed with a caustic solution. In some instances, nonspecifically adsorbed or precipitated material can be removed from the column with chaotropic agents such as urea or with nonionic detergents such as Triton® X-100. Sodium hydroxide effectively cleans and sanitizes ion exchange columns. Both Gram-negative and Gram-positive bacteria can be inactivated by sodium hydroxide under the appropriate conditions. Whitehouse and Clegg[10] reported that the inactivation of *Bacillus subtilis* spores is dependent on the concentration of sodium hydroxide used as well as the incubation time and temperature. Similarly, the inactivation times for *Pseudomonas aeruginosa* and *Escherichia coli* are both temperature and concentration dependent.[11] Therefore, the highest possible sodium hydroxide concentration should be used when sterilizing an ion exchange column.

III. AFFINITY CHROMATOGRAPHY

In affinity chromatography, a ligand is attached to an inert support and the product is specifically bound to the matrix during sample loading. Affinity chromatography systems can be divided into specific and general ligand types. Monoclonal antibodies and specific inhibitors produce affinity supports specific for one antigen or enzyme. The specific systems have great potential for one-step purification methods, but in practice some type of "cleanup" step is usually needed both before and after the affinity step. Specific affinity supports tend to be much more expensive than other chromatography resins, so long column lifetimes are important. Partial purification of a protein prior to the affinity step can reduce the total load of protein and other contaminants and facilitate the regeneration and reuse of the affinity column. Many specific affinity purification systems have been developed for both laboratory[12,13] and commercial applications.[14-16]

Immunoaffinity chromatography is used in the large-scale purification of recombinant leukocyte interferon A.[15,16] Crude *E. coli* extracts containing interferon are loaded onto the affinity column at pH 7.0. The column is then washed with pH 7.0 buffer, and the interferon is then eluted from the column with a low-pH buffer. The scaleup of this process to commercial levels required several changes in the original laboratory separation.[16]

The maximum flow rate achievable using an agarose-based immunosorbent was only 38 cm/h due to radial compression of the gel.[7] A comparison of alternative support matrices indicated that beaded cellulose or silica gel provided better mechanical support and permitted higher flow rates.[7] Silica gel was ultimately chosen as the support matrix because beaded cellulose disintegrated at pH values below 3.[7] Silica allowed fluxes ten times higher than those achieved on the agarose gel. Figure 10 shows the effect of flow rate on capture efficiency of recombinant interferon. As expected, capture decreased as flow rate increased. Based on this data and the desire to capture 80% of the interferon loaded, a linear flow rate of 720 cm/h was chosen for the large-scale purification of recombinant interferon.[7]

The high affinity of monoclonal antibodies for their antigen may make severe elution conditions necessary to recover the product. This may result in product loss by denaturation,

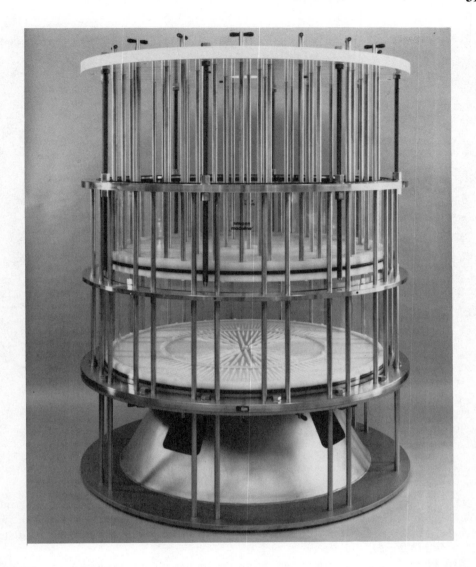

A

FIGURE 8. (A) A large-scale production column, showing materials of construction and overall dimensions; (B) a close-up of the flow distributor.

which has proven to be a particular problem.[17] This problem can be resolved using antibodies with minimal affinity for their antigen and by using tandem columns to rapidly neutralize the column effluent and stabilize the product.[17]

Another effective use of affinity chromatography is for removal of tiny amounts of contaminant proteins from an otherwise pure protein after all other methods of fractionation. Such "negative affinity" procedures are advantageous because only small quantities of adsorbent are needed and because conditions which might be detrimental to the product can be adopted to desorb unwanted contaminants during regeneration. For example, an immunoaffinity column is used to ensure the complete removal of the potentially deadly contaminant ricin B chain from the ricin A chain for use in immunotoxins.[18]

The growing use of monoclonal antibodies for *in vivo* imaging of human tumors and as therapeutic agents has focused attention on protein A affinity chromatography. The binding

FIGURE 8B.

of mouse immunoglobulins to protein A is pH dependent, with optimal binding occurring in the range of pH 6.0 to 8.0. At pH values below 6.0, this interaction is reversed. Immobilized protein A can therefore be used to purify antibodies produced either in ascites fluid or in *in vitro* cell culture.

Affinity purification of monoclonal antibodies on immobilized protein A has produced yields near 100%, with purity to match.[19-21] The scaleup potential of this technique is obvious, and it is currently being used on a commercial scale.[1] Although protein A affinity chromatography is the method of choice for antibody purification in some cases, the method is not applicable for all monoclonal antibody purifications for three reasons. First, some classes of antibody, such as IgM, do not bind to protein A. Second, protein A affinity resins are expensive and difficult to sanitize for reuse. Finally, leakage of protein from the column matrix constitutes a major regulatory concern.

Radial flow columns can also be used in affinity chromatography. As discussed above, these columns allow higher throughputs than traditional columns holding the same amount of resin. The extremely high binding capacity of affinity resins such as immobilized protein A reduces the material costs for this resin, but columns of the size required for this small amount of resin may have slower throughputs than larger columns of lower capacity resins. In this regard, radial flow columns offer the advantage of reducing material costs and at the same time increasing throughput. Figure 11 shows the purification of an anti-melanoma antibody on protein A agarose. Purified antibody was recovered from this column with >95% yield, and the entire separation was complete in less than 3 h.[22] Scaleup of separations

315

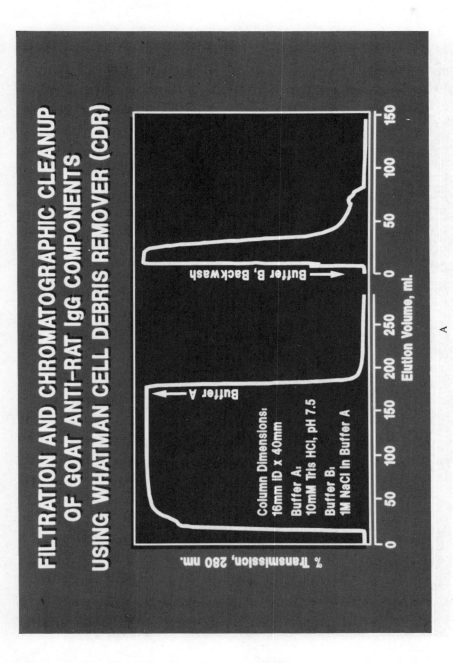

FIGURE 9. Use of a guard column to prevent fouling on an ion exchange column. (A) Heat-treated goat anti-rat antiserum was loaded onto a guard column of Whatman® CDR cellulose. After loading, the antibody-containing solution was washed through the column with buffer (10 m*M* Tris, pH 7.5). To clean the guard column, a solution of 1 *M* NaCl in 10 m*M* Tris, pH 7.5, was backflushed through the column. (B) Isoelectric focusing (IEF) analysis of the guard column separation shown in A. Lane 1: goat anti-rat antiserum; lane 2: guard column load; lane 3: unretained guard column fraction; lane 4: retained guard column fraction; lane 5: IEF markers.

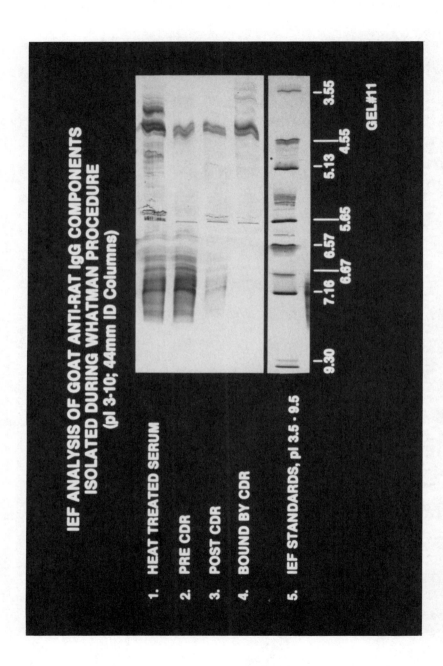

FIGURE 9B.

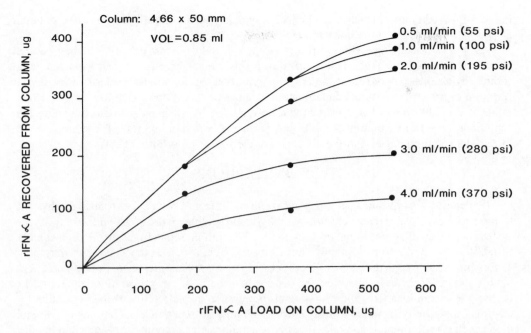

FIGURE 10. Effect of column flow rate on column capture efficiency. Purified recombinant leukocyte interferon A in 0.025 M ammonium acetate, pH 7.2 was loaded at various flow rates on a 4.66 × 50 mm immunoaffinity column equilibrated with buffer at pH 7.2. Elution was via a concave gradient from pH 7.2 to 2.8, at the same flow rate as loading.

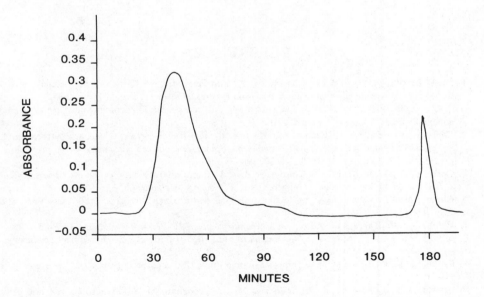

FIGURE 11. Purification of an anti-melanoma antibody on a radial flow column. Crude antibody (3 g) was loaded onto a 1.2-l radial flow column packed with protein A agarose. The sample was loaded at a flow rate of 104 ml/min and the resin then washed with buffer at a flow rate of 170 ml/min. Following washing, the antibody was eluted from the resin with a low-pH buffer at a flow rate of 92 ml/min.

in axial flow columns is easily accomplished by either increasing the length of the column or adding additional column units in parallel.

As with ion exchange resins, affinity resins can only be cost effective if they can be reused. Unlike ion exchange resins, affinity resins cannot be cleaned and sanitized with sodium hydroxide. Therefore, it is very important that the feedstock be clarified by sterile filtration to prevent microbial contamination of the affinity column. Also, impurities which might irreversibly bind to the resin should be removed by other means before loading the sample on an affinity column. As with ion exchange columns, the use of a guard column to remove such impurities can greatly increase the life of an affinity column.

IV. CONCLUSIONS

Preparative adsorption separations are so different from analytical chromatography that to refer to these operations as chromatography is misleading. Indeed, as we move to larger systems with smaller and smaller L:D ratios, adsorption separations begin to resemble filtration more than chromatography. Ion exchange filters such as the Zetaprep® may play a more important role in the future of bioprocessing than conventional resins, since theoretical plate measurements and separation factors are of little value in preparative situations. In large-scale separations, it is more important to optimize the selectivity of the separation by changing the affinity ligand in affinity chromatography or by altering the binding conditions in ion exchange chromatography. Preparative separations are very sample specific and require information on operating conditions, sample components, sample load, and the adsorbent to be fully optimized. As our understanding of proteins and the practical variables involved in their isolation grows, we can expect improved performance of adsorption separations.

REFERENCES

1. **Scott, R. W., Duffy, S. A., Moellering, B. J., and Prior, C.,** Purification of monoclonal antibodies from large-scale mammalian cell culture perfusion systems, *Biotechnol. Prog.,* 3, 49, 1987.
2. **Schwartz, W. E., Clark, F. M., and Sabran, I. B.,** Process-scale isolation and purification of immunoglobulin G, *LC-GC Mag.,* 4, 442, 1986.
3. **Carlsson, M., Hedin, A., Inganas, M., Harfast, B., and Blomberg, F.,** A two step purification of *in vitro* produced monoclonal antibodies. A two step procedure utilizing cation exchange chromatography and gel filtration, *J. Immunol. Methods,* 79, 89, 1985.
4. **Corthier, G., Boschetti, E., and Charley-Poulain, J.,** Improved method for IgG purification from various animal species by ion exchange chromatography, *J. Immunol. Methods,* 66, 75, 1984.
5. **Goding, J. W.,** in *Monoclonal Antibodies: Principles and Practice,* Academic Press, New York, 1983, chap. 4.
6. **Wankat, P. C.,** *Large Scale Adsorption Chromatography,* CRC Press, Boca Raton, FL, 1986.
7. **Roy, S. K., Weber, D. V., and McGregor, W. C.,** High performance immunosorbent purification of recombinant Leukocyte A interferon, *J. Chromatogr.,* 303, 225, 1984.
8. **Chase, H. A.,** Prediction of the performance of preparative affinity chromatography, *J. Chromatogr.,* 297, 179, 1984.
9. **Arnold, F. H., Chalmers, J. J., Saunders, M. S., Croughan, M. S., Blanch, H. W., and Wilke, C. R.,** *ACS Symp. Ser.,* 271, 113, 1985.
10. **Whitehouse, R. L. and Clegg, L. F. L.,** Destruction of *Bacillus subtilis* spores with solutions of sodium hydroxide, *J. Dairy Res.,* 30, 315, 1962.
11. **Anon.,** Process Hygiene in Industrial Chromatographic Processes, Downstream News and Views for Process Biotechnologists, No. 2, Pharmacia Fine Chemicals, Piscataway, NJ, 1987.
12. **Absolom, D. R.,** *Sep. Purif. Methods,* 10, 239, 1981.
13. **Cuatrecasas, P. and Anfinsen, C.,** Affinity chromatography, *Methods Enzymol.,* 22, 345, 1971.
14. **Chase, H. A.,** Affinity separations utilizing monoclonal antibodies — a new tool for the biochemical engineer, *Chem. Eng. Sci.,* 39, 1099, 1984.

15. **Staehlin, T., Hobbs, D. S., Kung, H. F., Lai, C. T., and Pestka, S.,** Purificaion and characterization of recombinant human leukocyte interferon with monoclonal antibodies, *J. Biol. Chem.,* 256, 9750, 1981.
16. **Tarvowski, S. J. and Liptak, R. A.,** in *Advances in Biotechnological Processes,* Vol. 2, Mizrake, A. and van Wezel, A. L., Eds., Alan R. Liss, New York, 1983, 271.
17. **van der Marel, P., van Wezel, A. L., Hazendonk, A. G., and Koovistra, A.,** Concentration and purification of poliovirus by immune adsorption on immobilized antibodies, *Dev. Biol. Stand.,* 41, 267, 1980.
18. **Fulton, R. J., Blakeley, D. C., Knowles, P. P., Uhr, J. W., Thorpe, P. E., and Vitetta, E. S.,** Isolation of pure ricin A_1, A_2, and B chains and characterization of their toxicity, *J. Biol. Chem.,* 261, 5314, 1986.
19. **Ey, P. L., Prowse, S. J., and Jenkins, C. R.,** Isolation of pure IgG_1, IgG_{2a} and IgG_{2b} immunoglobins from mouse serum using protein A Sepharose, *Immunochemistry,* 15, 429, 1978.
20. **Goding, J. W.,** Use of staphylococcal protein A as an immunological reagent, *J. Immunol. Methods,* 20, 241, 1978.
21. **Manil, L., Motte, P., Pernas, P., Troalen, F., Bognon, C., and Bellet, D.,** Evaluation of protocols for purification of mouse monoclonal antibodies, *J. Immunol. Methods,* 90, 25, 1986.
22. **Kawahate, R. T.,** The use of radial flow columns for rapid separations with soft gels, poster presentation of Cetus-UCLA Symp. Protein Purification: Micro to Macro, Frisco, Co, 1987.

Chapter 27

PREPARATIVE SCALE CHROMATOGRAPHIC PROCESSES: ION-EXCHANGE AND AFFINITY CHROMATOGRAPHIC SEPARATIONS

G. Mitra and M. H. Coan

TABLE OF CONTENTS

I. INTRODUCTION

The first steps of biochemical separations from natural and mammalian cell culture systems are usually based upon differences in physiochemical properties (size, solubility, charge, hydrophobic interactions) of the constituent macromolecules. The terminal step, however, is generally an affinity or immunoaffinity step specific for the particular macromolecule in question, ensuring a high degree of purity (from contaminant proteins and nucleic acids). This is particularly important for proteins derived from tissue culture systems and less of an issue for proteins derived from human plasma. For the former, the concerns are primarily toward residual levels of foreign proteins (antigenicity) and nucleic acids (oncogenes) in the purified product.[1] Contaminant proteins are homologous (not foreign proteins) in therapeutic entities isolated from human plasma and, hence, are not of similar concern. Of course, the criteria stated here are for therapeutic, intravenously administrable macromolecules; the human body is much more forgiving for orally administered dosage forms.

Anion-exchange chromatography in the pH range 6.5 to 8.0 (the pH of human plasma is 7.2 to 7.6; proteins of therapeutic importance are biologically active in this pH range) is commonly used at the front end of a purification process. Recent availability of highly cross-linked agarose matrices for industrial scale protein chromatography has resulted in the throughput limit being set not by flow resistance, but by the "on-off" rate of the protein in question. The so-called dynamic capacity for a particular protein is affected by a multitude of factors. The choice of the support matrix is based upon the charged groups on the matrix, particle diameter, pore size, and on-off rates of the proteins under the operating conditions. Unlike analytical scale applications, where approximately the top 10% or less of the column is utilized for protein adsorption, preparative scale ion-exchange chromatography uses almost all of the column capacity for adsorptive purposes. For a mixture of proteins, protein-protein interactions exert a significant effect on process chromatography. Adsorption-desorption behavior deviates significantly from that predicted by Langmuir-type isotherms commonly used to theoretically model chromatographic systems.

Adsorptive capacities of proteins on ion exchangers can be a function of the molecular weight of the proteins,[2] larger molecules having access only to the surface whereas smaller molecules can have access to the internal pore volume as well. Ion exchangers with higher exclusion limits (Q and S Sepharose® Fast Flow from Pharmacia [Piscataway, NJ]), exclusion limit 4×10^6 Da for globular proteins) obviate the accessibility problems of large molecular weight proteins. Uneven charge distributions over the protein surface can cause selective adsorptions not indicated by the net charge on the protein. For complex mixtures of proteins, stronger-binding proteins displace weaker-binding ones. The ensuing displacement effects can play a major role in selecting loading conditions favoring increased selectivity of the protein of choice.

Affinity/immunoaffinity chromatography, in contrast, utilizes specific ligand-macromolecule interactions which are reversible. The support matrix and the chemistry of attachment of the immobilized ligand have major influences on the selectivity and capacity of the support. Column configurations and operating conditions can be modified to minimize nonspecific interactions.

In this chapter we describe issues related to preparative scale purification of human alpha-1 proteinase inhibitor (alpha-1 PI) and human antithrombin III (AT III). Anion-exchange chromatography, as it relates to alpha-1 PI purification, has been optimized and scaled up to production scale. Heparin-Sepharose® chromatography for AT III purification has been optimized with respect to heparin immobilization methods and subsequently scaled up for preparative scale runs.

II. MATERIALS AND METHODS

A. ISOLATION AND CHARACTERIZATION OF ALPHA-1 PI

Alpha-1 PI was isolated from Cohn fraction IV-1[3] as described previously.[4] Alpha-1 PI was assayed by measuring elastase (porcine pancreatic elastase type III; Sigma, St. Louis, MO) inhibitory activity monitored during hydrolysis of chromogenic substrate *n*-succinyl-L-alanyl-L-alanyl-L-alanyl-*p*-nitroanilide (Sigma) at 405 nm and 37°C. The standard used was 100% pure alpha-1 PI. Slopes were determined for the hydrolysis reaction in the presence and absence of the inhibitor. The amount of alpha-1 PI in samples could be calculated by comparing the slopes, knowing that the stoichiometry is 1:1. The specific activity of pure alpha-1 PI, defined as milligrams of elastase inhibitory capacity per A_{280}, is 2.0.[4]

Purified alpha-1 PI in high-performance liquid chromatography (HPLC; TSK 3000 SW column, 30 cm × 7.5 mm) showed a major protein peak (93%) and some higher molecular weight material. DEAE-Sepharose® Fast Flow (Pharmacia) was the anion exchanger utilized in this study.

B. ISOLATION AND CHARACTERIZATION OF AT III

AT III was also isolated from Cohn fraction IV-1 via heparin-Sepharose® chromatography,[5] as described previously. AT III was assayed using the thrombin inactivation assay of Odegard,[6,7] utilizing chromogenic substrate S-2238 (Kabi Diagnostics, Stockholm) and bovine thrombin (Miles-Pentex, Kankakee, IL). Normal pooled plasma, by definition containing 1 U/ml of AT III, was used as a standard. This was equivalent to a plasma concentration of 125 µg/ml. Purified AT III has an extinction coefficient of 6.5 ± 0.1.[8]

Quantitation of bound heparin was based on the determination of the glucosamine concentration in the dried affinity matrices following acid hydrolysis, as described previously.[9] Heparin (porcine intestinal, U.S.P. grade, ≥150 U/mg) was obtained from Scientific Protein Products (Wisconsin).

Sepharose® CL4B and amino-hexyl (AH) Sepharose® were obtained from Pharmacia. Cyanogen bromide (CNBr) was obtained from Sigma. 1-Cyano-4-dimethylaminopyridinium tetrafluoroborate (CDAP) was synthesized[10] in-house.

III. RESULTS AND DISCUSSION

A. ANION-EXCHANGE CHROMATOGRAPHY FOR ALPHA-1 PI

For alpha-1 PI, levels of anion exchanger that had been used previously to purify the protein to homogeneity ranged from 1 to 5 ml of exchanger (column bed volume) per milligram alpha-1 PI.[11-14] This allowed for purification of small amounts of protein, but would be impractical for the production of large amounts of a therapeutic entity. This particular protein is quite abundant in human plasma,[15] and replacement therapy for an individual would require 4 g of alpha-1 PI per week.[16] Thus, a process has been developed which allows for the purification of grams and kilograms of the protein,[4] using anion-exchange chromatography as the major purification step.

In this purification, the feed material is only 15% pure; the major contaminant is albumin. A chromatography system was developed which would remove most of the albumin and other impurities and also allow for high loading of alpha-1 PI on the column. The principle of "displacement chromatography" was used.[17] This is based on overloading the column so that one protein is used to displace others.

The concept of displacement of albumin by alpha-1 PI is shown by the following model experiment. A column of DEAE Sepharose® (1 × 5 cm) was equilibrated in 0.025 *M* sodium phosphate, pH 6.5, at room temperature. Purified albumin (in the same buffer) was applied to the column until capacity was reached (the protein concentration in the flow-through was the same as in the feed). This is shown in Figure 1. After washing with buffer, the column was eluted with 0.1 *M* sodium phosphate at pH 6.5. The eluted albumin (representing the column capacity under these conditions) measured 335 mg. In a second run, alpha-1 PI was loaded onto the same column under identical conditions (Figure 2). The capacity for this protein was found to be 380 mg. Then, in a third column run (Figure 3), albumin was loaded onto the column to capacity. After washing with buffer, alpha-1 PI was applied to this preloaded column. The protein in the effluent peak contained both proteins; however, the initial fractions were pure albumin, and the latter fractions were mostly alpha-1 PI. Enough alpha-1 PI was applied to the DEAE column to saturate it with this protein. Then, after washing with equilibration buffer, the DEAE column was eluted with 1 *M* sodium chloride in 0.1 *M* sodium phosphate at pH 6.5. This eluate contained the same amount of alpha-1 PI as in the previous experiment, and the purity was the same. Thus, preloading the column with albumin did not affect the column capacity for alpha-1 PI. All of the albumin was displaced.

In the early developmental stages of the alpha-1 PI purification project, a polyethylene glycol (PEG)-containing solution (11.5% w/v) was applied to the DEAE-Sepharose® column. Approximately 10 mg of the protein of interest could be bound to each milliliter of ion exchanger. It was found that under the conditions being used (pH 6.5, 0.04 *M* sodium phosphate buffer), no more could be applied without losses occurring in the breakthrough fraction and in the subsequent wash. Elution with 0.10 *M* sodium phosphate recovered the alpha-1 PI with good yield and a purity of 60 to 70%. The major problem with this method was the flow rate during column loading. The PEG caused the ion-exchanger bed to compact, thereby increasing the pressure and slowing the flow rate to intolerably low levels. The purification was good; alpha-1 PI displaced many of the impurities.

The problem, then, was to remove or lower the PEG concentration. The method chosen was further precipitation with PEG. The proteins could be quantitatively precipitated with more PEG (17%). The resulting precipitate, when dissolved in seven volumes of buffer, contained 3% PEG, and this did not affect the column flow rate.

In the first experiment, a quantity of dissolved Cohn fraction IV-1 (the source of alpha-1 PI)[4] was fractionated with PEG. The resulting precipitate was dissolved and applied to the DEAE-Sepharose® column. A 50-ml bed volume was used, which was sufficient (from previous experience) for 450 to 500 mg of alpha-1 PI. The resulting eluate, which was produced very quickly due to the increased flow rate now possible during loading, was surprisingly less pure than expected. Also, a lower than expected yield was obtained. In order to examine this phenomenon, side-by-side comparison of the two methods was made. On one column, the 11.5% PEG solution was applied directly to the DEAE Sepharose®. On the second, an equal portion of the same PEG-containing solution was further fractionated, and the resulting precipitate obtained at 28.5% PEG (pH 6.5, +5°C) was dissolved in buffer (0.025 *M* phosphate, pH 6.5) and then applied to the column. Again, the yield and purity were lower than expected. This experiment is summarized in Table 1.

The obvious question, then, is why the purity is lower in column 2. During the loading of column 1 there was some loss of alpha-1 PI in the flow-through and in the wash fractions, whereas in column 2 there was none. Thus, the answer may be that purity would have been improved if *more* protein had been loaded. PEG apparently affects the column capacity, perhaps by blocking sites. If the column capacity were now higher, then loading more protein would allow the alpha-1 PI to displace some further amount of the more weakly binding impurities, just as albumin was displaced in the previously described model system.

This hypothesis was tested. At slightly lower ionic strength than had been used earlier (0.025 *M* in place of 0.03 *M* sodium phosphate), 1500 mg of alpha-1 PI (as the redissolved precipitate) were applied to 50 ml DEAE Sepharose®. (This is 30 mg/ml of exchanger, or about threefold more than loaded previously.) All the protein of interest bound. Washing

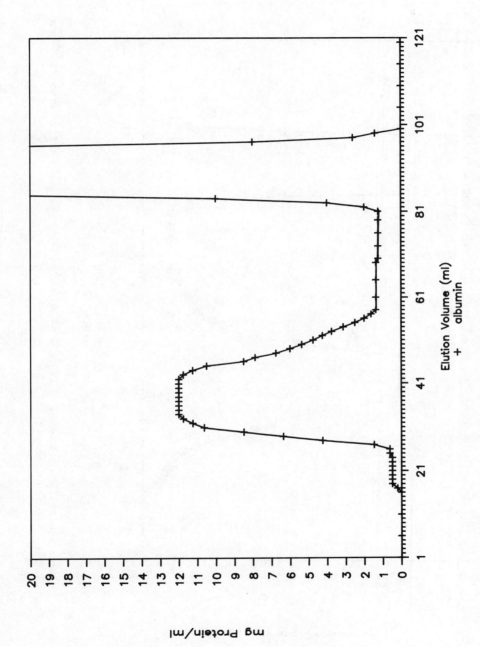

FIGURE 1. Elution of albumin (column not preloaded). DEAE Sepharose® (1 × 5 cm) equilibrated in 0.025 *M* sodium phosphate, pH 6.5; 1-ml fractions collected; 615 mg albumin applied. At 38 ml, wash started. Elution with 0.1 *M* buffer began at 77 ml. Eluate contained 335 mg albumin.

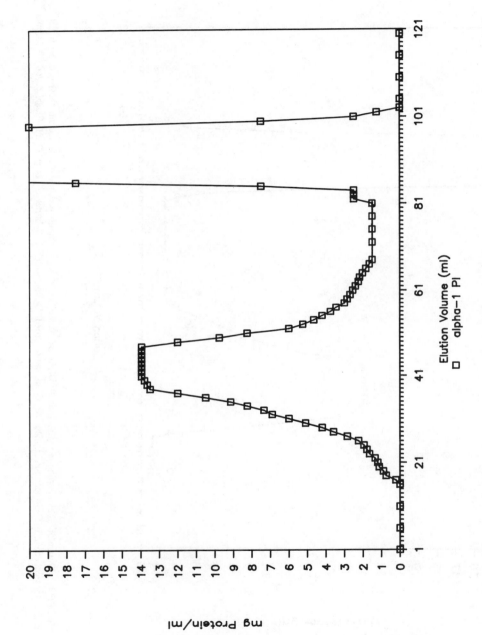

FIGURE 2. Elution of alpha-1 PI (column not preloaded). Column as in Figure 1. Alpha-1 PI (736 mg) applied. At 43 ml, wash started. Elution began at 77 ml. Eluate contained 380 mg alpha-1 PI.

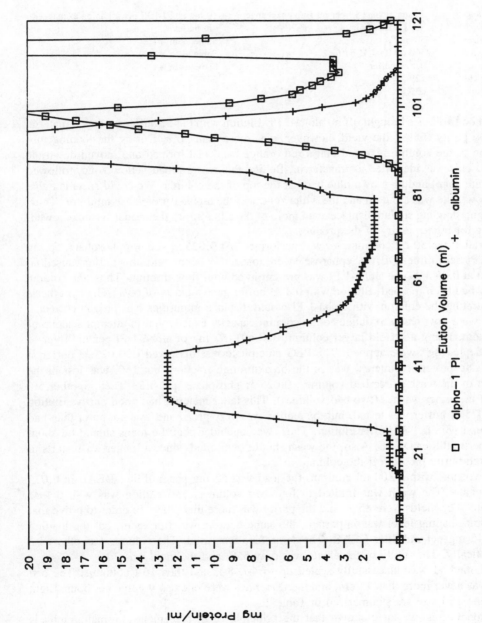

FIGURE 3. Elution of alpha-1 PI (column preloaded with albumin). Column as in Figure 1. Albumin applied as in Figure 1. Washed with 40 ml buffer, then applied 570 mg alpha-1 PI; washed with 25 ml buffer, then eluted. Eluate contained 380 mg alpha-1 PI.

TABLE 1
Comparison of Feed Materials

	Experiment	
Step	**#1**	**#2**
Material loaded	10 mg alpha-1 PI per ml anion exchanger in 11.5% PEG	10 mg alpha-1 PI per ml anion exchanger in buffer
Column size	50 ml	Same as in #1
Column wash	500 ml 0.04 M buffer	Same as in #1
Elution	0.1 M buffer	Same as in #1
Process yield	72%	68%
Product purity	63%	51%

with ten bed volumes brought off no alpha-1 PI. Elution with 0.1 M buffer produced material with low purity (55%); the yield, however, was more than 90%. During the loading, observation of the column revealed that a red-orange band had formed and moved down the column. This was identified as transferrin. Behind this was a broad yellow zone, followed by a white zone, and then by a blue band at the top of the column. We could speculate that the yellow zone was albumin and the white zone was the alpha-1 proteinase inhibitor. Thus, even higher loading should displace even more of the albumin, as the model system showed, allowing for higher purity of the product.

A trial at the 50-ml column size was performed. At 0.025 M sodium phosphate, 53 mg alpha-1 PI per milliter DEAE Sepharose® were loaded. Washing with this buffer caused no loss, but at 0.03 M some alpha-1 PI was present in column flow-through. Thus, the column capacity had been reached. Elution with 0.1 M buffer gave a purity of 63%. The hypothesis that saturating the column with alpha-1 PI would displace impurities was proved correct.

We were now ready to delineate the column capacity exactly and to attempt scaleup of this system. Using a sixfold larger column (300 ml), 57 mg of alpha-1 PI per milliliter of DEAE Sepharose® were applied. The PEG precipitate was dissolved in 0.025 M buffer at pH 6.5. No protein of interest was in the breakthrough fraction, but 15% was lost in the first part of the wash (three bed volumes, 0.025 M phosphate, pH 6.5). Then, another 5% was lost in further washes (two bed volumes). This latter material had good purity. Elution with 0.075 M buffer was tested, but the elution was very broad and was not purer than had been found with 0.1 M buffer elution. Thus, we concluded that the loads should be about 50 mg per milliliter DEAE. Also, the wash should be limited, since a longer wash starts to move high-purity protein off the column.

In two runs with a 300-ml column, the load was 53 mg per milliliter DEAE in 0.025 M phosphate. The wash was limited to four bed volumes, and elution was with 0.1 M phosphate. The yields were 85%, and the purity was more than 70%. In order to prove that the column loading level was important, the same column was run again, but the loading was at 45 mg per milliliter DEAE. The same feed was used. The column was otherwise run identically. The purity of the eluate was only 55% and the yield was 70%.

This method was successfully scaled up to 4-l, 8-l, and then 16-l columns. The bed height was never more than 15 cm, and the flow rates were one bed volume per hour. Eight runs at the 16-l size are summarized in Table 2.

In conclusion, we have shown that the principle of displacement chromatography is clearly applicable to large-scale systems. The methods described allow for the purification of enough alpha-1 PI for therapeutic use.

B. HEPARIN-SEPHAROSE® CHROMATOGRAPHY FOR AT III

AT III capacity of each affinity matrix was tested with 10 g of resin packed into a 1.6-cm-diameter column and equilibrated with 0.05 M Tris-0.15 M sodium chloride at pH 7.4

TABLE 2
Purification on 16-L Columns

| | Feed material | | Eluate | | |
Experiment number	Protein (g)	Alpha-1 PI (g)	Alpha-1 PI (g)	Purity (%)	Yield (%)
1[a]	3573	630	553	50	88
2	5632	756	525	60	69
3	4320	808	656	59	81
4	5472	792	739	67	93
5	4276	787	700	63	89
6	4174	768	712	66	93
7	4272	810	739	70	91
8	4271	809	675	70	83

[a] Number 1 was performed at reduced load (39 mg alpha-1 PI per ml DEAE). The purity was therefore low. The average for numbers 2 through 7 is 85% yield and 65% purity.

and 22°C. Affinity-purified AT III (8 U/mg) was applied at a flow rate of 1 ml/min until protein was detected in the column flow-through fractions. The column was washed with the equilibration buffer until the outlet A_{280} returned to baseline. A second wash with 0.05 M Tris-0.3 M sodium chloride at pH 7.4 elicited a second peak; this was followed by elution with 0.05 M Tris-2.0 M sodium chloride at pH 7.4. Each resin was run in triplicate, and the average values of amounts of At III eluted were correlated with amounts of bound heparin (Table 3).

In general, a positive correlation was obtained between amounts of bound heparin and AT III capacity for a particular affinity matrix. Lower binding efficiencies were obtained with CNBr activation of heparin (1 to 3, Table 3) compared to cyanate ester derivatives (resin–O–C=N) of the support matrix (4 and 5, Table 3). Modifications of the hydroxyl groups in the reactive tetrasaccharide portions of the heparin molecule during CNBr activation are the likely cause for the reduced AT III affinity of these supports.

For a given affinity matrix, purity of the feed solution and the temperature of operation had significant effects on capacity and nonspecific binding of the support (Table 4). For affinity-purified AT III (#1, Table 4), an inverse relationship between temperature and gel capacity was noted, which agrees well with the literature report of an inverse relationship between K_{dissoc} of the heparin-AT III complex and temperature.[18] This effect was less pronounced for less pure feed solutions (#2 and #3, Table 4).

Negative charges on the heparin molecule promote electrostatic interactions, and the support has discernible ion-exchange properties when the feed contains plasma proteins other than AT III (#2 and #3, Table 4). Colder temperatures increased nonspecific binding, as reflected in lower specific activity of AT III. At any given temperature, decreased feed purity resulted in decreased AT III capacity — since the negatively charged sulfate groups also reside within the reactive tetrasaccharide portions of the heparin molecule (responsible for AT III binding), the ion-exchange properties compete with affinity binding of AT III in these cases.

Heparin-Sepharose® chromatography has been scaled up to 20-l columns prepared in-house. Predictable column operations have been obtained on a reproducible basis for the preparation of preclinical and clinical lots.

In conclusion, the following points have been identified in this chapter regarding their importance in process scale chromatography:

TABLE 3
AT III Capacity of Different Heparin-Sepharose® Supports

#	Support	Bound heparin ([mmol/g support] $\times 10^{-5}$)	AT III capacity ([mmol/g support] $\times 10^{-5}$)	$\dfrac{\text{mol AT III}^a}{\text{mol heparin}}$
1	CNBr-activated heparin-AH Sepharose®	11.8	1.9	0.16
2	CNBr-activated heparin-AH Sepharose®	23.7	3.7	0.16
3	CNBr-activated heparin-AH Sepharose®	47.3	9.2	0.19
4	CNBr-activated Sepharose® CL 4B	2.9	1.2	0.41
5	CDAP-activated Sepharose® CL 4B	9.7	3.2	0.33

[a] Average mol wt of heparin = 15,000 Da, mol wt of AT III = 58,000 Da.

TABLE 4
AT III Capacity of CDAP-Activated Heparin-Sepharose® as a Function of Temperature and Feed Solution Purity

#	Feed Solution	Temperature (°C)	AT III capacity ([mmol/gm support] × 10^{-5})	Eluted AT III specific activity (U/m)
1	Affinity-purified AT III (sp act 8U/mg)	5	4.03	—
		22	3.14	—
		37	2.38	—
2	Dissolved IV-1 (sp act 0.05 U/mg)	5	1.69	3.9
		22	1.72	6.1
3	Effluent II + III (sp act 0.02 U/mg)	5	0.95	3.1
		22	0.91	3.8

1. Preparative scale ion-exchange strategy has been one of increased selectivity at high protein loading.
2. Ligand-binding chemistry can have a significant effect on column capacity and, hence, column size.
3. Nonspecific interactions can influence affinity adsorption/elution processes. These interactions need to be minimized through processing conditions.

REFERENCES

1. **Hopps, H. E. and Petricciani, J. C., Eds.**, Abnormal Cells, New Products and Risk; Proceedings of a Workshop held July 30—31, 1984 at NIH, Bethesda, MD, Tissue Culture Association, Gaithersburg, MD, 1985.
2. **Scopes, R. K.**, Quantitative studies in ion exchange and affinity elution chromatography of enzymes, *Anal. Biochem.*, 114, 8, 1981.
3. **Cohn, E J., Strong, L. E., Hughes, W. L., Mulford, D. J., Ashworth, J. N., Melin, M., and Taylor, H. L.**, Preparation and properties of serum and plasma proteins. IV. A system for the separation into fractions of the protein and lipoprotein components of biological tissues and fluids, *J. Am. Chem. Soc.*, 68, 459, 1946.
4. **Coan, M. H., Brockway, W. J., Eguizabal, H., Krieg, T., and Fournel, M.**, Preparation and properties of alpha₁-proteinase inhibitor concentrate from human plasma, *Vox Sang.*, 48, 333, 1985.
5. **Miller-Andersson, M., Borg, H., and Andersson, L.-O.**, Purification of antithrombin III by affinity chromatography, *Thromb. Res.*, 5, 439, 1974.
6. **Odegard, O. R., Lie, M., and Abildgaard, U.**, Heparin cofactor activity measured with an amidolytic method, *Thromb. Res.*, 6, 287, 1975.
7. **Odegard, O. R.**, Evaluation of an amidolytic heparin cofactor assay method, *Thromb. Res.*, 7, 351, 1975.
8. **Nordenman, B., Nyström, C., and Björk, I.**, The size and shape of human and bovine antithrombin III, *Eur. J. Biochem.*, 78(1), 195, 1977.
9. **Mitra, G., Hall, E., and Mitra, I.**, Application of immobilized heparins for isolation of human antithrombin III, *Biotechnol. Bioeng.*, 28, 217, 1986.
10. **Wakselman, M. E., Guibe-Jampel, E., Raoult, A., and Busse, W. D.**, 1-Cyano-4-dimethylamino-pyridinium salts: new water-soluble reagents for the cyanylation of protein sulfhydryl groups, *J. Chem. Soc. Chem. Commun.*, p. 21, 1976.
11. **Crawford, I. P.**, Purification and properties of normal human alpha₁-antitrypsin, *Arch. Biochem. Biophys.*, 156, 215, 1973.
12. **Glaser, C. B., Karic, L., and Fallat, R.**, Isolation and characterization of alpha-1-antitrypsin from the Cohn fraction IV-I of human plasma, *Prep. Biochem.*, 5(4), 333, 1975.
13. **Kress, L. M. and Lakowski, M.**, Large scale purification of alpha-1-trypsin inhibitor from human plasma, *Prep. Biochem.*, 3, 541, 1973.
14. **Pannell, R., Johnson, D., and Travis, J.**, Isolation and properties of human α-1-proteinase inhibitor, *Biochemistry*, 13, 5439, 1974.

15. **Travis, J. and Salveson, G.,** Human plasma proteinase inhibitors, *Annu. Rev. Biochem.,* 52, 655, 1983.
16. **Gadek, J. E. and Crystal, R. G.,** α_1-Antitrypsin deficiency, in *The Metabolic Basis of Inherited Disease,* Stanbury, J. B., Wyngaarden, J. B., Fredrickson, D. S. et al., Eds., McGraw-Hill, New York, 1983, 1450.
17. **Janson, J. and Hedman, P.,** On the optimization of process chromatography of proteins, *Biotechnol. Prog.,* 3, 9, 1987.
18. **Jordan, R. E., Beeler, D., and Rosenberg, R.,** Fractionation of low molecular weight heparin species and their interaction with antithrombin, *J. Biol. Chem.,* 254, 2902, 1979.

Chapter 28

ADSORPTION CHROMATOGRAPHY: APPROACHES TO COMPUTER-BASED OPTIMIZATION AND MODELING OF COLUMN PROCESSES

Per Hedman, Jan-Gunnar Gustafsson, Kristina Wiberg, and Christine Markeland-Johansson

TABLE OF CONTENTS

I. INTRODUCTION

Chromatography packings are chosen primarily according to their selectivities.* The choice is also influenced by kinetic limitations which are inherent in the packing material. In our development of new packing materials, we are trying new methods for the evaluation of kinetic parameters. This evaluation can be done using zone spreading measurements or batch adsorption rate determinations. Using batch adsorption experiments, it is possible to measure the rate of solute adsorption under conditions which are easier to control than the conditions affecting column experiments. The time course of the batch adsorption experiment will depend on the initial conditions as well as the adsorption kinetics. Our aim was to study the kinetic limitations inherent in the chromatographic packing. For this reason we were not satisfied with earlier models which resulted, for example, in mass transfer coefficients that were dependent on volume or mass ratios (i.e., the initial conditions). In these earlier models, the different possible mass transfer resistances were lumped together with a reaction rate constant (characterizing the solute-ligand complex-forming reaction). It follows that if one wants to distinguish between diffusional resistances and sorption reaction kinetics, etc., the model will necessarily be more complicated.

II. COLUMN EXPERIMENTS

In the van Deemter equation, where H is the height equivalent of a theoretical plate and u represents the linear mobile-phase velocity in a packed column, the parameters A, B, and C represent zone-broadening contributions from eddy diffusion, longitudinal molecular diffusion, and mass transfer resistance, respectively.[1]

$$H = A + B/u + C \times u \tag{1}$$

Thus, the kinetic resistances are lumped together in the coefficient C. When peak widths are measured at different flow rates in isocratic runs using a series of buffers with decreasing buffer strength, one may separately estimate the influence of sorption kinetics as well as external and internal diffusional resistances.

Using buffers of increasing elution strength, it should be possible to reduce the influence of the sorption kinetics keeping the diffusional resistance approximately constant. For such experiments it is desirable to obtain a measure of the elution strength of the buffer rather than merely the composition. A correlation between the retention (k') and the elution strength has been suggested by Rounds and Regnier.[2] However, their retention model was unable to predict the behavior of bovine serum albumin (BSA) using S Sepharose® High Performance, a new 30-μm agarose-based cation exchanger. On the contrary, with chymotrypsinogen A, we were able to use this model in order to select suitable buffer compositions. An aberrant behavior of BSA was also noticed by Kopaciewicz et al.,[3] although they were using Mono S™, a 10-μm cation exchanger.

III. CALCULATION OF KINETIC PARAMETERS FROM PEAK WIDTHS

Horvath and Lin[4] presented a theoretical method for the estimation of axial dispersion as well as film diffusion and pore diffusion resistance. We were unable to detect a sufficient deviation from quasilinearity in order to use Horvath's equations for the determination of

---

* The chromatographic packings mentioned in this work are products of Pharmacia LKB Biotechnology AB. Mono S™, Mono Q™, and Sepharose® are trademarks of Pharmacia.

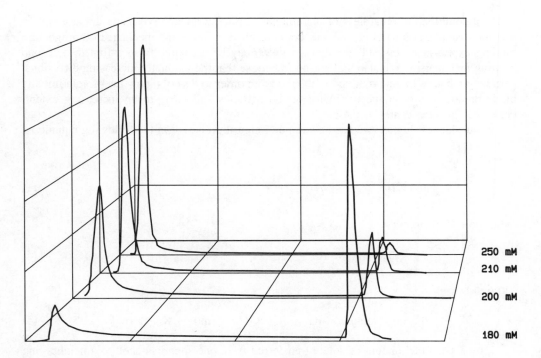

FIGURE 1. Chromatograms obtained from isocratic runs in 50 mM sodium acetate, pH 4.5 containing different sodium chloride concentrations. After approximately ten column volumes, a step gradient to 1 M sodium chloride was used to elute BSA remaining on the column. The peaks to the right occurred only after this step gradient.

the film diffusion resistance. However, Arnold and co-workers[5] have suggested the correlation

$$Sh = 2 + 1.45 \times Re^{0.5} \times Sc^{0.33} \qquad (2)$$

where Sh is the Sherwood number, Re the Reynolds number, and Sc the Schmidt number, for the estimation of the film diffusion resistance. Although this correlation may be useful, we are reluctant to rely upon peak width measurements. Corrections for extra column effects, sample variance, and zone compression effects may introduce substantial errors. Nevertheless, we found Arnold's determination of BSA diffusivity in pores of Sepharose® CL-4B, 1.8×10^{-7} cm^2/s, interesting as an independent measurement. Due to our difficulties in controlling the retention volumes by changing the buffer composition (see Figure 1), we could not make a comparison using BSA. With chymotrypsinogen A this should have been possible (data not shown). Still, we hesitated to use data obtained with underivatized packings for prediction of diffusivities in derivatized packings such as ion exchangers because the pores may be altered during the derivatization procedure. This is probably not a significant problem when data obtained with underivatized packings are used for the prediction of diffusivities in immunoaffinity packings because of the lower substitution degrees obtained in the latter case. Consequently, we found no reason not to accept Arnold's data.

IV. BATCH ADSORPTION EXPERIMENTS

In previous work in our laboratories we tried less complicated models, lumping several mass transfer resistances together as described by Chase.[6] These modeling experiments did not satify our need to find parameter values that would be characteristic of the solute-packing material combination regardless of the amount of protein added. We made the assumption that the overall mass transfer rate was not controlled by the same mechanism throughout a binding experiment. Instead, we thought there may be a change from a first to a second limiting mechanism. We also had reason to believe that the gradually increasing diffusional path length had to be incorporated in the model in order to describe how the protein initially binds predominantly at the perimeter of the particles and later, to an increasing extent, penetrates into the center of the particles.

In our data evaluation we used a computer algorithm based on the following equations:

film diffusion

$$J_0 = Kf \times A_0 \times (C_0 - C_1) \tag{3}$$

radial pore diffusion

$$J_i = Ds \times A_i \times (C_i - C_{i+1})/r_i \tag{4}$$

sorption rate

$$dC_i/dt = qmax \times Kl \times [C_i \times (1 - rm_i) - Kd \times rm_i] \tag{5}$$

J_i represents the amount of solute transferred from one spherical shell to a neighboring shell. These imaginary shells are radial subdivisions of the column packing particles. C_i represents the liquid concentration of the solute in the respective shell. C_0 represents the bulk liquid concentration, and A_0 is the area of the boundary film layer. Kf is the film mass transfer coefficient, and Ds is the stationary phase (pore) diffusivity. The thickness of the ith shell is denoted by r_i, and the protein binding capacity per unit volume of the packing is denoted by qmax. K1 is the forward binding rate constant, whereas Kd is the solute-immobilized ligand complex dissociation constant. The fraction of occupied binding sites within the ith shell is denoted by rm_i.

The algorithm was written in C and used as a binary program together with a Hewlett Packard® Technical Basic user interface program on an Integral PC (M68000 processor). Estimated values of Kf, Ds, and K1 were entered until the computer model generated a satisfactory concentration vs. time plot as compared to our measurements. We used Kf values according to Arnold's correlation, knowing that our experiments (which were carried out at considerably higher Reynolds numbers than Arnold's) were less dependent on film diffusion resistance. Admittedly, our algorithm should be less efficient than the numerical methods used by Arve and Liapis.[7] Our algorithm is, in principle, equivalent to the numerical solution presented by Nigam and Wang.[8] Unfortunately, we are not aware of any results from evaluations of experimental data using either of these numerical methods. Figures 2 through 5 show computer-generated predictions of batch adsorption experiments. These adsorption rate predictions were compared to actual measurements such as those in Figure 6.

Our model agreed well with experiments using S Sepharose® High Performance at low protein loadings (40% of the available protein loading capacity), as shown in Figure 7. We were unable, however, to obtain good agreement at higher protein loadings (80% of the total binding capacity). At 40% loading, we obtained from the best fit a BSA diffusivity value of 8.33×10^{-8} cm²/s, which is 2.2 times lower than the value obtained by Arnold. On the other hand, our value is higher than the value (3.5×10^{-8} cm²/s) reported by Gibbs and Lightfoot[9] for a 40,000-mol-wt enzyme with Mono Q™, a 10-μm anion exchanger. Blanch et al.[10] previously reported a BSA diffusivity value of 5.2×10^{-8} cm²/s, which is closer to our value. If these differences are not due to systematic errors in any of the

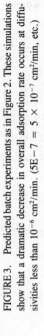

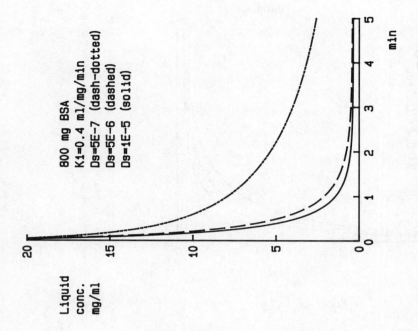

FIGURE 3. Predicted batch experiments as in Figure 2. These simulations show that a dramatic decrease in overall adsorption rate occurs at diffusivities less than 10^{-6} cm²/min. ($5E - 7 = 5 \times 10^{-7}$ cm²/min, etc.)

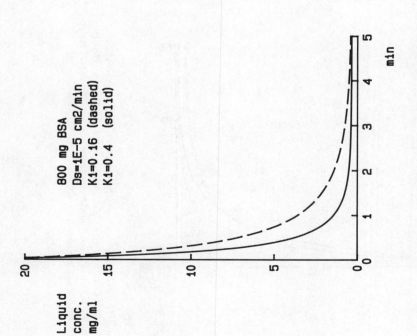

FIGURE 2. Predicted time course of a batch adsorption experiment. The total binding capacity and the dissociation constant were obtained from adsorption isotherms determined in separate experiments. (Pore diffusivity parameter, DS, of $1E - 5$ cm2/min $= 1 \times 10^{-5}$ cm²/min.)

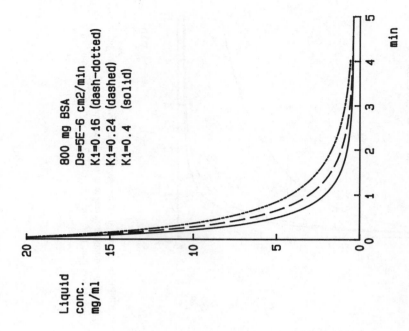

FIGURE 5.   Predicted batch experiments as in Figure 2. These (and other) simulations did not fit experimental data, although they were carried out using the diffusivity value which resulted in the best fit at 40% loading. (5E − 6 cm2/min = 5 × 10⁻⁶ cm²/min.)

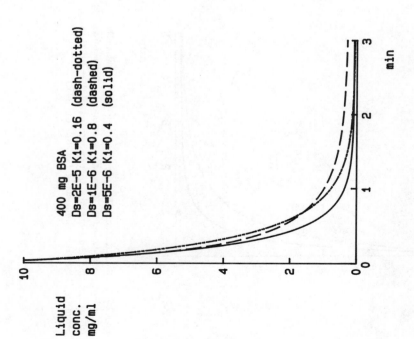

FIGURE 4.   Predicted batch experiments as in Figure 2. These simulations show that a combination of high diffusivity and low binding rate, although initially slower, will reach equilibrium earlier than a combination of high binding rate and low diffusivity. (2E − 5 = 2 × 10⁻⁵ cm²/min, etc.)

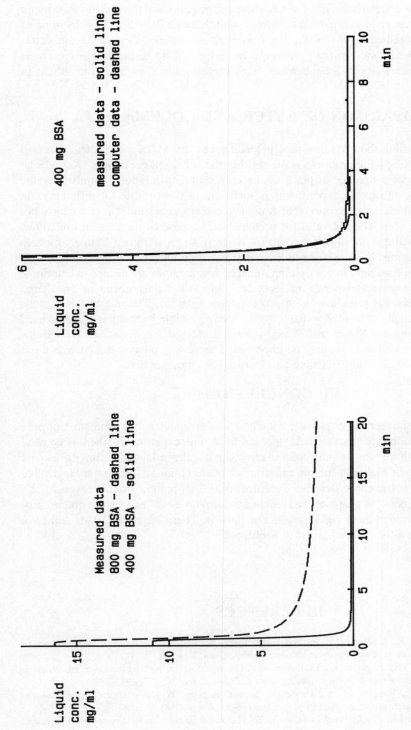

FIGURE 6. Measured BSA concentration vs. time in batch experiments. A sample flow was pumped through a filter and a UV 1 detector and back to the stirred batch adsorption reactor. A_{280} readings were digitized using an LCC-500 liquid chromatography controller. Data was converted to concentration units and stored using an HP® 86 microcomputer.

FIGURE 7. Comparison of experimental data (solid line) obtained at 40% (400 mg) loading with the computer model prediction. The calculated curve was obtained with a diffusivity of 8.33×10^{-8} cm²/s, a forward reaction rate constant of 0.4 ml/mg/min, and a film mass transfer coefficient of 2.2×10^{-3} cm/s.

determinations, then these figures may indicate a more optimal pore structure in Sepharose® CL-4B as compared to the more extensively cross-linked 30-μm agarose or Mono Q™. Unfortunately, these comparisons involve two different proteins and three different column packings. Still, we chose to compare our findings with those of Blanch and Gibbs in order to verify that we obtained results of the right order of magnitude. Comparing our determination with Blanch's and Arnold's findings, we believe that, despite extensive cross-linking, the BSA diffusivity in S Sepharose® High Performance is as favorable as it is in Sepharose® CL-4B.

V. COMPARISON OF BATCH AND COLUMN DATA

Our model algorithm should (at low protein loadings) be able to describe protein transport rates in a column using an appropriate value for the film diffusion coefficient. Knowledge of the column Reynolds number in place of the corresponding Reynolds number for the stirred batch reactor adsorption experiments is sufficient to compensate for differences in the boundary layer thickness, provided that Arnold's correlation is valid. Since we have not seen proof that this correlation is valid for proteins and because of the present limitations of our algorithm (loadings not higher than 40%), we have so far refrained from attempts to calculate chromatograms. We are eager to find an explanation for the slow approach toward equilibrium at higher loadings. In our opinion, independent methods are needed to determine whether this slow approach depends on secondary interaction phenomena or on altered diffusivities due to bound protein or high protein concentrations. The numerical methods used by Arve and Liapis[7] also rely on the second-order reversible interaction equation used in our algorithm, whereas Nigam and Wang[8] used a simplified rate equation. We believe that their models, as well as ours, may be improved if secondary phenomena such as those described by Hearn et al.[11] and Hethcote and DeLisi[12] are incorporated.

VI. CONCLUSIONS

Batch adsorption experiments provide a method for the simultaneous estimation of pore diffusivity and sorption rate constants. As opposed to column experiments, there is no need to account for axial dispersion or extra column effects. Also, the influence of film diffusional resistance is likely to be higher in column experiments. Assumptions of elementary reversible interactions are not sufficient to describe experiments at high protein-loading levels.

Recent developments in agarose cross-linking technology have made it possible to produce beads that are much more rigid (results not shown). These improved beads still have the same good pore diffusion properties as Sepharose®.

REFERENCES

1. **van Deemter, J. J., Zuiderweg, F. J., and Klinkenberg, A.,** Longitudinal diffusion and resistance to mass transfer as causes of nonideality in chromatography, *Chem. Eng. Sci.,* 5, 71, 1956.
2. **Rounds, M. A. and Regnier, F.,** Evaluation of retention model for high-performance ion-exchange chromatography using two different displacing salts, *J. Chromatogr.,* 283, 37, 1984.
3. **Kopaciewicz, W., Rounds, M. A., Fasnaugh, J., and Regnier, F. E.,** Retention model for high-performance ion-exchange chromatography, *J. Chromatogr.,* 266, 3, 1983.
4. **Horvath, C. and Lin, H.-J.,** Band spreading in liquid chromatography. General plate height equations and a method for the evaluation of the individual plate height contributions, *J. Chromatogr.,* 149, 43, 1978.
5. **Arnold, F. H., Blanch, H. W., and Wilke, C. R.,** Analysis of affinity separations. I. The characterization of affinity columns by pulse techniques, *Chem. Eng. J.,* 30, B25, 1985.

6. **Chase, H. A.,** Prediction of the performance of preparative affinity chromatography, *J. Chromatogr., 297,* 179, 1984.
7. **Arve, B. H. and Liapis, A. I.,** Modeling and analysis of biospecific adsorption in a finite bath, *AIChE J.,* 33(2), 179, 1987.
8. **Nigam, S. C. and Wang, H. Y.,** Mathematical modeling of bioproduct adsorption using immobilized affinity absorbents, in *ACS Symp. Ser. Separation, Recovery and Purification Mathematical Modeling,* Vol. 314, American Chemical Society, Washington, D.C., 1986, 53.
9. **Gibbs, S. J. and Lightfoot, E. N.,** Scaling up gradient elution chromatography, *Ind. Eng. Chem. Fundam.,* 25, 490, 1986.
10. **Blanch, H. W., Arnold, F. H., and Wilke, C. R.,** Mass transfer effects in affinity chromatography, in *Third European Congress on Biotechnology,* Vol. 1 (abstr.), Verlag Chemie, Weinheim, West Germany, 1984, I-613.
11. **Hearn, M. T. W., Hodder, A. N., and Aguilar, M. I.,** High performance liquid chromatography of amino acids, peptides and proteins. LXVI. Investigations on the effects of chromatographic swell in the reversed-phase high-performance liquid chromatographic separation of proteins, *J. Chromatogr.,* 327, 47, 1985.
12. **Hethcote, H. W. and DeLisi, C.,** Determination of equilibrium and rate constants by affinity chromatography, *J. Chromatogr.,* 248, 183, 1982.

Chapter 29

IMMOBILIZED METAL ION AFFINITY CHROMATOGRAPHY OF TRANSFERRINS

Eugene Sulkowski

TABLE OF CONTENTS

I. INTRODUCTION

Immobilized metal ion affinity chromatography (IMAC)[1,2] has been exploited for the purification of many proteins.[3-5] Porath et al. were the first to apply IMAC to the isolation of human serum transferrin[1] and the purification of human milk lactoferrin[6] on iminodiacetate (IDA)-Zn^{2+} and IDA-Cu^{2+}, respectively.

Studies with model proteins have revealed that a single histidine residue on the surface of a protein may suffice for its retention on IDA-Cu^{2+}.[4,7,8]

What does constitute a binding site for IDA-Zn^{2+} and IDA-Co^{2+} on the surface of a protein? An inspection of model proteins analyzed to date[4,8] has revealed a common structural feature: two proximal histidines in an α-helix, separated by two or three other residues. This type of putative binding site has been identified in the following proteins: sperm whale myoglobin (His 113, His 116), human fibroblast interferon (His 93, His 97), and human lymphotoxin (His 17, His 21, His 24 and His 131, His 135). All of those proteins bind to IDA-Zn^{2+} and IDA-Co^{2+}.[4,8]

An appropriate folding of a polypeptide may also bring two histidines, distant in the primary structure, to close proximity on the surface of a protein. Plausibly, such a protein could display an affinity for IDA-Zn^{2+} (IDA-Co^{2+}). If so, then one could distinguish "sequential" and "conformational" types of binding sites for IDA-Zn^{2+} (IDA-Co^{2+}).

Human serum transferrin, human lactoferrin, and chicken ovotransferrin afford excellent opportunities to probe for the existence of the "conformational" type of IDA-Zn^{2+} (IDA-Co^{2+}) binding sites. Both human serum transferrin and chicken ovotransferrin do not have clusters of two (or more) histidines in their primary structures,[9,10] and, therefore, the putative binding sites for IDA-Zn^{2+} (IDA-Co^{2+}) of the "sequential" type are absent. Each transferrin, however, has two Fe (III) binding sites. Each Fe (III) binding site is postulated to have two histidines participating in the coordination with iron.[11-15] If so, then those histidines may be available for coordination with IDA-Zn^{2+} (IDA-Co^{2+}) and represent an example of the sought-after "conformational" type of IDA-Zn^{2+} (IDA-Co^{2+}) binding site.

II. MATERIALS AND METHODS

Human serum transferrin (Fe saturated, lot 134601), human serum transferrin (holo, lot 138601), and human serum transferrin (apo, lot 134601) were purchased from Boehringer Mannheim Biochemicals, Indianapolis, IN. Human lactoferrin L-8010 (0.8 Fe/mol, lot 15F-0089), ovotransferrin (conalbumin) from chicken egg white, type II (iron complex, Fe, 0.22%, lot 104F-8065), and ovotransferrin (conalbumin), type I (substantially iron-free, Fe, 0.02%, lot 104F-8060) were purchased from Sigma, St. Louis, MO. Chelating Sepharose® 6B was purchased from Pharmacia, Piscataway, NJ.

A. PREPARATION OF IDA-Me²⁺ COLUMNS

Chelating Sepharose® 6B was washed with water, degassed under vacuum, and equilibrated with 0.1 M sodium acetate (1 M sodium chloride), pH 4.0. The gel was poured into a chromatographic column with a bed volume of approximately 5 ml (0.9 × 8 cm) and charged with 50 ml of a solution containing 5 mg/ml of metal salt in 0.1 M sodium acetate (1 M sodium chloride), pH 4.0. Excess metal was washed out of the column with metal-free 0.1 M sodium acetate (1 M sodium chloride), pH 4.0. In the case of Co^{2+}, the column charging and washing were performed at pH 6.0. Finally, all columns (Zn^{2+}, Co^{2+}, Ni^{2+}, and Cu^{2+}) were equilibrated with 0.02 M sodium phosphate (1 M sodium chloride), pH 7.0. When IDA-Me²⁺ columns were developed with imidazole, all columns, after an initial equilibration with 0.02 M sodium phosphate (1 M sodium chloride), pH 7.0, were saturated with 50 ml of 10 mM imidazole in 0.02 M sodium phosphate (1 M NaCl), pH 7.0. The columns were then emptied into 10 mM imidazole solution, repacked, and equilibrated with 50 ml of 2 mM imidazole in 0.02 M sodium phosphate (1 M sodium chloride), pH 7.0.

B. CHROMATOGRAPHY ON IDA-Me^{2+} (FALLING pH PROTOCOL)

A sample of protein, 10 mg in 5 ml of column equilibrating buffer (0.02 M sodium phosphate, 1 M sodium chloride, pH 7.0), was applied on an IDA-Me^{2+} column, and the column was washed with 20 ml of column equilibrating buffer, pH 7.0 (arrow 1). Each column was then developed with 25 ml of 0.1 M sodium acetate (1 M sodium chloride), pH 6.0 (arrow 2). The pH gradient was then developed by mixing 25 ml of 0.1 M sodium acetate (1 M sodium chloride), pH 6.0, and 25 ml of 0.1 M sodium acetate (1 M sodium chloride), pH 4.0 (arrow 3). Finally, a column was developed with 0.1 M sodium acetate (1 M sodium chloride), pH 4.0 (arrow 4). This protocol applies to Figures 1, 2, 3, and 7.

In one case, Figure 6, the falling pH gradient (arrow 2) was developed by mixing 25 ml of 0.02 M sodium phosphate (1 M sodium chloride), pH 7.0, with 25 ml of 0.02 M sodium phosphate (1 M sodium chloride), pH 6.0. The final wash (arrow 3) was done with 0.02 M sodium phosphate (1 M sodium chloride), pH 6.0.

C. CHROMATOGRAPHY ON IDA-Me^{2+} (AFFINITY ELUTION PROTOCOL)

A sample of protein, 10 mg in 5 ml of column equilibrating buffer (2 mM imidazole in 0.02 M sodium phosphate, 1 M sodium chloride, final pH of 7.0 after an adjustment with hydrochloric acid), was applied to an appropriately equilibrated IDA-Me^{2+} column; the column was then washed (arrow 1) with 20 ml of 2 mM imidazole (column equilibrating buffer). The columns were then developed with linear concentration gradients of imidazole at pH 7.0 (arrow 2) and with a finite eluant containing imidazole (arrow 3). Details of the gradients are illustrated in Figures 4, 5, 8, and 9.

All IDA-Me^{2+} columns were developed at room temperature. The flow rates were maintained at 1 ml/5 min by means of a peristaltic pump. Fractions of 1 ml were collected, and their protein content was measured by spectrophotometry at 280 nm.

III. RESULTS

All transferrins employed in this study were obtained from commercial sources and were used without further treatment.

A. CHROMATOGRAPHY OF TRANSFERRINS ON IDA-Zn^{2+}

Human serum transferrin (apo), human lactoferrin (0.8 Fe per mole), and ovotransferrin type I (substantially iron free) were chromatographed on IDA-Zn^{2+} columns. The results are illustrated in Figure 1. Both ovotransferrin and lactoferrin did not bind to IDA-Zn^{2+}. Human serum transferrin was retained on IDA-Zn^{2+} at pH 7 and was subsequently displaced from the column by lowering the pH of the eluant from pH 7 downward. A small portion of the protein was recovered at pH 6 and the major portion at a still lower pH value. Thus, human serum transferrin and ovotransferrin, both essentially iron free, displayed an affinity for IDA-Zn^{2+} and the lack of it, respectively.

The lack of retention of human lactoferrin on IDA-Zn^{2+} may result from its partial loading with iron. Whether human lactoferrin in its apo form will bind to IDA-Zn^{2+} still remains an open question at this time, although this is not likely to occur.

B. CHROMATOGRAPHY OF TRANSFERRINS ON IDA-Co^{2+}

All three transferrins, i.e., human serum transferrin (apo), human lactoferrin (0.8 Fe per mole), and ovotransferrin type I (substantially iron free), were chromatographed on IDA-Co^{2+} columns. The results are illustrated in Figure 2. Chicken ovotransferrin and human

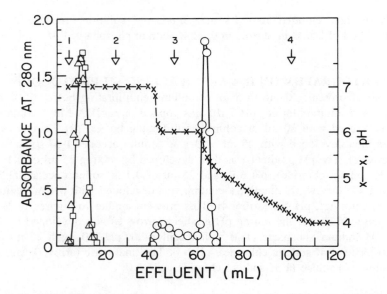

FIGURE 1. Chromatography of human serum transferrin, human lactoferrin, and chicken ovotransferrin on IDA-Zn^{2+}. Ovotransferrin, △; lactoferrin, □; serum transferrin, ○.

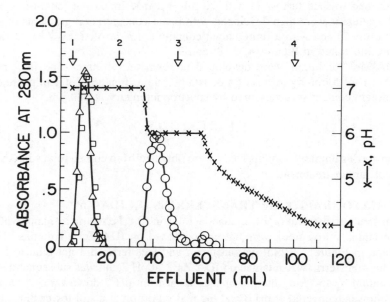

FIGURE 2. Chromatography of human serum transferrin, human lactoferrin, and chicken ovotransferrin on IDA-Co^{2+}. Ovotransferrin, △; lactoferrin, □; serum transferrin, ○.

lactoferrin were not retained on IDA-Co^{2+}. Human lactoferrin may not bind to IDA-Co^{2+} for the same reason for which it did not bind on IDA-Zn^{2+}, i.e., because of partial loading with iron. The outcome of this experiment must be reexamined with genuine apolactoferrin. Human serum transferrin was retained on IDA-Co^{2+} at pH 7; practically all of the protein was then recovered by lowering the pH of the eluant down to 6. Therefore, the elution of human serum transferrin from the IDA-Co^{2+} column occurred at a somewhat higher pH value than that from IDA-Zn^{2+}, although the difference was not a large one.

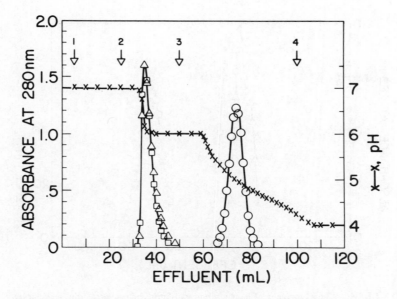

FIGURE 3.   Chromatography of human serum transferrin, human lactoferrin, and chicken ovotransferrin on IDA-Ni^{2+}. Ovotransferrin, $\triangle$; lactoferrin, $\square$; serum transferrin, $\bigcirc$.

C. CHROMATOGRAPHY OF TRANSFERRINS ON IDA-Ni^{2+}

In view of the binding of human serum transferrin (apo) to both IDA-Zn^{2+} and IDA-Co^{2+}, one could anticipate its binding to IDA-Ni^{2+}, and with a stronger retention as well. The results illustrated in Figure 3 confirm both of those expectations. However, whether chicken ovotransferrin (apo) and human lactoferrin would bind to IDA-Ni^{2+} or not remained to be established. In fact, as shown in Figure 3, both of those transferrins were retained on IDA-Ni^{2+} at pH 7 and were then readily displaced (sharp elution profiles) by a stepwise lowering of the pH of the eluant from 7 to 6.

Thus, all three transferrins displayed an affinity for IDA-Ni^{2+} at pH 7. The strongest retention on IDA-Ni^{2+} was observed for human serum transferrin; it was recovered from the column at the lowest pH value.

D. CHROMATOGRAPHY OF TRANSFERRINS ON IDA-Cu^{2+}

All three transferrins were applied to IDA-Cu^{2+} columns. All were retained, and none was displaced in a falling pH protocol employed for the elution, as described above (Figures 1, 2, and 3). Those results are not illustrated. No attempt at their recovery was made by a prolonged elution of the columns at pH 4 or an even lower pH.

In order to assess the affinity of individual transferrins for IDA-Cu^{2+}, their elution was accomplished by the affinity elution protocol with imidazole. To this end, all transferrins bound to IDA-Cu^{2+} columns at a low concentration of imidazole (2 mM) were displaced with a linear concentration gradient of imidazole. The results are illustrated in Figure 4. As could be anticipated, human serum transferrin (apo) was retained the strongest vis-à-vis chicken ovotransferrin (apo) and human lactoferrin (partially charged with iron). Again, the affinity of human lactoferrin for IDA-Cu^{2+} must be reexamined with a sample of apolactoferrin.

E. CHROMATOGRAPHY OF HUMAN SERUM TRANSFERRIN ON IDA-Me^{2+}

In order to better assess the relative affinity of human serum transferrin (apo) for various IDA-Me^{2+} sorbents, a series of experiments were performed utilizing an affinity elution protocol on all IDA-Me^{2+} columns. The results are given in Figure 5.

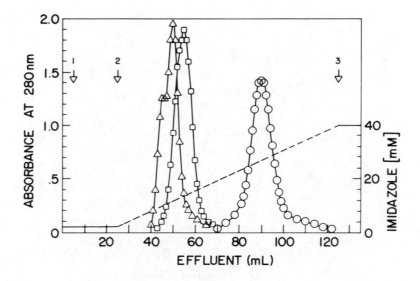

FIGURE 4.   Chromatography of human serum transferrin, human lactoferrin, and chicken ovotransferrin on IDA-Cu^{2+}. Ovotransferrin, $\triangle$; lactoferrin, $\square$; serum transferrin, $\bigcirc$.

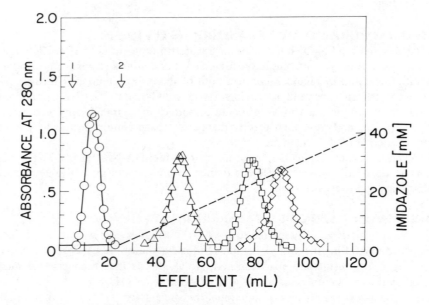

FIGURE 5.   Chromatography of human serum transferrin (apo) on IDA-Me^{2+}. Zn, $\bigcirc$; Co, $\triangle$; Ni, $\square$; Cu, $\diamond$.

Human transferrin (apo) was not retained on an IDA-Zn^{2+} column equilibrated at 2 mM imidazole; however, it was retained on IDA-Co^{2+}, IDA-Ni^{2+}, and IDA-Cu^{2+}. The retention on IDA-Cu^{2+} was the strongest.

The chromatographic behavior of human serum transferrin (apo) on IDA-Zn^{2+} and IDA-Co^{2+} columns equilibrated with imidazole (2mM) is at variance with the behavior of this transferrin on IDA-Zn^{2+} and IDA-Co^{2+} columns which had not been equilibrated with imidazole (Figures 1 and 2). Human serum transferrin (apo) binds more strongly to IDA-Zn^{2+} (no imidazole) than to IDA-Co^{2+} (no imidazole); the reverse is true in the presence of imidazole (Figure 5). The reason for this disparate behavior is not clear at present. Perhaps

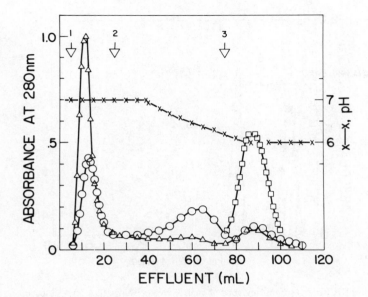

FIGURE 6. Chromatography of human serum transferrin on IDA-Zn^{2+}. Fe saturated, $\triangle$; holo, $\bigcirc$; apo, $\square$.

the IDA-Co^{2+} column itself is being saturated with imidazole at a somewhat higher concentration than is IDA-Zn^{2+}.

It should be observed that human serum transferrin could be recovered from IDA-Ni^{2+} at pH 5, but still remained bound to IDA-Cu^{2+} at pH 4, in a falling pH protocol. In the presence of imidazole, when an affinity elution protocol is employed (Figure 5), human serum transferrin is still retained more strongly on IDA-Cu^{2+} than on IDA-Ni^{2+}, but the effective concentration of imidazole displacing this transferrin is rather similar (overlapping elution profiles).

F. CHROMATOGRAPHY OF HUMAN SERUM TRANSFERRIN — APO, HOLO, AND Fe SATURATED — ON IDA-Zn^{2+}

Figure 6 illustrates the chromatographic behavior of human serum transferrin on IDA-Zn^{2+} as a function of its saturation with iron.

Apotransferrin was completely retained at pH 7 and was subsequently displaced from the column at pH 6. Holotransferrin, however, displayed significant heterogeneity: one portion was found in the breakthrough fractions, another portion was retained and subsequently eluted at about pH 6.5, and a small portion was eluted in the area corresponding to apotransferrin.

Human serum transferrin, Fe saturated, also displayed chromatographic heterogeneity, qualitatively comparable to that of holotransferrin. The major portion of the sample applied to the IDA-Zn^{2+} column was found, however, in the breakthrough fractions.

It seems plausible to interpret the chromatographic behavior of different forms of human serum transferrin as a function of iron residency (apoferric, monoferric, diferric). Presumably, diferric transferrin is not recognized by IDA-Zn^{2+}, whereas apotransferrin binds the most strongly to IDA-Zn^{2+}. The chromatographic fate of two possible variants of monoferric transferrin, if present in the sample, is not clear; they may display different affinity for IDA-Zn^{2+}. It is clear, however, that at least one of them is retained (Figure 6, portion eluted at about pH 6.5).

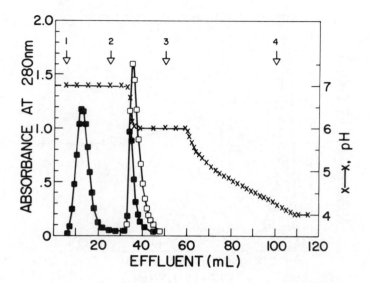

FIGURE 7. Chromatography of chicken ovotransferrin IDA-Ni^{2+}. Substantially iron free, □; iron complex, ■.

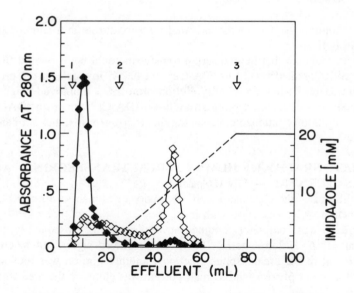

FIGURE 8. Chromatography of chicken ovotransferrin on IDA-Ni^{2+}. Substantially iron free, ◇; iron complex, ◆.

G. CHROMATOGRAPHY OF OVOTRANSFERRIN, "SUBSTANTIALLY IRON FREE" AND "IRON COMPLEX", ON IDA-Ni^{2+}

Figure 7 illustrates the chromatography of "substantically iron-free" and "iron-complex" preparations of chicken ovotransferrin on IDA-Ni^{2+}. The column was developed by a falling pH protocol. "Iron-free" ovotransferrin was retained on IDA-Ni^{2+} at pH 7 and was subsequently displaced from the column by lowering the pH of the eluant to 6. The chromatographic behavior of the "iron-complex" ovotransferrin indicated its heterogeneity: the major portion of the protein was not retained on IDA-Ni^{2+} at pH 7, while a small portion of the protein was retained and was subsequently displaced at pH 6.

Figure 8 again illustrates the chromatography of "iron-complex" and "substantially

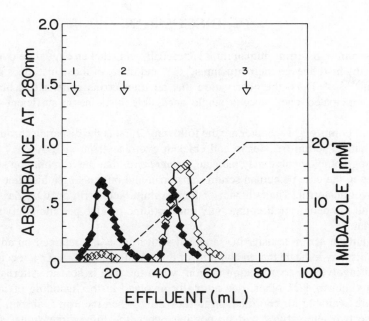

FIGURE 9.   Chromatography of chicken ovotransferrin on IDA-Cu²⁺. Substantially iron free, ◇; iron complex, ◆.

iron-free'' preparations of chicken ovotransferrin on IDA-Ni²⁺; the column was developed this time by the affinity elution protocol using imidazole. The "iron-complex" ovotransferrin was recovered almost quantitatively in the breakthrough fractions; this was in contrast to the chromatographic behavior in the falling pH elution protocol, where a portion of the protein was retained (Figure 7). The "iron-free" ovotransferrin displayed some heterogeneity in the imidazole elution protocol: a small portion of the protein was not retained on IDA-Ni²⁺, while a major portion of the protein was retained and was subsequently displaced from the column at 10 mM imidazole.

The data of Figures 7 and 8 can be rationalized if one assumes that "iron-complex" ovotransferrin is a mixture of diferric and monoferric forms with a trace of apotransferrin present. A "substantially iron-free" ovotransferrin preparation could contain a trace of diferric transferrin and some monoferric, with the major component being apotransferrin.

H. CHROMATOGRAPHY OF OVOTRANSFERRIN, "SUBSTANTIALLY IRON FREE" AND "IRON COMPLEX", ON IDA-Cu²⁺

Figure 9 illustrates the chromatography of "iron-free" and "iron-complex" preparations of chicken ovotransferrin on IDA-Cu²⁺. The columns were developed by affinity elution with imidazole. The "iron-free" ovotransferrin was retained on IDA-Cu²⁺ and was subsequently displaced from the column at 10mM imidazole. The "iron-complex" ovotransferrin displayed chromatographic heterogeneity: a major portion of the protein was found in the breakthrough fractions, while the remainder of the protein, retained on IDA-Cu²⁺, was subsequently displaced at 8 mM imidazole.

It is plausible that the portion of the protein binding to IDA-Cu²⁺ represents the monoferric component of the "iron-complex" ovotransferrin, whereas the portion found in the breakthrough fractions is the diferric component. However, all experiments depicted in Figures 6 through 9 must await a more detailed interpretation until an electrophoretic analysis of the chromatographic fractions has been performed.

IV. DISCUSSION

Human serum transferrin, human milk lactoferrin, and chicken egg white ovotransferrin all contain many histidines in their structures,[9,10,16] and many of those histidines are exposed to the solvent.[11,13,16] Thus, the observation that all transferrins under study bind to IDA-Cu^{2+} is an anticipated one, since a single accessible histidine is sufficient to retain a protein.[4,7,8]

The salient points of this chapter are the following. First is the disparate chromatographic behavior of human serum transferrin and chicken ovotransferrin on IDA-Zn^{2+} (Figure 1). This chromatographic outcome may indicate, *prima facie,* that the iron binding site contains two histidines in the case of human serum transferrin and only a single histidine in the case of chicken ovotransferrin. The behavior of human milk lactoferrin on IDA-Zn^{2+} would be consistent with the demonstration that only one histidine residue per site is involved in the binding of iron.[17-19]

Second, human serum transferrin, saturated with iron, is not retained on an IDA-Zn^{2+} (Figure 6). This may indicate that the binding of this transferrin to IDA-Zn^{2+} occurs, indeed, via histidines involved in coordination to iron, when the iron is absent. Alternatively, two histidines proximal to each other, one normally involved in the liganding of iron and the other not,[18] are available for coordination to IDA-Zn^{2+} when the iron is absent. The choice between these two alternatives will be possible when the three-dimensional structure of human serum transferrin has been determined. Whether or not both iron binding sites of human serum transferrin can be involved in the recognition of IDA-Zn^{2+} must await further study, perhaps with proteolytically derived fragments.[20,21]

Third, IMAC may be useful in the resolution of human serum transferrin into its iron-content-varying components: apo, monoferric, and diferric. However, in order to stabilize the binding of iron to serum transferrin, chromatographic conditions should be further modified — phosphate buffer replaced, bicarbonate included, pH increased, etc.

Finally, IMAC should provide the means for the advanced purification of various transferrins.

This initial foray into IMAC of transferrins has proved to be of some value both for further development of IMAC itself and for the characterization of iron binding sites on transferrins. Clearly, additional and more analytical studies are very much in demand.

ACKNOWLEDGMENT

I am grateful to Marcia Held for her excellent assistance in preparing this manuscript.

REFERENCES

1. **Porath, J., Carlsson, J., Olsson, I., and Belfrage, G.,** Metal chelate affinity chromatography, a new approach to protein fractionation, *Nature (London),* 258, 598, 1975.
2. **Porath, J. and Olin, B.,** Immobilized metal ion affinity adsorption and immobilized metal ion affinity chromatography of biomaterials. Serum protein affinities for gel-immobilized iron and nickel ions, *Biochemistry,* 22, 1621, 1983.
3. **Lonnerdal, B. and Keen, C. L.,** Metal chelate affinity chromatography of proteins, *J. Appl. Biochem.,* 4, 203, 1982.
4. **Sulkowski, E.,** Purification of proteins by IMAC, *Trends Biotechnol.,* 3, 1, 1985.
5. **Fatiadi, A. J.,** Affinity chromatography and metal chelate affinity chromatography, *Crit. Rev. Anal. Chem.,* 18, 1, 1987.
6. **Lonnerdal, B., Carlsson, J., and Porath, J.,** Isolation of lactoferrin from human milk by metal-chelate affinity chromatography, *FEBS Lett.,* 75, 89, 1977.

7. **Sulkowski, E., Vastola, K., Oleszek, D., and von Muenchhausen, W.,** Surface topography of interferons: a probe by metal chelate chromatography, *Anal. Chem. Symp. Ser.,* 9, 313, 1982.
8. **Sulkowski, E.,** Immobilized metal ion affinity chromatography of proteins, in *Protein Purification: Micro to Macro,* Burgess, R., Ed., Alan R. Liss, New York, 1987, 149.
9. **MacGillivray, R. T. A., Mendez, E., Sinha, S. K., Sutton, M. R., Lineback-Zins, J., and Brew, K.,** The complete amino acid sequence of human serum transferrin, *Proc. Natl. Acad. Sci. U.S.A.,* 79, 2504, 1982.
10. **Jeltsch, J.-M. and Chambon, P.,** The complete nucleotide sequence of the chicken ovotransferrin mRNA, *Eur. Biochem.,* 122, 291, 1982.
11. **Rogers, T. B., Gold, R. A., and Feeney, R. E.,** Ethoxyformylation and photooxidation of histidines in transferrins, *Biochemistry,* 16, 2299, 1977.
12. **Chasteen, N. D.,** The identification of the probable locus of iron and anion binding in the transferrins, *Trends Biochem. Sci.,* 8, 272, 1983.
13. **Thompson, C. P., Grady, J. K., and Chasteen, N. D.,** The influence of uncoordinated histidines on iron release from transferrin, *J. Biol. Chem.,* 261, 13128, 1986.
14. **Penner, M. H., Osuga, D. T., Meares, C. F., and Feeney, R. E.,** The interaction of anions with native and phenylglyoxal-modified human serum transferrin, *Arch. Biochem. Biophys.,* 252, 7, 1987.
15. **Baldwin, D. A. and Egan, T. J.,** An inorganic perspective of human serum transferrin, *S. Afr. J. Sci.,* 83, 22, 1987.
16. **Mazurier, J., Metz-Boutigue, M.-H., Jolles, J., Spik, G., Montreuil, J., and Jolles, P.,** Human lactotransferrin: molecular, functional and evolutionary comparisons with human serum transferrin and hen ovotransferrin, *Experientia,* 39, 135, 1983.
17. **Mazurier, J., Leger, D., Tordera, V., Montreuil, J., and Spik, G.,** Comparative study of the iron-binding properties of transferrins. Differences in the involvement of histidine residues as revealed by carboethoxylation, *Eur. J. Biochem.,* 119, 537, 1981.
18. **Anderson, B. F., Baker, H. M., Dodson, E. J., Norris, G. E., Rumball, S. V., Waters, J. M., and Baker, E. N.,** Structure of human lactoferrin at 3.2-Å resolution, *Proc. Natl. Acad. Sci. U.S.A.,* 84, 1769, 1987.
19. **Baker, E. N., Rumball, S. V., and Anderson, B. F.,** Transferrins: insights into structure and function from studies on lactoferrin, *Trends Biochem. Sci.,* 12, 350, 1987.
20. **Brock, J. H. and Arzabe, F. R.,** Cleavage of diferric bovine transferrin into two monoferric fragments, *FEBS Lett.,* 69, 63, 1976.
21. **Lineback-Zins, J. and Brew, K.,** Preparation and chracterization of an NH_2-terminal fragment of human serum transferrin containing a single iron-binding site, *J. Biol. Chem.,* 255, 708, 1980.

Chapter 30

HPLC SCALEUP FOR URACIL-DNA GLYCOSYLASE PURIFICATION

G. Barrie Kitto and Lois Davidson

TABLE OF CONTENTS

I. INTRODUCTION

Now that the first products of biotechnology, including both those from recombinant DNA technologies and those relating to monoclonal antibodies, are assuming major importance in the marketplace, the problems and economics of protein and peptide purification are gaining increased importance. The focus of the Bioseparations Program, a part of the Separations Research Program at The University of Texas at Austin, is the development of effective generic scaleup methods for taking protein products of biotechnology from the laboratory to pilot and industrial scales. Particular emphasis is being placed on the use of high-performance liquid chromatography (HPLC) in downstream processing.

Scaleup of the purification of the enzyme uracil-DNA glycosylase, derived from the fermentation of *Escherichia coli,* has been chosen as a model system. There were several reasons for this choice. First, the enzyme could serve as a model for recombinant-DNA-produced proteins derived from fermentation. The enzyme has been cloned in *E. coli,* and several high-producing strains have been developed.[1] Second, the scaleup could serve as a model for the transition of processing from the laboratory to the pilot scale. A well-documented laboratory-scale procedure for purification of uracil-DNA glycosylase has been published,[2] but larger scale processing has not been reported. Third, successful development of scaleup procedures for the enzyme could potentially lead to a viable commercial product. Uracil-DNA glycosylase is an indispensable element in a new and very effective method for site-directed mutagenesis.[3] Full utilization of this method is hampered by ready availability of the enzyme, and development of scaleup procedures would allow for the widespread adoption of this new route for site-specific modification of proteins.

Uracil-DNA glycosylase had been characterized extensively prior to its use for site-directed mutagenesis. This enzyme plays a role in the normal process of DNA repair. DNA does not usually contain uracil. In part, this is because of uracil removal by uracil-DNA glycosylase. This enzyme catalyzes the cleavage of uracil-deoxyribose bonds, yielding free uracil and an unbroken DNA with an apyrimidinic site. The uracil-DNA glycosylase from *E. coli* is a single subunit, globular protein with a molecular weight of 25,000 Da. The enzyme is moderately stable, retaining activity in solution up to 45°C.[4-7]

II. MATERIALS AND METHODS

E. coli D110 was obtained from Dr. Ian Molineux, Department of Microbiology, The University of Texas at Austin. ^{3}H-Uracil DNA was synthesized and provided by Dr. Dale Mosbaugh and his colleagues of the Department of Chemistry, The University of Texas at Austin. Dithiothreitol, HEPES, bovine serum albumin (BSA), and Trizma® base were purchased from Sigma Chemical Company, St. Louis, MO. Ultrapure ammonium sulfate was obtained from Schwartz Mann, Spring Valley, NY. Hydroxyapatite and a Bio-Gel® HPHT column were from Bio-Rad Laboratories, Richmond, CA. Sephadex® G-75 and DNA-Agarose were purchased from Pharmacia, Piscataway, NJ. QMA-Accell and an I-125 gel filtration column were obtained from Waters Chromatography Associates, Milford, MA, and the CM-300 column was from Synchrom, Inc., Linden, IN. HPLC-grade water was obtained from a Continental Type I system (Continental Water Company, Austin, TX). All other chemicals were reagent grade.

Uracil-DNA glycosylase was assayed according to the procedure of Lindahl et al.;[2] 20 μl of BSA (1 mg/ml) were used as a carrier instead of calf thymus DNA. For the enzyme, 9.2 nmol of calf thymus ^{3}H-uracil DNA (212 to 375 cpm/pmol uracil) were used as substrate.

TABLE 1
Uracil-DNA Glycosylase Purification: Comparison of Original and Modified Procedures

Sample	Volume (ml)	Protein		Enzyme activity		Recovery (%)	Specific activity (E.U./ mg)	Fold purification
		mg/ml	Total mg	E.U./ml	E.U. total			
Original (Conventional Chromatography)								
Crude sample	100	39.2	3920	0.95	95.0	100	0.024	0
Streptomycin	190	10.6	2021	0.54	102.6	108	0.189	7.9
Ammonium sulfate (35—65%)	19	28.5	541	2.8	53.2	56	0.098	4.1
Sephadex® G-75	13.5	9.2	125	1.4	19.0	20	0.152	6.3
Hydroxyapatite	4.4	4.6	20	2.9	12.7	13.3	0.641	26.7
DNA agarose	2.0	0.94	1.9	14.3	28.5	30	15.0	625.0
Modified (HPLC Chromatography)								
Crude sample	27	47	1270	0.88	24	100	0.02	0
Streptomycin	56	9.3	523	0.56	31	132	0.06	3.2
30% PEG supernatant	140	1.42	199	0.03	4.3	18	0.02	1.1
5.75—30% PEG pellet	8.4	10.9	91	1.8	15	64	0.17	8.7
QMA-Accell HPLC (fractions 70—85)	4.8	4.25	20.4	2.2	11	45	0.52	27.3
Hydroxyapatite HPLC (fractions 16—71)	5.0	2.7	13.4	2.16	11	45	0.81	42.4
CM-300 HPLC (fractions 46—56)	15	0.03	0.51	0.61	9.2	39	18	949
I-125 gel filtration HPLC (fractions 31—40)	16	0.012	0.20	0.27	4.3	18	21.4	1126

III. RESULTS

A. LABORATORY-SCALE PURIFICATION

An initial objective in this project was to carry out the purification of uracil-DNA glycosylase according to the published procedure of Lindahl et al.[2] Our aims in doing so were to provide an appreciation of the overall difficulties that might be encountered in scaleup and to provide benchmarks for the development of scaleup procedures. The types of purification steps involved are very typical of bench-scale methodologies; they include precipitation, salt fractionation, gel permeation, adsorption chromatography, and affinity chromatography. As adjunct methods to these separatory steps, a number of centrifugation and dialysis procedures are necessary. The laboratory-scale purification, according to the original procedure, proved reasonably straightforward and gave stepwise and overall yields and degrees of purification comparable to the published data.

Analysis of the purification steps of the original procedure indicated several steps that would pose difficulties for scaleup. These included centrifugation and dialysis steps and, in particular, the use of DNA-agarose affinity chromatography. For large-scale applications, the DNA-agarose would be expensive and would have poor physical properties.

Because of the variety of modes in which it can operate and because of the relative ease with which it can be scaled up,[8] HPLC was explored in detail as a substitute for the affinity chromatography procedure and for other steps. A series of trials led to the modified laboratory-scale purification shown in Table 1, where it is compared with the original procedure.

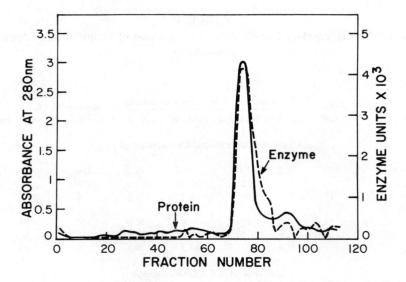

FIGURE 1.   Analytical-scale HPLC separation of uracil-DNA glycosylase on QMA-Accell anion-exchange packing. The injected sample was 18.0 ml of dialyzed 5.7 to 30% PEG fraction. Fractions were collected at 1.0-min intervals. Elution conditions were as follows:

Time (min)	Flow (ml/min)	% A	% B	% C
0	1.5	100	0	0
20	1.5	100	0	0
30	1.5	0	100	0
35	1.5	0	100	0
55	1.5	0	0	100
60	1.5	0	0	100
65	1.5	100	0	0
70	1.5	100	0	0

Eluants: (A) 0.05 M Tris-acetate buffer, pH 7.8, containing 0.001 M EDTA and 0.001 M dithiothreitol; (B) 0.05 M Tris-acetate buffer, pH 7.8, containing 0.001 M EDTA, 0.001 M dithiothreitol, and 1.0 M sodium acetate; (C) 0.05 M Tris-acetate buffer, pH 7.8, containing 0.001 M EDTA, 0.001 M dithiothreitol, and 2.0 M sodium acetate.

Substitution with the HPLC procedures speeds up the overall process while allowing for comparable yields of product and being readily amenable to scaleup. Examples of the efficacy of the four new HPLC steps are shown in Figures 1 to 4. Figure 1 shows the separation achieved in the first HPLC ion-exchange step using QMA-Accell as the packing material. A 60 cm × 7.8 mm column was loaded with 18.2 ml (84.5 mg) of dialyzed 5.75 to 30% polyethylene glycol (PEG) sample. A very simple elution system was developed, using an initial Tris-acetate buffer and effecting desorption of the enzyme by increasing concentrations of sodium acetate (see figure legend for details). One major enzyme peak was eluted. The enzyme-containing fractions were combined, concentrated, and pumped onto a hydroxy-apatite column (HPHT, 10 cm × 7.8 mm). The gradient conditions and chromatogram are shown in Figure 2. Peak enzyme fractions were combined and concentrated before being pumped onto a CM-300 column. Gradient conditions and the elution profile are shown in Figure 3. The major enzyme peak (fractions 46 to 56) was combined separately from the

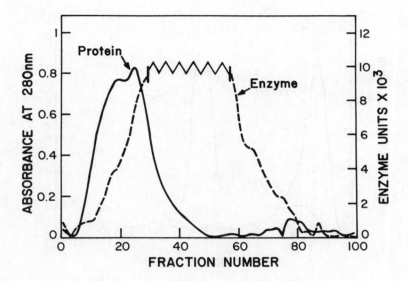

FIGURE 2.   Analytical-scale HPLC separation of uracil-DNA glycosylase on hydroxyapatite packing. The sample was the combined and concentrated fractions 70
to 85 from the QMA-Accell column. Fractions were collected at 2.0-min intervals.
Elution conditions were as follows:

Time (min)	Flow (ml/min)	% A	% B
0	0.6	100	0
40	0.6	100	0
70	1.0	0	100
90	1.0	0	100
100	1.0	100	0
115	0.6	100	0

Eluants: (A) 0.01 M potassium phosphate buffer, pH 7.4, containing 0.001 M
dithiothreitol; (B) 0.2 M KCl in 0.01 M potassium phosphate buffer, pH
7.4, containing 0.001 M dithiothreitol.

minor peak (fractions 14 to 45). The major peak was concentrated and then injected onto a
gel filtration column (Waters I-125, 60 cm × 7.8 mm). Results and eluting conditions are
shown in Figure 4.

One of the particular advantages of substituting the HPLC steps in the modified purification procedure is that it avoids the necessity for several time-consuming dialysis steps.
Recent studies have shown that samples from one HPLC step can be diluted with the initial
eluant buffer from the next step and pumped directly onto the column. Only for the gel
filtration step does the sample need to be concentrated.

B. SCALEUP TRIALS

Following development of the modified laboratory-scale purification of uracil-DNA
glycosylase, scaleup parameters for each of the HPLC steps were examined. A Waters Delta
Prep 3000 chromatography system proved very convenient for these scaleup studies. This
instrument is specifically designed for such purposes and has flow-rate capabilities from 1
to 180 ml/min, with column sizes from 3.9 mm × 15 cm in the analytical range to 57 mm
× 30 cm and larger in the preparative mode. Sample capacity runs from microgram to
multigram loads. A typical scaleup run for ion-exchange chromatography by HPLC is shown

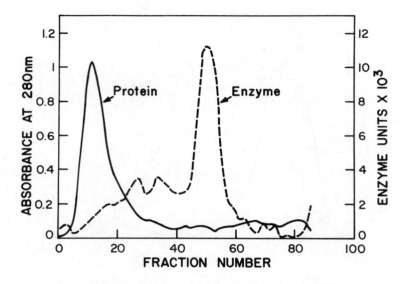

FIGURE 3.   Analytical-scale HPLC separation of uracil-DNA glycosylase on CM-300 cation-exchange packing. The applied sample was fractions 16 to 71 from the hydroxyapatite column. Fractions were collected at 1.0-min intervals. Elution conditions were as follows:

Time (min)	Flow (ml/min)	% A	% B	% C
0	1.0	100	0	0
30	1.0	100	0	0
50	1.0	0	100	0
55	1.0	0	100	0
60	1.0	0	0	100
65	1.0	0	0	100
70	1.0	100	0	0
75	1.0	100	0	0

Eluants: (A) 0.05 M Tris-acetate buffer, pH 6.8, containing 0.001 M EDTA and 0.001 M dithiothreitol; (B) 0.05 M Tris-acetate buffer, pH 7.8, containing 0.001 M EDTA, 0.001 M dithiothreitol, and 1.0 M sodium acetate; (C) 0.05 M Tris-acetate buffer, pH 7.8, containing 0.001 M EDTA, 0.001 M dithiothreitol, and 2.0 M sodium acetate.

in Figure 5. At the laboratory scale, a typical protein loading of 85 mg gave an overall enzyme yield of 70%, with a 22-fold purification. This can be compared with a pilot-scale run using 2.65 g of protein, where an 80% yield with a 30.6-fold purification was achieved. At the larger scale, separation of the enzyme was much cleaner than at the analytical level, with the bulk of the enzyme being eluted as a single peak. Scaleups with hydroxyapatite and cation-exchange packing materials have also been completed. Excellent enzyme resolution and very satisfactory yields of enzymatic activity were obtained.

   Among other HPLC packings tested for protein separations were a variety of reverse-phase C-18 materials from several manufacturers. These were examined because such C-18-type packings are typically quite stable and of relatively low cost, and noncomplicated eluant conditions can be used. Although resolution of protein peaks was often excellent, the use of this class of materials for uracil-DNA glycosylase purification was only marginally successful. Often enzymatic activity was either lost or spread throughout a relatively larger

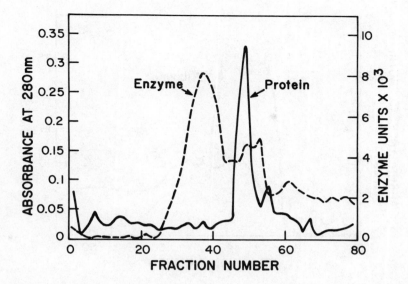

FIGURE 4. Analytical-scale HPLC separation of uracil-DNA glycosylase by gel filtration using a Waters I-125 column. The injected sample was 0.75 ml of the combined and concentrated peak samples (fractions 46 to 56) from the CM-300 column. Fractions were collected at 0.5-min intervals, using a 1.1-ml/min flow rate. Elution was isocratic, using a 0.05 M Tris-acetate buffer (pH 7.5) containing 0.001 M EDTA and 0.001 M dithiothreitol.

number of fractions of eluant. This was true over a wide range of eluant conditions and at both analytical and preparative scales. In some cases, mechanical strength of the particles was problematical when large-scale columns were used.

IV. DISCUSSION

Purification of uracil-DNA glycosylase according to the originally published procedure gave results which closely paralleled the reported values. Analysis of this procedure enabled us to identify several steps which could profitably be replaced, for scaleup, by HPLC techniques. Laboratory-scale enzyme purification incorporating these modifications gave an overall yield comparable to the original procedure, but with an improved specific activity of the final product.

Scaleup of several of the steps involving HPLC has been carried out with excellent results, including gains both in yield and in degree of purification, while providing greatly increased throughput compared with more conventional chromatographic procedures. At the present time, we are approximately two thirds of the way through a pilot-level scaleup of the total purification procedure. Our only unsatisfactory experience has come in the attempted use of reverse-phase adsorption chromatography packings. In this case it was the physical failure of the packing materials themselves, rather than an inability to predict scaleup parameters, that caused difficulties.

Overall, our work with the purification of uracil-DNA glycosylase has provided us with strong evidence that, with appropriate packing materials, HPLC techniques offer a highly predictable and effective means of scaling up the production of protein products from biotechnology.

ACKNOWLEDGMENTS

This work was supported, in part, by funds from the Separations Research Program,

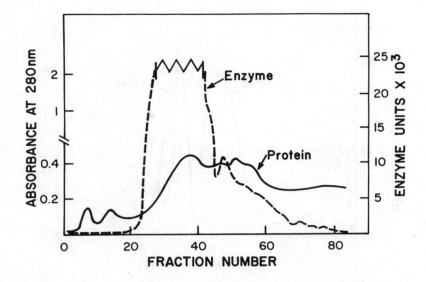

FIGURE 5.   Preparative-scale HPLC separation of uracil-DNA glycosylase on QMA-Accell anion-exchange packing. The applied sample was 5.5 ml of 35 to 65% ammonium sulfate fraction, dialyzed against buffer A, containing 2.4 g of protein. Elution conditions were as follows:

Time (min)	Flow (ml/min)	% A	% B	% C
0	10	100	0	0
20	10	50	50	0
30	10	0	100	0
35	10	0	100	0
36	10	0	0	100
40	10	0	0	100
45	10	100	0	0

Eluants: (A) 0.05 M Tris-acetate buffer, pH 7.8, containing 0.001 M EDTA and 0.001 M dithiothreitol; (B) 0.05 M Tris-acetate buffer, pH 7.8, containing 0.001 M EDTA, 0.001 M dithiothreitol, and 1.0 M sodium acetate; (C) 0.05 M Tris-acetate buffer, pH 7.8, containing 0.001 M EDTA, 0.001 M dithiothreitol, and 2.0 M sodium acetate.

The University of Texas at Austin, and by the Clayton Foundation for Research. We thank Dr. Larry Stepp, Jill Bedgood, Nancy Waidelich, and Raquelle Keegan for their assistance.

REFERENCES

1. **Duncan, B. K. and Chambers, J. A.,** The cloning and overproduction of *Escherichia coli* uracil-DNA glycosylase, *Gene,* 28, 211, 1984.
2. **Lindahl, T., Ljungquist, S., Siegert, W., Nyberg, B., and Sperens, B.,** DNA *N*-glycosidases: properties of uracil-DNA glycosidase from *Escherichia coli, J. Biol. Chem.,* 252, 3286, 1977.
3. **Kunkel, T. A.,** Rapid and efficient site-specific mutagenesis without phenotypic selection, *Proc. Natl. Acad. Sci. U.S.A.,* 82, 488, 1985.
4. **Lindahl, T.,** DNA repair enzymes, *Annu. Rev. Biochem.,* 51, 61, 1982.

5. **Lindahl, T.,** DNA glycosylases, endonucleases for apurinic/apyrimidinic sites, and base excision-repair, *Prog. Nucleic Acid Res. Mol. Biol.,* 22, 135, 1979.
6. **Sekiguchi, M., Hayakawa, H., Makino, F., Tanaka, K., and Okada, Y.,** A human enzyme that liberates uracil from DNA, *Biochem. Biophys. Res. Commun.,* 73, 293, 1976.
7. **Duncan, B. K.,** DNA glycosylases, in *The Enzymes,* Vol. 14, Boyer, P., Ed., Academic Press, New York, 1981, 565.
8. **Rahn, P., Joyce, W., and Schratter, P.,** Scalability: the challenge to chromatography, *Am. Biotechnol. Lab.,* 4, 34, 1986.

Chapter 31

BIOSEP: A NOVEL APPROACH TO PRECOMPETITIVE RESEARCH IN DOWNSTREAM PROCESSING

Michael G. Norton

TABLE OF CONTENTS

I. BACKGROUND

The U.K. has maintained a high level of interest in biotechnology and its industrial applications for many years. For instance, British companies were among the first to employ large-scale continuous aseptic fermentation techniques for single-cell protein manufacture in the early 1970s, and the U.K. was also the source of one of the two basic discoveries (hybridoma technology) which underpin the whole of modern biotechnology. Despite this strong early position, however, the commercialization of recombinant DNA and hybridoma technologies proceeded more slowly than in the U.S. and was the subject of a study in 1980 by the U.K. Advisory Committee on Applied Research and Development.[2]

As a result of this review, the U.K. Department of Trade and Industry (DTI) assigned a high priority to creating a framework within which biotechnology could grow and flourish in the U.K. Actions taken included encouraging the private sector to fund new companies (e.g., Celltech and Agricultural Genetics) and helping to establish the necessary infrastructure for a successful biotechnology industry.

In the latter context, reviews were conducted to identify any rate-limiting steps in the commercialization of biotechnology which might be addressed through cooperative research, and it was concluded that one of these was in the area of downstream processing (DSP). Costs of separation and purification could account for up to 70% of the production cost of biological products; thus, the economic success of many products could ultimately depend on reducing these costs. Early discussions with biotechnology companies suggested that there could be scope for significant improvements in the efficiency of primary cell/debris recovery, in the use of membranes, and in the large-scale application of adsorption and chromatography. As a result, one of the industrial research laboratories of the DTI (Warren Spring Laboratory) and the Harwell Laboratory of the U.K. Atomic Energy Authority cooperated to set up the Biotechnological Separations Project (BIOSEP) in 1983. The aim was to build on existing research expertise at the two laboratories to tackle generic research problems in DSP, which would benefit a wide range of companies.[3]

II. HOW BIOSEP OPERATES

BIOSEP is a multiclient cooperative research enterprise which is funded by a grant from the U.K. DTI and by annual subscriptions from the participating companies. At present, its 50 member companies all have equal access to the results of a program of work costing about $2 million (1988, U.S.) per annum. While most of this is conducted at the Harwell and Warren Spring laboratories, the program also funds work at several universities and involves academic specialists in DSP in reviewing results of the BIOSEP program. The membership is representative of the different types of companies in the industry (equipment suppliers, chemical, pharmaceutical, and new biotechnology companies, contractors, etc.) and has become increasingly international, with members from five countries in Europe as well as from North America and Australia.

The BIOSEP program is decided by members and has included a number of state-of-the-art reports (SARs), which comprise in-depth critical technical reviews of all data available on the selected subject, using both published and unpublished (e.g., industrial) sources. Such reviews are not only valuable products for members, but also allow the rate-limiting steps for that technology to be identified as priority areas for future research. A total of 14 SARs have been completed or are in preparation, covering the major unit operations in separation and more specific subjects such as inorganic membranes, flocculants as an aid to separation, and membrane-cleaning technologies. Early reviews confirmed the research priorities of the membership as belonging to the three main areas of primary separations, membrane processes, and adsorption/chromatography. Active research programs are underway in all three areas under the direction of technical panels attended by members every 6 months. These research activities (described in Section III) produce data on which companies can base process or equipment improvements in-house, and they also provide the basis for design reports (DRs) which make recommendations on the best currently available methods for selecting, designing, and using separation equipment. The first two DRs cover cell recovery efficiency improvements for centrifugation and the computer-aided optimization of packed-bed adsorption columns.

III. RESEARCH AND DEVELOPMENT ACTIVITIES

SARs and DRs are confidential to members. However, while members have immediate access to all results of the research program, some results are also published in the open literature after a delay period (e.g., References 4 to 12).

The main elements of the BIOSEP research program are summarized in Table 1, where the rationale and the objectives of each research project are listed. By way of illustration, further details of three of the projects (including hitherto unpublished results) are given in the remainder of this section.

A. APPLICATION OF FLOCCULANTS TO CENTRIFUGAL SEPARATION
1. Rationale

Separation of bacterial cells can involve considerable difficulties due to their small size (1 to 3 μm for cells, <1 μm for debris) and densities close to that of broth. Disk stack centrifugation is a common separation method, but relatively low flow rates are obtained where good cell recoveries are required. Stoke's law applied to centrifugation[8] shows that improvements in the efficiency of separation can be achieved by

- Increasing cell density
- Reducing broth viscosity
- Increasing particle diameter

Flucculants are widely used in the water industry to increase the particle size of bacterial flocs in order to aid settlement, but have not been widely used in the biotechnology industry. This research project has evaluated the effectiveness of flocculants in increasing the particle size in fermenter broths and quantifying the beneficial effect on centrifuge performance.

2. Methods

A model fermentation was used, based on the production of α-amylase by *Bacillus amyloliquefaciens*. Samples of fermentation broth were mixed with flocculant solutions and aggregation behavior examined by settlement tests, sedimentation tests, and particle size analysis. The effect of increased particle size on the efficiency of centrifugal separation was determined by passing the fermenter broth through a disk stack centrifuge (Alfa-Laval® LAPX) at a variety of flow rates. The selected flocculant was added immediately before entry to the centrifuge, and recovery rates were compared with and without addition of flocculant.

3. Results

A number of flocculants have been evaluated for their effectiveness in achieving cell aggregation. Results of particle-size analysis for one typical flocculant (polyacrylamide cationic of 15,000 mol wt) are shown in Figure 1. The effects of this flocculant on the efficiency of centrifugal recovery at two flocculant concentrations are shown in Figure 2.

TABLE 1
A Summary of BIOSEP Research Projects

Technical panel area	Problem	Laboratory approach	Objective
Primary separation	Efficiency of centrifugal recovery of small cells/debris is poor	Quantify potential of flocculation to increase particle size	Design recommendation for selection and optimization of flocculant use
	Physical models of particle-particle interactions not applicable to biological particles	Synthesize model "cell" of latex bead and cell surface molecules and study	Insight into mechanisms of bacterial aggregation at molecular level
	Selective separation or detection of cells difficult	Develop lab-scale dielectrophoretic separator cell	Evaluate potential of dielectrophoresis
Membrane processes	Protein adsorption causes extensive fouling	Measure adsorption isotherms of proteins on membranes, and study clean and fouled membrane structure via scanning electron microscopy	Insight into process of fouling to assist in membrane design and cleaning
	Transmission of soluble proteins through MF membranes is poor in the presence of cells/debris	Identify key factors affecting transmission	Design recommendations on membrane selection and configuration
	Antifoams foul membranes in downstream processing	Investigate the surface chemistry of the membrane-antifoam interaction and the effects of antifoam solution behavior	Selection guide for membrane and antifoam compatability
Adsorption and chromatography	There is a lack of comparative physical data on adsorbents	Assemble a standard test rig for measuring pressure drop and mass transfer properties	Performance characteristics of commercial products quantified
	Mathematical models for affinity adsorption are complex or inaccurate	Develop lab-scale measurements allowing full-scale predictions via model	User-friendly mathematical model (for IBM®-compatible PC) based on minimum input data
	Purification processes are multistep	Study potential of adsorbents to recover product directly from fermenter broth	Selection and design recommendations on product adsorption

4. Significance

These initial results demonstrate that flocculants are effective in significantly increasing the particle size of bacterial suspensions in broth and that this can lead to enhanced efficiency of centrifugal separation. Simple laboratory-based selection and optimization procedures are now under development to identify the most effective flocculant, determine its optimum concentration, and define the best conditions of mixing and conditioning to provide a floc which does not break up in the centrifuge. Laboratory predictions are also being correlated with performance at the pilot scale to generate design recommendations on selection, optimization, and full-scale use of flocculants in centrifugal cell recovery.

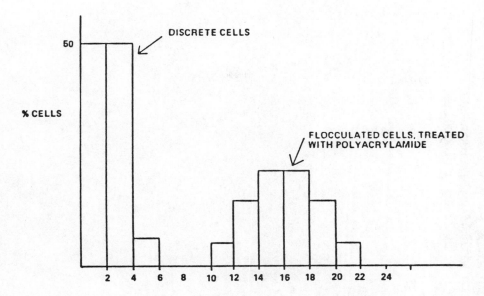

FIGURE 1. Increase in particle size distribution of *B. amyloliquefaciens* cells due to polyacrylamide flocculation.

B. MEMBRANE PROCESSES

1. Rationale

The highest priority in BIOSEP membrane research is the problem of flux decline and its avoidance.[9,10] The reversible buildup of solutes at the membrane surface (concentration polarization) and the irreversible adsorption of solutes such as proteins to the membrane material (fouling) cause a decrease in throughput. The latter is a major limitation to the use of both ultrafiltration (UF) and microfiltration (MF) membranes in biotechnological applications and has therefore been studied in the BIOSEP research program. One component of this program has quantified adsorption isotherms for proteins on different membrane materials.

2. Methods

The adsorption of bovine serum albumin (BSA) to a range of membranes was studied at Harwell using radioactively labeled BSA. Membrane disks were exposed to a given concentration of BSA solution for different lengths of time, washed in phosphate buffer to remove unbound protein, and the adsorbed protein determined by its level of radioactivity. By comparing the amounts adsorbed by different membranes under the same conditions, the relative affinity of different materials for BSA could be determined.

3. Results

Typical results comparing the adsorption of BSA to membranes of differing composition are shown in Table 2, illustrating the major differences in affinity that occur with different membrane materials.

4. Significance

The results typified by the above emphasize the importance of the interaction between protein and membrane in determining fouling potential. Other components of the membrane research program have quantified the various contributions to membrane fouling and have investigated the mechanism of fouling. One approach to the latter employs protein staining and scanning electron microscope techniques to locate protein on and in fouled UF mem-

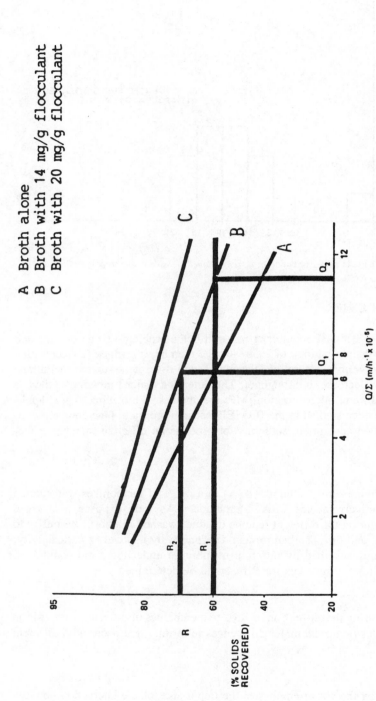

A Broth alone
B Broth with 14 mg/g flocculant
C Broth with 20 mg/g flocculant

FIGURE 2.   Recovery rate (R) of *B. amyloliquefaciens* cells from broth treated with polyacrylamides, at various flow rates (Q) in a disk stack centrifuge with equivalent settling area.

TABLE 2
BSA Adsorption to Various Membrane Materials

Membrane type	Membrane material	Nominal molecular weight cutoff	Adsorption (mg/m²)	Sample standard deviation (mg/m²)
3038	Acrylic	20 kDa	63.6	7.3
XM50	Substituted polyolefin	50 kDa	498	39
FS61	Fluoro polymer	20 kDa	1020	110
Anotec	Alumina	0.025 μm[a]	74.7	13
GR61	Polysulfone	20 kDa	18.8	4.9
SM14549	Cellulose acetate	20 kDa	60.3	4.7

Note: 1 h contact with 0.5% BSA solution in 0.05 M phosphate buffer (pH 7) at 25°C.

[a] Pore size.

branes. Gold immunostaining has shown that BSA can enter the body of some membranes with nominal molecular weight cutoff sizes of only 10,000 Da.

The interaction between protein and membrane is only one of several interactions which may contribute to fouling. Another important source of fouling may arise from the use of antifoams and other components of fermenter broth, and work is going on to elucidate the surface chemistry of the interaction between these substances and the membrane.

C. MODELING TO AID SCALEUP OF LIQUID CHROMATOGRAPHY
1. Rationale

Adsorption and chromatographic techniques for recovery and purification are generally developed at small bench scale, requiring scaleup calculations to be conducted in order to achieve optimal performance at production scale. A requirement therefore exists for computer methods for the design, performance prediction, and optimization of adsorption equipment. This project develops the necessary models, together with the required small-scale experiments, to provide the adsorption isotherm and kinetic data for the model.

2. Methods
a. Modeling

The system is modeled as a reversible mass transfer reaction in which the free adsorbate becomes bound to the adsorbent. The forward rate of the reaction is proportional to the amount of free adsorbate multiplied by the amount of adsorbent free to bind with the adsorbate. The reverse rate is proportional to the amount of adsorbate bound to the adsorbent, i.e.,

$$C \underset{K_2}{\overset{K_1(Q_m-q)}{\rightleftharpoons}} Q$$

where C is the free adsorbate, Q is the bound adsorbate, Q_m is the maximum capacity of the adsorbent, and K_1 and K_2 are constants. Further details of the development of the model have been published.[12] The differential equations involved in the solution of the problem are solved using a proprietary language developed by Harwell (FACSIMILE) and allow the adsorption, washing, and elution phases to be solved speedily on a PC.

b. Experimental

Two laboratory experiments are required to give the required constants: measurement

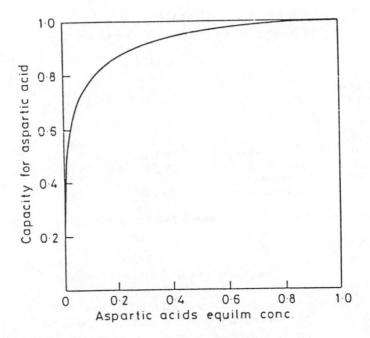

FIGURE 3.   Adsorption isotherm (dimensionless).

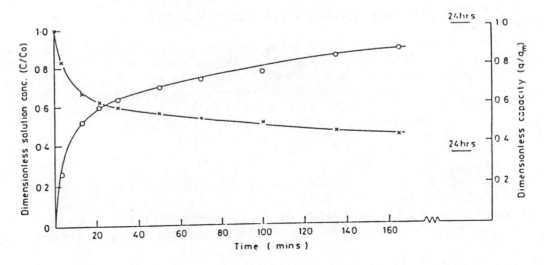

FIGURE 4.   Kinetics of aspartic acid adsorption in a stirred cell; Co = initial solution concentration, q_m = capacity at equilibrium. Adsorbent uptake: o——o; solution concentration: x——x.

of the adsorption isotherm and determining the kinetics of adsorption. The adsorption isotherm may be generated by contacting known concentrations of the adsorbate with known weights of adsorbent at a constant temperature until equilibrium is obtained. The kinetics can be measured most readily by monitoring the uptake of adsorbate in a stirred cell, giving a plot of uptake vs. time. These two experiments provide the necessary data to run the model.

3. Results

A typical validation run is shown in Figures 3 to 5. Figure 3 is the adsorption isotherm

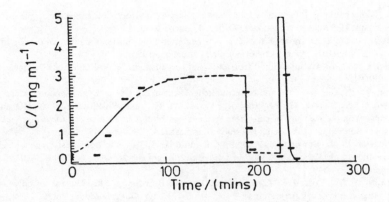

FIGURE 5.   Experimental data and simulation for loading, washing, and elution of aspartic acid onto Duolite® A162.

for the sytem of aspartic acid onto Duolite® A162. Figure 4 shows the kinetics of aspartic acid adsorption in a stirred cell. These curves are used to derive values for constants K_1 and K_2 and for Q_m which can then be used in the model to predict the behavior of the aspartic acid in a column of Duolite® A162. Figure 5 shows the results of a typical validation run for the adsorption, washing, and elution stages, and it demonstrates the closeness of the theoretical fit with the experimental data.

4. Significance

The validation of the above model allows the prediction of the performance of packed-bed adsorption columns from constants derived from simple small-scale experiments. This enables cost-effective optimization and scaleup of adsorption chromatography equipment. Recent expansions of the basic model described above enable more complex systems to be modeled, including the consideration of liquid film mass transfer rates, pore diffusion limitations, and competitive adsorption in multicomponent mixtures.

IV. CONCLUSIONS

BIOSEP has established itself over the last 4 years as an effective means of organizing precompetitive research in DSP. The research and technology transfer activities under BIO-SEP have succeeded in generating a consensus on research priorities between the various participants in the biotechnology industry, including suppliers, equipment users, contractors, and consultants.

Members set the priorities at the outset of each research project, provide guidance and midterm corrections during the project, and can also assist in final validation of design recommendations at industrial scale. The BIOSEP structure thus encourages a continuing dialogue between researchers and industrial customers which ensures that research remains relevant to real industrial needs.

REFERENCES

1. **Senior, P.,** Scale-up of a fermentation process, in *Biotechnology in Food Processing,* Harlander, S. K. and Labuza, T. P., Eds., Noyes, Park Ridge, NJ, 1986, 249.

2. **Anon.**, Biotechnology: Report of a Joint Working Party, Advisory Committee for Applied Research and Development, Her Majesty's Stationery Office, London, 1980, 1.

3. **Thomson, A. R.**, Downstream processing — a neglected field, in *Bioprocessing in the Eighties*, Institute of Chemical Engineers Symp., Chameleon Press, London, 1982.

4. **Bowden, C. P.**, Primary solid/liquid separation, in *Proc. Biotech '84*, Vol. 2, Online Publications, London, 1984, 139.

5. **Bowden, C. P.**, Cell recovery and biotechnological separation, *Chem. Eng. (London)*, March, 50, 1985.

6. **Bowden, C. P., Leaver, G., Melling, J., Norton, M. G., and Whittington, P. N.**, Recent and novel developments in the recovery of cells from fermentation broths, in *Separations for Biotechnology*, Verall, M. and Hudson, M. J., Eds., Ellis Horwood, Chichester, U.K., 1987, 49.

7. **Whittington, P. N., Bowden, C. P., Pethig, R., and Burt, J.**, Dielectrophoresis; a novel technology for selective bioparticle separations, in *Proc. 4th Eur. Congr. Biotechnology*, Vol. 2, TUP-231, Elsevier, Amsterdam, 1987, 540.

8. **Salusbury, T. T., Cosgrove, T., Langford, P., and Bowden, C. P.**, The use of polyelectrolytes as conditioning agents in cell harvesting, in *Proc. 4th Eur. Congr. Biotechnology*, Vol. 2, TUP-230, Elsevier, Amsterdam, 1987, 539.

9. **Leaver, G. and Newdick, P. C.**, Concentration polarisation and retention properties of microporous membranes, in *Proc. 4th Eur. Congr. Biotechnology*, Vol. 2, TUP-229, Elsevier, Amsterdam, 1987, 535.

10. **Gutman, R. G. and Leaver, G.**, Membranes separate big from small in biotechnology, *Process Eng.*, 65(6), 37, 1984.

11. **Caughlin, R. A., Cowan, G. H., Laws, G. H., and Ottoway, J.**, Preliminary results for the pressure drop characteristics of inorganic adsorbents from packed bed experiments, paper to the 1st Meet. Int. Biotechnol. Network, Universite Parl Sabatier, Toulouse, France, February 19 to 21, 1985.

12. **Cowan, G. H., Gosling, I. S., Laws, J. F., and Sweetenham, W. P.**, Physical and mathematical modelling to aid scale-up of liquid chromatography, *J. Chromatogr.*, 363, 37, 1986.

Part 6. Emerging Technologies in Bioprocessing

The work of some laboratories can be recognized as conspicuously progressive, inconspicuously promising, or dedicated to long-term, highly rewarding goals. A few projects in these categories have been singled out for emphasis in the following section, which is devoted to "emerging technologies". In the chapters that follow, Runstadler and Young present the Verax Process, which recognizes the real needs of mammalian cells and adapts reactor engineering technology to these needs; Matson and Lopez discuss means of approaching continuous production and processing using enzymes and membranes in an extractive reactor; and Ulmer describes the engineering of specific protein structures using chemical and genetic manipulations, which requires good three-dimensional structure information, which in turn requires good crystals.

Chapter 32

LARGE-SCALE FLUIDIZED-BED BIOREACTOR SYSTEMS

Peter W. Runstadler, Jr. and Michael W. Young

TABLE OF CONTENTS

I. INTRODUCTION

The biotechnology industry is rapidly approaching the introduction to the commercial market of high-valued, recombinant-engineered, therapeutic medical proteins. After years of research and development efforts, products now under development hold the promise of making a significant impact upon such scourges of mankind as heart disease and cancer. Most of these products are based upon naturally occurring biomolecules that are produced by genetically engineered cells using recombinant DNA techniques. Early in the development of the industry, bacteria (especially *Escherichia coli*) and yeast were the organisms of choice for biosynthesis. However, for the large-sized, complex protein molecules that are the focus of interest, bacteria and yeast are not capable of the complex posttranslational folding and glycosylation of the protein product necessary to produce effective biological activity. The mammalian cell has therefore emerged as the vehicle of choice for the manufacture of these products.[1] The new process technology that has evolved to produce these products requires a large-scale mammalian cell culturing sytem, followed by downstream processing for purification to final product form.

This chapter describes an immobilized, continuous culturing process developed by Verax Corporation (Lebanon, NH) for the low-cost, large-scale, reliable production of efficacious high-value biochemicals. The process is based on the immobilization of cells in a particulate matrix which is maintained as an optimum culture environment in a fluidized-bed bioreactor. The fundamental elements of the process are described. Data are presented to display the operation of the bioreactor systems at large production scale. The advantages of the fluidized-bed bioreactor process for continuous mammalian cell culture are described. Typical examples based on culturing data are provided to demonstrate the economics of these culturing systems for the future production of valuable health products.

II. THE CONTINUOUS, IMMOBILIZED-CELL, FLUIDIZED-BED BIOREACTOR SYSTEM

The technical objectives for an effective mammalian cell culturing process encompass the following:

- Predictable scalability
- Continuous immobilized-cell process
- Ease of process optimization
- Maximum cost-effective product yield

A. PREDICTABLE SCALABILITY

The need to bring commercial products to market on a timely schedule with confidence demands process scalability. The very nature of the development process usually implies operation of the culturing system at small scale with an evolution to ever larger scales as development and market demands dictate. Reliability and cost effectiveness are important parameters in scalability considerations. Reliability encompasses the ability of a given process to function well at larger scales as well as the ability to predict properly sized equipment as production demands increase. Cost considerations can become predominant when the process becomes immensely costly, both in capital and operating costs, in order to substantially increase volume output. The ability to reliably scale up the culturing process is one of the most important factors in bringing a product to market.

B. CONTINUOUS IMMOBILIZED-CELL PROCESS

A continuous immobilized-cell process permits the attainment of steady-state operating conditions. In contrast to a batch or semicontinuous batch process, the continuous process permits the realization of sophisticated optimization. One parameter at a time can be varied to truly understand the way to achieve optimal process operation.

C. PROCESS OPTIMIZATION

The continuous immobilized-cell process offers the advantage of decoupling the feed rate of the medium from the cell-specific growth rate. The limitation of washout, such as occurs in a chemostat suspension cell system, is no longer an issue, and a truly optimum culture environment can be produced by specifying the levels of medium substrate, the concentration of inhibitory (waste) products, and overall physical and chemical culture conditions. Cell metabolic energy, normally directed toward cell growth and division, can instead be directed toward increased cell-specific productivity (grams of product produced per viable cell per hour).

D. MAXIMUM COST-EFFECTIVE YIELD

The continuous immobilized process and the optimization of the bioreactor system, properly implemented, imply maximum cost-effective yield, i.e., maximum grams of product produced per liter of medium consumed. The continuous process has the distinct advantage that the cell colony need not reproduce periodically, as it must in a batch process. The savings of repeated downtime is valuable in the overall cost analysis of producing a product, but even more so are the advantages of a steady and reliable product quality output. If the culturing system and downstream processing are developed from a systems approach, significant savings in product cost and increases in product yield can be realized; as much as 80% of the cost of goods in manufacturing lies in downstream processing costs. The use of high medium-dilution rates, realized by the decoupling of medium feed rate from cell-specific growth rate by immobilizing the cells, also allows flexibility in specifying the time that the secreted product spends in the culture milieu.[2] For some products this can have a significant impact on downstream processing costs and, hence, on overall economics of the entire processing system.

On the basis of these arguments we developed a process based upon continuous, immobilized, long-term culture of mammalian cells for the mass production of biochemicals.

III. THE FLUIDIZED-BED BIOREACTOR

The technical objectives discussed above were achieved in our hands by immobilizing mammalian cells in a three-dimensional collagen matrix of spherical particles, each about 500 μm in diameter.[3] These collagen microspheres are fluidized in a bioreactor, thus forming the continuously operating, immobilized-cell, fluidized-bed bioreactor system. Figure 1 is a schematic illustration of the basic bioreactor system.

On the left is the fluidized-bed bioreactor. The basic element in the bioreactor is the 500- to 600-μm-diameter collagen microsphere (which is weighted with small, heavy particles to increase the specific gravity) in which cells are immobilized. This fundamental element, the microsphere, serves several basic purposes. It is the immobilization matrix in which the cells populate to extremely high densities, thus enabling the reactor system to be run as an immobilized perfusion-flow system. Second, the millions of microspheres distribute uniformly throughout the bioreactor volume, thereby ensuring uniformity of cell density and culture environment. Third, the characteristics of a properly fluidized bed provide the high rates of mass transfer required to deliver oxygen to the dense populations of cells in the microspheres. Fluidized-bed technology is well understood, and the use of a purely solid-liquid fluidized-bed configuration ensures good flow stability performance and confidence in scalability. It is well known that fluidized beds scale at constant bed depth.[4] Therefore, proper design of even the smallest reactors permits orders of magnitude scaling of reactor performance with excellent confidence.

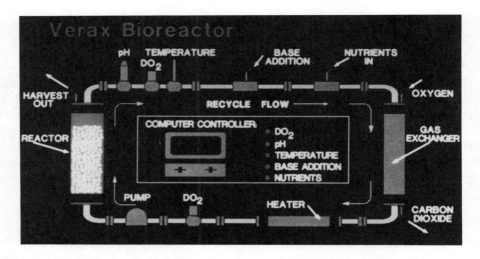

FIGURE 1.   Schematic of basic bioreactor system. (From Runstadler, P. W. and Cernek, S. R., in *Animal Cell Biotechnology,* Vol. 3, Spier, R. E. and Griffiths, J. B., Eds., Academic Press, London, 1988, chap. 13. With permission.)

The remainder of the piping and components shown in Figure 1 illustrate the recycle flow portions of the system and the flow paths for the addition of nutrients and the extraction of harvest from the continuously operating system. The pump in the recycle loop provides the continuous circulation of culture liquor to support the fluidized bed. Because of the characteristics of a fluidized bed, the culture liquor separates from the cell culturing microspheres at a clear separation horizon and continues out of the top of the reactor. In order to ensure good mass transfer of oxygen and nutrients, vertical superficial velocities in the reactor are on the order of 100 cm/min.[5]

Leaving the reactor, the recycled culture liquor passes pH-, dissolved oxygen-, and temperature-measuring instruments before entering a membrane gas exchanger (shown on the right in Figure 1). The gas exchanger is of the shell and tube type. The culture liquor flows internally in the silicone tubes of the gas exchanger, while oxygen is passed through the shell side. Silicone membranes are permeable to oxygen and carbon dioxide. The culture liquor, depleted of oxygen after passing through the reactor, now has the partial pressure of oxygen increased by permeation of oxygen from the shell side of the exchanger through the silicone membrane to the culture liquor. In the reverse direction, carbon dioxide permeates to the shell of the gas exchanger and is removed. Culture liquor thus "recharged" to a high partial pressure of oxygen leaves the gas exchanger and passes through a heater and a dissolved oxygen probe before reentering the pump for return to the reactor.

Conditions are such that the recycle flow circulates culture liquor rapidly through the bioreactor. Recycle flow rates are sufficient to exchange a bioreactor volume every several minutes (recycle dilution rate = recycle flow rate/bioreactor volume = 30 h^{-1}).

At the same time, nutrients are added to the bioreactor by injecting medium from a reservoir into the recycle loop. Because the bioreactor and recycle loop contain only liquids, an equal volume of harvest is always extracted from the bioreactor for every input of medium.

Dissolved oxygen levels, pH, temperature, and requirements for base addition and nutrients (medium flow rate) are all controlled by a computer. Sensors and the various components shown in the schematic are coupled to and are under monitor and feedback control of the computer.

In addition to the requirements of high mass transfer rates for oxygen and nutrients and the requirement of control of the bioreactor environment to achieve a homogeneous, uniform distribution of cells throughout the bioreactor, good, reliable aseptic design is an imperative

TABLE 1
Verax Microsphere Design
Specifications

- Diameter 500 μm
- Pore size, 20—40 μm
- Wet specific gravity, 1.6—1.7
- Biocompatible
- Postinoculate

for a mammalian culture system that is expected to run continuously for many months. This requires careful attention to engineering details so that not a single renegade organism is permitted to enter the aseptic envelope over the time of a reactor campaign.

Other features of the hardware design that demand careful engineering attention are the selection and design of components such as the feed and recycle pumps, valves, reactor, and the gas exchanger. Depending on cell type, 1 to 5% of the total cells in the bioreactor will be in the recycled culture liquor. It is important to maintain minimum levels of fluid dynamic shear throughout the entire system in order to prevent degradation of products produced by the cells or damage and death to the cells themselves.

IV. MICROSPHERE TECHNOLOGY

The heart of the system that enables the extremely high-density, immobilized-cell, fluidized bed is the weighted microsphere. Table 1 displays typical microsphere specifications. The microspheres are spherical, sponge-like balls of native collagen, each approximately 500 μm in diameter. The sponge-like character is provided through pores and interconnecting channels on the order of 20 to 40 μm in size. This pore size allows cells to easily enter and populate the entire internal volume of each microsphere. Both anchorage-dependent and anchorage-independent cells easily do this, with the result that the microspheres rapidly become populated to cell densities in excess of 10^8 viable cells per milliliter of microsphere volume.

Another essential feature of the microsphere is its high specific gravity, required to achieve the high mass transfer rates in the rapidly upward-flowing culture liquor that passes through the bioreactor. Wet specific gravities of 1.6 to 1.7 are required to maintain a fluidized bed state of the microspheres in the 100 cm/s upward flow through the bioreactor. Otherwise, the microspheres would be washed right out of the reactor. The collagen matrix is highly biocompatible. Collagen is the major constituent of the extracellular structural matrix in higher animals, and it is immunologically benign. Native collagen is a natural substrate for cell adhesion in anchorage-dependent cells.

The open structure of the microspheres also allows the easy postinoculation of each microsphere. Therefore, the microspheres are made prior to cell inoculation and are stored in a ready-to-use state.

Figure 2 is a schematic of the slurry of microspheres in the fluidized bed. Culture liquor circulates through and between the microspheres, while nutrients diffuse from the surface of the microsphere into the cells. Products and by-products are diffused outward and into the culture liquor.

Processing techniques for manufacturing the microspheres permit consistent control of the morphology, porosity, pore size, and fluidization characteristics. Figure 3 is a scanning electron micrograph of a portion of the collagen microsphere without cells. The best morphology is "leafy like", in which the cells can find optimum surface area for mechanical or physical attachment to the collagen. Although difficult to measure, estimates of the matrix surface area available for cells in the microspheres (estimated by calculation using the known

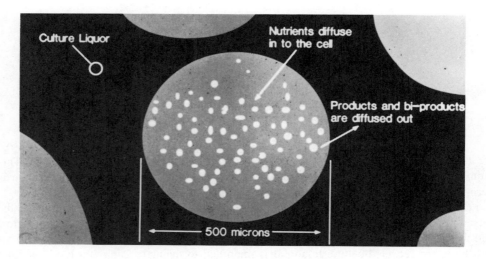

FIGURE 2.   Schematic of fluidized bed with microspheres.

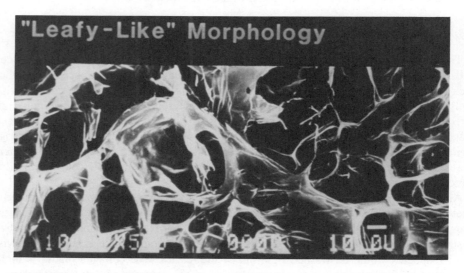

FIGURE 3.   Scanning electron micrograph of microsphere morphology. (From Runstadler, P. W., Jr. and Cernek, S. R., in *Animal Cell Biotechnology*, Vol. 3, Spier, R. E. and Griffiths, J. B., Eds., Academic Press, London, 1988, chap. 13. With permission.)

amount of collagen in the microspheres and knowledge of the conformation and average thickness of the sheets of collagen) yield values approaching 30 m²/l of packed microspheres. Figure 4 is a microphotograph of individual microspheres. The open pores and sponge-like character of the microspheres are readily apparent. The shiny particles seen in these unpopulated microspheres are small amounts of benign weighting materials that are added to the microspheres to obtain high specific gravities.

Figure 5 is a scanning electron micrograph of densely packed hybridoma cells populating the internal structure of a microsphere. Figure 6 is a scanning electron micrograph showing anchorage-preferred cells covering both the weighting materials and the collagen matrix.

V. PERFORMANCE

A variety of cell types, both anchorage-dependent and suspension-type cells, have been

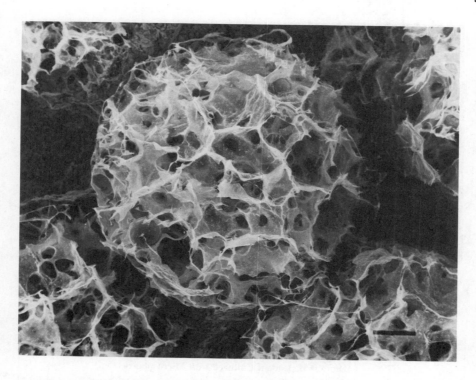

FIGURE 4. Microphotograph of individual microspheres. (From Runstadler, P. W., Jr. and Cernek, S. R., in *Animal Cell Biotechnology*, Vol. 3, Spier, R. E. and Griffiths, J. B., Eds., Academic Press, London, 1988, chap. 13. With permission.)

FIGURE 5. Hybridoma cells growing in collagen microspheres.

cultured to cell densities $>10^8$ viable cells per milliliter of matrix in 100-ml to 1-l systems. Table 2 shows these different cell types. Among the anchorage-dependent genetically engineered cells that have been cultured are Chinese hamster ovary (CHO) cells (making five different genetically engineered products), human tumor cells, African green monkey kidney cells, human embryonic kidney cells, and mouse mammary tumor cells. Nonengineered

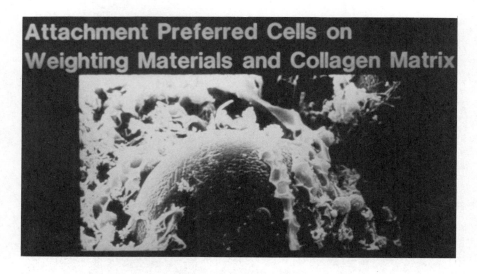

FIGURE 6.   Attachment-preferred cells growing in microspheres.

TABLE 2
Cell Types

Anchorage Dependent Cultured with Microspheres
    Engineered
        rDNA-CHO, human tumor, AGMK, human embryonic kidney, mouse mammary tumor
    Nonengineered
        Rat kidney, transformed rat kidney, COS 7, human hepatoma
Hybridoma Cell Types Cultured in Microspheres
    30 hybridomas
    Mouse/mouse
    Mouse/rat
    Mouse/human
    Human/human
    IgM, IgG, IgG2a, IgG2b, IgG3

cells have included rat kidney cells, transformed rat kidney cells, and human hepatoma cells. Over 30 different hybridomas have also been cultured successfully, including mouse/mouse, mouse/rat, mouse/human, and human/human, making a wide range of immunoglobulin products. Figure 7 illustrates monoclonal antibody production using the fluidized-bed process, and it shows typical start-up conditions leading to steady-state antibody production. For this hybridoma making an IgG product, steady-state conditions were reached in approximately 22 d, yielding harvest concentrations of approximately 90 µg/ml and total product production on the order of 400 mg/d from an 800-ml reactor (0.5 g/l of reactor per day).

Figure 8 displays the performance of a 5-l fluidized-bed bioreactor using a recombinantly engineered CHO cell producing a lymphokine. Steady-state performance with a cell-specific productivity equal to cell performance in a batch reactor was reached after approximately 15 d. However, continued operation of the system at a steady-state medium feed rate of 25 l/d induced a further increase in cell-specific productivity on the order of 1.5 times equivalent semicontinuous batch performance at the end of 30 d.

A recombinant CHO cell line making tissue plasminogen activator (tPA) was cultured for more than 60 d in a small 150-ml fluidized-bed bioreactor, and the same cell line was

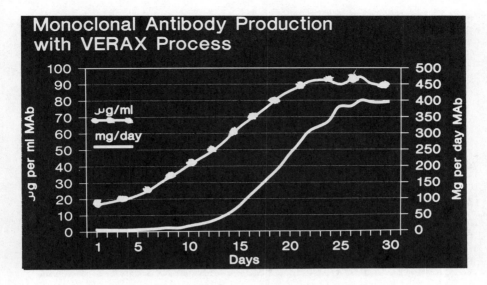

FIGURE 7. Monoclonal antibody production with Verax process.

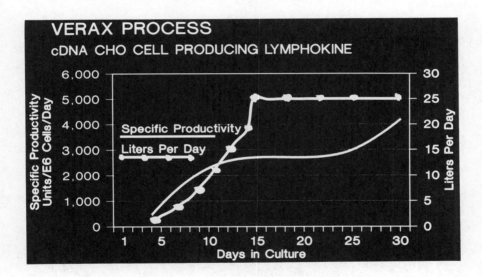

FIGURE 8. Verax process — cDNA CHO cell producing lymphokine.

cultured for 30 d in a larger 24-l system. In both cases, cell densities greater than 10^8 viable cells per milliliter of matrix at viabilities of approximately 80% were obtained at steady-state conditions at the end of each run. Figure 9 shows the concentration of tPA in the harvest and the medium feed rate during the run in the large 24-l fluidized-bed system. Figure 10 shows the daily tPA output in grams during this run. During the steady-state portion of the run (day 24 on), greater than 20 g of tPA were produced per day by the system.

VI. COST-EFFECTIVE PERFORMANCE

Thus, the fluidized-bed bioreactor using the collagen microsphere matrix has been shown to be useful for many, if not all, cell types, both anchorage dependent and anchorage

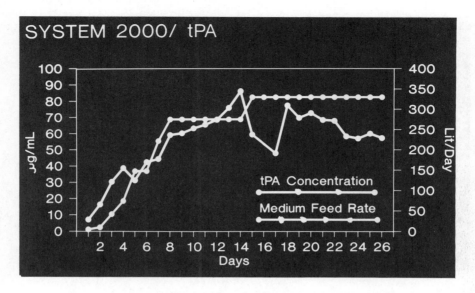

FIGURE 9.   Concentration of tPA in harvest and medium feed rate during run in 24-l fluidized-bed System 2000.

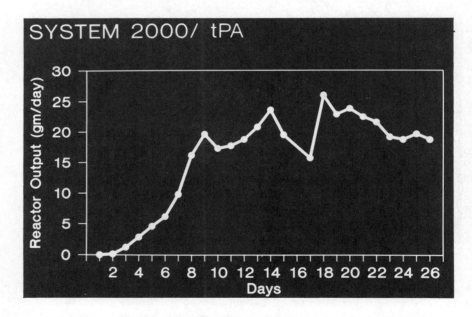

FIGURE 10.   Daily tPA output in grams during run with 24-l System 2000 bioreactor.

independent. Operation under continuous culture conditions provides optimization of the culture environment through the understanding of the kinetics of growth and product formation. The process yields a product with excellent consistency, purity, and biological activity. These factors have important implications for downstream purification, leading to reduced costs, if a total systems approach is used to optimize overall performance.

The entire fluidized-bed bioreactor process is inherently scalable.[4,5] Properly operated systems are easily scaled over many orders of magnitude in reactor volume size. The smallest reactors run to date are 150-ml fluidized beds, while the largest reactors encompass a bed

FIGURE 11.   Verax® System 2000 bioreactor in operation.

of 24 l. The compatibility of cells to the native collagen and the morphology and structure of the microspheres ensure maximum cell densities. All of these factors lead to an extremely cost-effective bioreactor system. An additional feature is the stability of cultures in the fluidized-bed system. In addition to the stabilization of the cells as an inherent result of immobilization, these reactor systems become essentially immune to genetic mutations by isolating such mutations to one or, at the very most, a few microspheres within the millions of microspheres that comprise the reactor bed.

VII. LARGE-SCALE PRODUCTION SYSTEMS

Large-scale systems for commercial production are available using this technology. The largest of these systems is a 24-l fluidized-bed reactor that is computer controlled and is cleaned and sterilized in place. The entire system is approximately $3.5 \times 2 \times 1.2$ m high. Figures 11 and 12 are photographs of this system in operation. Normal recycle flow rates are on the order of 12 l/min, and medium and harvest rates vary from 200 to 900 l/d (actual values being dependent upon the particular cell-line requirements and optimum culturing conditions). Thus, the relatively small 24-l fluidized-bed system is equivalent in product output to 1000-l and more conventional stirred-tank fermentation systems.

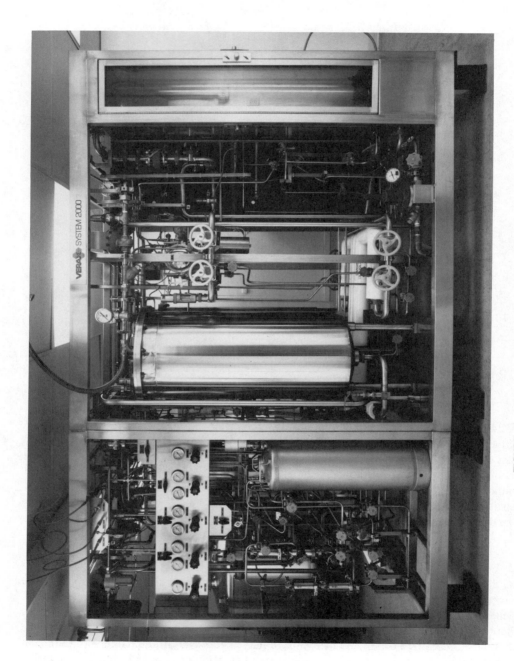

FIGURE 12.   Side view of Verax® System 2000 bioreactor.

TABLE 3
Estimate of Market Economics with
Verax Process

An Estimate of Profitability — A Blood Protein

Market size	55,800 g/year
Selling price	$14,000/g
Market potential	$781 million
Market share	$148 million revenue
Cost of goods	$10 million
Gross margin	$138 million (94%)

Note: All monetary values are 1987 U.S. dollars.

VIII. PROCESS ECONOMICS

The bottom line in the selection of a process for commercial production is the economics of the process, although factors related to timing and reliability of market entry, etc., as discussed above, must also be considered. A valid economic analysis should consider a thorough treatment of market factors as well as the evaluation of the process economics. An example of this type of analysis is presented for a recombinant blood protein, tPA, produced using the fluidized-bed process.

Table 3 shows an estimate of profitability for a typical therapeutic blood protein product, based upon an analysis of

- Total market size of 55,800 g/year, estimated from dose level per treatment and number of patients per year
- Selling price of $14,000/g*
- Market potential of $781 million per year
- Market share (gross revenues) of $148 million per year (= 19% of total market)
- Cost of goods of $10 million per year
- Gross margin** of $138 million per year or 94%

The cost of goods estimate is obtained from an analysis of the production process and includes details of

- Cell culture specifics
- Medium costs
- Costs for labor and overhead
- Materials and supplies
- Bioreactor costs (including matrix, license, royalty, and maintenance costs)
- Separation and purification costs and yields

Table 4 displays the essential information on the cell culture specifics used for this cost analysis. Costs for a defined medium (serum free) were estimated at $1.00/l. Almost 80% of the production costs for this particular product were estimated to be associated with separation and purification, assuming a 30% yield of purified to raw product from the culture harvest. The final calculated cost of raw product in the culture harvest is less than $100/g,

---

* All monetary figures are 1987 U.S. dollars.
** Gross margin = gross revenues − cost of goods; % gross margin = gross margin/gross revenues × 100.

TABLE 4
Cell Culture Specifics Using Verax Process

Cell Culture Specifics — A Blood Protein

Productivity	1.8 μg/10^6 viable cells/h
Density	1 $\times$ 10^8 viable cells/ml of matrix
Viability	80%
Generation time	19 h (growth), 30 h (production)
Medium feed rate	0.02 ml/10^6 cells/h
Yield	94 mg/l
Medium cost[a]	$1/l

[a]   In 1987 U.S. dollars.

and seven System 2000-type production systems could meet the annual required production of 64 kg of raw product harvest.

These data are typical of the performance obtained from the fluidized-bed bioreactor process and are the type of data needed to make appropriate decisions on manufacturing strategies based on accurate process economics.

IX. CONCLUSION

The immobilized, continuous, fluidized-bed bioreactor culturing process provides the following benefits for large-scale production of mammalian cell bioproducts:

1.   It is suitable for both anchorage-dependent and anchorage-independent cells.
2.   It is easily scaled from research development systems to the large systems required for commercial production.
3.   The process provides maximum cell densities for cost-effective volumetric output.
4.   Combined with a systems approach to process design, the process provides for reduced downstream purification costs and increased purification yields.
5.   It provides long-term culture stability.
6.   Product consistency, purity, and biological activity are also provided.

Because of the ability to precisely control the bioreactor environment, the kinetics of growth and product formation enable the user to achieve optimum cost-effective performance.

REFERENCES

1. **Arathoon, W. R. and Birch, J. R.,** Large-scale cell culture in biotechnology, *Science,* 232, 1390, 1986.
2. **Vieth, W. R. and Venkatasubramanian, K.,** in *Immobilized Microbial Cells,* Am. Chem. Soc. Symp. Ser. 106, Venkatasubramanian, K., Ed., American Chemical Society, Washington, D.C., 1979, 1.
3. **Dean, R. C., Jr., Karkare, S. B., Phillips, P. G., Ray, N. G., and Runstadler, P. W., Jr.,** Continuous cell culture with fluidized sponge beads, in *Large-Scale Cell Culture Technology,* Lyderson, B. K., Ed., Hanser Publications, New York, 1987, 145.
4. **Young, M. W. and Dean, R. C., Jr.,** Optimization of mammalian cell bioreactors, *Biotechnology,* 5, 835, 1987.
5. **DeLucia, D. E. and Chow, T.,** Design and scale-up of fluidized bed bioreactors, presented at the American Society of Mechanical Engineers Winter Annu. Meet., Boston, December 13 to 18, 1987.

Chapter 33

MULTIPHASE MEMBRANE REACTORS FOR ENZYMATIC RESOLUTION: DIFFUSIONAL EFFECTS ON STEREOSELECTIVITY

Stephen L. Matson and Jorge L. Lopez

TABLE OF CONTENTS

I. INTRODUCTION

Fulfillment of the commercial potential of biotechnology requires solutions to several demanding bioprocessing problems that pose unique challenges to chemical engineers. One response to this has been the development and emergence of novel bioreactors employing immobilized or entrapped enzymes and cells. Within this field of biochemical engineering, process integration — more specifically, the advantageous coupling of bioconversion and product recovery operations — is a prevalent theme; another is exploration of bioreactor configurations that operate efficiently in multiphase reaction systems. Significantly, at the same time that biotechnology scaleup is making these aggressive demands on process technology, some of the products of that industry (e.g., protein-engineered enzymes and monoclonal antibodies) are promising to serve as new process tools for the transformation and purification of high-value species of chemical and biological origin.

Such classical membrane separation processes as microfiltration, ultrafiltration, and reverse osmosis successfully address a number of important downstream product-recovery problems.[1] Moreover, membrane scientists have recently become fascinated with the idea of "activating" membranes by incorporating catalytically reactive or adsorbent species within them. Typically, the purpose in doing this is to render the membrane capable either of performing a separation more selectively or of performing a chemical conversion simultaneously with a separation.

In this chapter we discuss a class of membrane reactors wherein a membrane-contained biocatalyst converts a water-insoluble reactant to product in a multiphase reaction system. The particular reactor described here is representative of a family of related enzyme membrane reactors that have in common the fact that some nontrivial attribute of the membrane support is exploited to endow the membrane reactor with capabilities that are without counterpart in more conventional immobilized-enzyme reactors.[2-4] Finally, we examine the application of these multiphase membrane bioreactors to the important problem of isolating individual stereoisomers at high optical purity from racemic mixtures.

II. BIOCONVERSIONS OF SPARINGLY SOLUBLE SUBSTRATES

Many chiral resolution opportunities involve organic compounds that are only slightly soluble in water. Clearly, one problem with enzyme-based resolution methods is finding an enzyme that exhibits adequate activity and the desired stereoselectivity toward what is usually an unnatural substrate. Many enzymes have been identified, however, that have utility in the production or resolution of a number of chiral carboxylic acids, esters, and alcohols. The problem that remains is to perfect bioreactor and bioprocess designs that make such enzyme-based resolution schemes economical at the commercial scale.

The problem with existing dispersed-phase reactor systems can be appreciated by close examination of the mass transfer and reaction steps. Consider the generalized case in which it is desired to produce a resolved carboxylic acid. A preferred enzymatic approach is to use a hydrolytic enzyme, such as a lipase or carboxyl esterase, to stereoselectively convert a simple ester derivative to its corresponding acid and alcohol:

$$R'CO_2R'' + H_2O \rightarrow R'CO_2H + R''OH$$

The acyl moiety, R', in many of the chiral pharmaceuticals and pesticides of commercial interest is composed of substituted benzyl, naphthyl, and/or aryloxy functionalities that make the ester poorly soluble in water. The resulting acids, however, are frequently water soluble, since most often the optimum pH at which the enzymatic reaction is conducted is above the pK_a value of the acidic product.

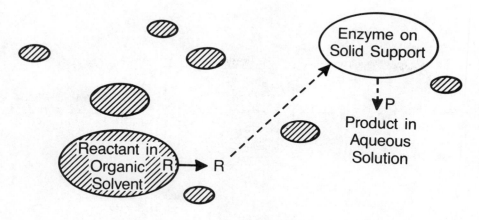

FIGURE 1.   Conventional dispersed-phase reaction system.

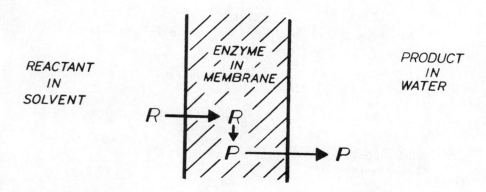

FIGURE 2.   Schematic of multiphase enzyme-membrane reactor operation.

Such reactions are conventionally carried out in a multiphase stirred tank reactor or in a catalytic packed-bed configuration. In either case, reactor productivity can be poor. The nature of the productivity limitation is shown schematically in Figure 1. The organic substrate is typically dispersed in a continuous aqueous phase, either as the "neat" organic liquid or dissolved in an appropriate water-immiscible solvent. The enzyme catalyst is immobilized on, or contained within, a porous support of some type. In order for the reaction to proceed, the substrate must first partition into the aqueous phase and be transported to the supported catalyst. However, its low water solubility limits the diffusive flux of substrate to the catalyst; consequently, reactor productivity suffers. Additional problems associated with dispersed-phase reaction systems relate to difficulties in phase separation and product recovery.

III. MULTIPHASE MEMBRANE BIOREACTORS

Figure 2 shows a cross-sectional view of the enzyme membrane in a multiphase membrane reactor designed to address many of these limitations of dispersed-phase reactors. A hydrophilic, microporous membrane — suitably activated by incorporation of an enzyme within its pores — is disposed at the interface between aqueous and organic process streams. In operation, the reactant partitions from the organic-phase feed stream into the water-wet enzymatic membrane, where it is subsequently converted to the water-soluble product. The product then diffuses out into the aqueous process stream.

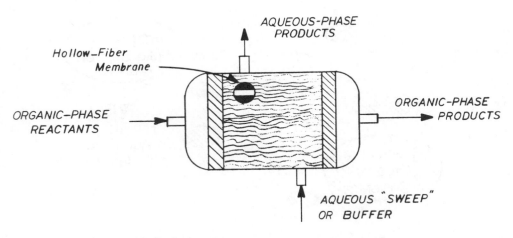

FIGURE 3.   Hollow-fiber multiphase membrane reactor.

TABLE 1
Batchwise Hydrolysis of Ethyl Butyrate in a Multiphase Membrane Reactor

Module	Solvent-resistant PAN module (1.0 m^2)
Enzyme	Pig liver esterase
Enzyme loading	150 mg
Intrinsic enzyme activity	200 μmol EtOBu hydrolyzed/min-mg
Organic phase	Ethyl butyrate (100%)
Aqueous phase	Phosphate buffer, 0.2 M, pH 8.0
Reactor productivity	9000 μmol/min; 416 kg/year-m^2
Enzyme effectiveness factor	30%

The membrane in a multiphase membrane reactor thus serves three functions. First, it provides high-surface-area contact between the two immiscible process streams on either side of it. Second, the membrane serves at the same time to separate the bulk phases, thus avoiding the need to disperse one phase within the other. Finally, the enzyme-activated membrane also functions as an interfacial catalyst, putting the supported enzyme in direct contact with the organic phase containing the reactant. In this manner, it is possible to minimize the diffusive limitations associated with the intervening bulk aqueous phase that is characteristic of dispersed-phase bioreactors. Hollow-fiber membrane modules are particularly attractive as multiphase membrane reactors (see Figure 3).

Table 1 shows experimental results for the enzymatic hydrolysis of ethyl butyrate (EtOBu), a model substrate, fed as the neat organic liquid to a multiphase membrane bioreactor. Noteworthy is the enzyme effectiveness factor of 30%, which is calculated as the ratio of the observed activity of 9,000 units to the intrinsic enzyme activity of the 30,000 units initially loaded onto the membrane. The productivity of the reactor in this case exceeds 400 kg/year-m^2 of active membrane area. Such productivities make multiphase membrane bioreactors quite feasible from an economic viewpoint when applied, for example, to the optical resolution of high-value compounds.

Significantly, the enzyme was not covalently attached to the polyacrylonitrile copolymer membrane in this experiment; rather, it was reversibly entrapped within the membrane in a manner that facilitates replacement of deactivated biocatalyst. Figure 4 shows a cross section of the type of asymmetric, hydrophilic hollow-fiber membrane used. The two features most critical to its enzyme entrapment capabilities are the fine microporous fiber walls and the existence of a "skin" layer at one of the membrane surfaces. This asymmetric membrane "immobilizes" the enzyme by entrapping it between two barriers: (1) the enzyme-imperme-

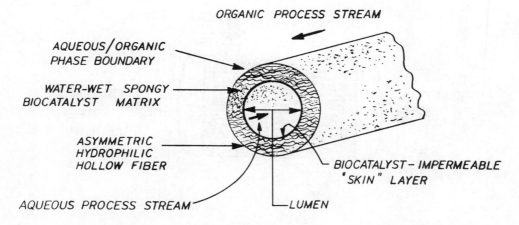

FIGURE 4. Reversible enzyme containment in an asymmetric, hydrophilic hollow fiber.

able skin layer and (2) the aqueous/organic interface maintained at the opposite membrane surface. The large size of the enzyme (125 kDa) relative to the molecular weight cutoff ($MWCO_{50\%}$) of 30 kDa of the surface pores prevents enzyme from diffusing across the skin layer of the membrane, while the poor solubility of the enzyme in organic solvents prevents it from partitioning across the aqueous/organic phase boundary.

Membrane modules containing hollow fibers of the type shown are first loaded with enzyme by charging an aqueous solution of the biocatalyst to the shell side of the module and ultrafiltering the solvent (i.e., water) through the fiber wall. In the process, enzyme accumulates within the porous fiber wall. Excess enzyme solution is then displaced from the fiber bundle by flushing with air and/or a water-immiscible organic solvent. Deactivated enzyme can be removed and replaced as required by means of a simple back-flushing and reloading procedure.

IV. OPTICAL RESOLUTION OF D,L-BTEE

The application of a multiphase membrane reactor to the enzymatic resolution of racemic amino acids has been demonstrated by feeding it a racemic mixture of *N*-benzoyl tyrosine ethyl ester (D,L-BTEE). Moreover, this substrate is only sparingly water soluble (the saturation concentration of racemic BTEE is approximately 0.2 m*M*) so that this substrate is also representative of several commercially important chiral compounds that exhibit low water solubilities.

The 1.0-m² multiphase membrane reactor employed for BTEE resolution consisted of polyacrylonitrile-based hollow-fiber membranes activated by incorporation of 4 g of chymotrypsin. This enzyme exhibits essentially perfect stereoselectivity for the hydrolysis of L-BTEE to the corresponding acid, *N*-benzoyl-L-tyrosine (L-BT). In this particular case, the enzyme was first adsorbed to the micropore walls; subsequently, the adsorbed protein layer was stabilized by covalently cross-linking it with a 2.5% solution of glutaraldehyde. The poorly water-soluble, racemic ester was then fed to the membrane reactor as a 30 m*M* solution in 1.1 l of octanol; the opposite surface of the membrane (see Figure 5) was contacted with 0.2 l of aqueous phosphate buffer (pH 7.8) to maintain reaction pH and provide a reservoir for the water-soluble reaction product. Both the organic and aqueous phases were then recycled to the multiphase membrane reactor, with the batchwise reaction essentially proceeding to completion overnight.

Table 2 summarizes concentrations and quantities of each of the species present in the reaction system before and after the run; both organic- and aqueous-phase concentrations

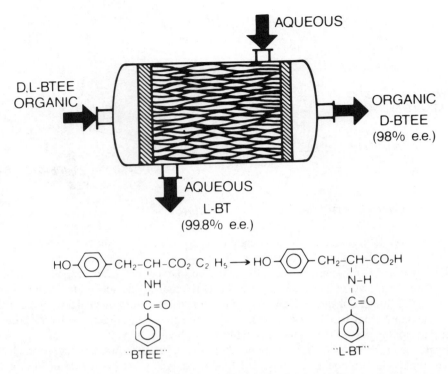

FIGURE 5.   Enzymatic resolution of D,L-BTEE in a multiphase membrane bioreactor.

TABLE 2
Enzymatic Resolution of D,L-BTEE in a Multiphase Membrane Reactor

Species	Stream	Initial		Final	
		Conc. (mM)	mmol	Conc. (mM)	mmol
D-BTEE	Organic	14.9	16.8	14.9	16.79
L-BTEE	Organic	14.9	16.8	0.0	0.0
L-BT	Organic	0.0	0.0	0.12	0.14
D-BTEE	Aqueous	0.0	0.0	0.07	0.01
L-BTEE	Aqueous	0.0	0.0	0.0	0.0
L-BT	Aqueous	0.0	0.0	80.0	16.7

Note: Membrane: 4 g chymotrypsin on 1-m^2 PAN hollow-fiber module; organic phase: 1.13 l 1-octanol; aqueous phase: 0.21 l phosphate buffer, pH 7.8; organic-phase enantiomeric excess: 98%; aqueous-phase enantiomeric excess: 99.8%.

are tabulated. Nearly all of the relatively water-soluble reaction product, the L-BT acid, was found in the aqueous process stream, where it was enriched approximately fivefold over the initial concentration of L-ester in the feed stream. In contrast, the inert D-BTEE ester was recovered in high yield from the organic stream. A final accounting of the concentrations of the various enantiomers yielded enantiomeric excess values of 0.98 (i.e., [D − L]/[D + L]) and 0.998 (i.e., [L − D]/[L + D]) in the organic and aqueous phases, respectively.

This example serves to illustrate the ability of single-layer, hollow-fiber-based multiphase membrane reactors to separate and enrich reaction products in systems involving the organic-phase feed of water-insoluble reactants. Moreover, it highlights an important area for commercial application of the technology, namely, the enzymatic resolution of sparingly water-soluble chiral pharmaceuticals and agricultural chemicals.

V. DIFFUSIONAL EFFECTS ON STEREOSELECTIVITY

The enantioselectivity exhibited by chymotrypsin toward the two isomers of BTEE is, for all practical purposes, absolute. However, in many other cases, enzymes that show stereospecificity do so only to a certain extent. The practical use of these enzymes in the large-scale resolution of enantiomers will depend on the degree of specificity that the enzyme exhibits toward a racemic mixture of the isomers. Chen and co-workers have introduced a concept which establishes a simple relationship between conversion, optical purity, and stereospecificity.[5] The derivation of their relationship for the case of simple, homogeneous enzyme kinetics is reproduced below.

Consider the following reactions, where A_R and A_S refer to the R- and S-enantiomers of a chiral compound:

$$A_R + Ez \longleftrightarrow EzA_R \xrightarrow{k_{2R}} Ez + B_R \tag{1}$$

$$A_S + Ez \longleftrightarrow EzA_S \xrightarrow{k_{2S}} Ez + B_S \tag{2}$$

Assuming that the net reaction is irreversible and not inhibited by product, one has the following reaction rate expressions for the two isomers:

$$v_R = -(k_2/K_m)_R[Ez][A_R] \tag{3}$$

$$v_S = -(k_2/K_m)_S[Ez][A_S] \tag{4}$$

$$= -(k_2/K_m)_R[Ez][A_S]/E \tag{5}$$

where

$$E = \frac{(k_2/K_m)_R}{(k_2/K_m)_S} \tag{6}$$

and the conversion, c, is given by

$$c = 1 - \frac{[A_R] + [A_S]}{[A_R]^0 + [A_S]^0} \tag{7}$$

The enantiomeric excess (ee), a measure of optical purity, is given by the following expressions for unconverted reactant and product, respectively:

$$ee(A) = \frac{[A_S] - [A_R]}{[A_S] + [A_R]} \tag{8}$$

and

$$ee(B) = \frac{[B_R] - [B_S]}{[B_R] + [B_S]} \tag{9}$$

For a racemic mixture, $[A_R]^0 = [A_S]^0$, and the equations describing enantiomer concentrations as a function of time are

$$\frac{d[A_R]}{dt} = -v_R \tag{10}$$

and

$$\frac{d[A_S]}{dt} = -v_S \tag{11}$$

By taking the ratio of Equations 10 and 11, the time variable can be eliminated:

$$\frac{d[A_R]}{d[A_S]} = \frac{v_R}{v_S} \tag{12}$$

Finally, integration of Equation 12 yields the following relationships between conversion, enantiomeric excess, and the enzyme stereospecificity parameter E:

$$\frac{\ln\{(1 - c)[1 - ee(A)]\}}{\ln\{(1 - c)[1 + ee(A)]\}} = E \tag{13}$$

and

$$\frac{\ln\{1 - c[1 + ee(B)]\}}{\ln\{1 - c[1 - ee(B)]\}} = E \tag{14}$$

The higher the value of E, the more optically pure the product will be at a specified conversion.

The above relationships between conversion, enzyme stereospecificity, and optical purity are presented graphically in Figures 6 and 7 for a few values of E. Figure 6 pertains to the situation where it is the optical purity of unconverted reactant that is of interest, whereas Figure 7 summarizes the situation wherein it is the optical purity of the reaction product that is important.

The above derivation is valid only for the case of homogeneous enzyme kinetics; it is implicitly assumed that diffusional limitations between the reactant and the immobilized enzyme are insignificant. In most large-scale, practical applications of enzymatic resolution, however, the enzyme will be immobilized on a solid support in order to permit its reuse, and the attendant diffusional resistance will give rise to concentration gradients in the reacting species within the solid-phase biocatalyst. This is particularly true where high enzyme loadings are used to maximize reactor productivity, for example, in order to minimize the required membrane area in a multiphase membrane reactor.

Such diffusional resistance can lead to a reduction in the effective stereospecificity exhibited by an immobilized enzyme. Assume that for a particular enzyme the stereospecificity value E, as measured in a homogeneous system, is greater than 1 and that the enzyme has been immobilized within a porous spherical particle without affecting its intrinsic stereospecificity. Imagine further that the particle is now immersed in a solution of racemic reactant A. Barring any external mass transfer resistances, the rate of diffusion of each isomer into the particle will be proportional to its rate of reaction inside the particle. Since a value of E greater than 1 means that one of the isomers is consumed faster than the other,

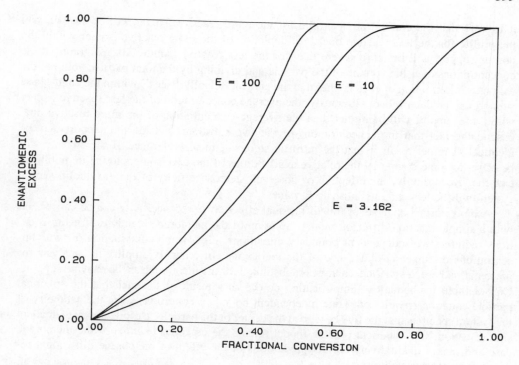

FIGURE 6. Enantiomeric excess of reactant A vs. conversion for several values of E in a homogeneous reaction system.

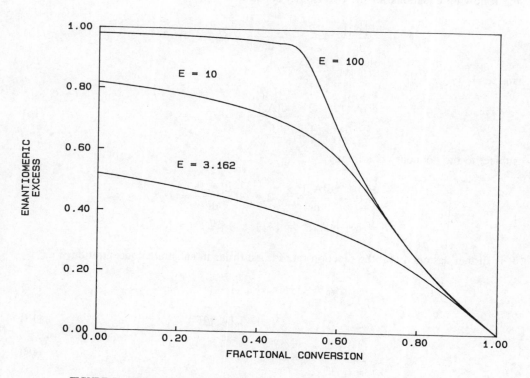

FIGURE 7. Enantiomeric excess of product B vs. conversion for several values of E.

it follows that the concentration profile for the more reactive isomer will be steeper than the profile for the less reactive one; i.e., depletion of the more reactive isomer within the porous support will be more severe than for the less reactive isomer. By integrating these concentrations over the volume of the particle and dividing by the total particle volume, the average concentration of each isomer that the enzyme actually "sees" within the solid-phase support can be determined. Obviously, the average concentration of the less reactive isomer within the support will be higher than the average concentration of the more reactive one, despite the fact that the concentrations of the two substrates outside the support may be identical. The net result is that the intrinsic stereochemical preference of the enzyme will be offset to some degree by the relative availabilities of the two isomers to the immobilized enzyme. Accordingly, the effective or observed stereospecificity of the enzyme in such a system will be less than its intrinsic E value.

A more formal analysis of this diffusional effect on stereospecificity can be carried out with a simple reaction/diffusion model. Solution of the membrane reaction/diffusion equations requires two independent boundary conditions, e.g., some combination of two concentrations or fluxes at either side of the membrane. In order to simplify the analysis for presentation here, a spherical catalyst is substituted for the membrane as the enzyme support. Considerable mathematical simplification comes as a result of the fact that the spherical particle equations require only one independent boundary condition, since the geometry of the system specifies that the flux be zero at the center of the particle. Despite the mathematical simplification, this spherical particle model retains the essential features of the membrane case and makes qualitatively correct predictions of the effect of membrane diffusional resistance on stereospecificity.

The pertinent reaction/diffusion equations for a spherical catalyst particle are given by the following equations for the two isomer substrates:

$$\frac{d^2[A_R]^*}{dr^2} + \frac{2}{r} \cdot \frac{d[A_R]^*}{dr} + v_R/D_{eff} = 0 \tag{15}$$

and

$$\frac{d^2[A_S]^*}{dr^2} + \frac{2}{r} \cdot \frac{d[A_S]^*}{dr} + v_S/D_{eff} = 0 \tag{16}$$

subject to the boundary conditions

$$r = 0 \rightarrow \frac{d[A_R]^*}{dr} = 0, \qquad \frac{d[A_S]^*}{dr} = 0$$

$$r = \delta \rightarrow [A_R]^* = [A_R], \qquad [A_S]^* = [A_S]$$

It is further assumed that the reaction rate is first order in substrate concentration, i.e.,

$$v_R = -(V_m/K_m)_R[A_R] \tag{17}$$

$$v_S = -(V_m/K_m)_R[A_S]/E \tag{18}$$

$$V_m = k_2[Ez^0] \tag{19}$$

Solution of these differential equations results in the following equations for the reactant fluxes into the supported enzyme particle:

$$N_R = (D_{eff}/\delta)[A_R]\phi(1/\tanh\phi - 1/\phi) \tag{20}$$

$$N_S = (D_{eff}/\delta)[A_S](\phi/E^{1/2})[1/\tanh(\phi/E^{1/2}) - 1/(\phi/E^{1/2})] \tag{21}$$

where the parameter ϕ is the Thiele modulus, defined as

$$\phi = \delta[(V_m/K_m)_R/D_{eff}]^{1/2} \tag{22}$$

and

$$E = \frac{(V_m/K_m)_R}{(V_m/K_m)_S} \tag{23}$$

The Thiele modulus is a measure of whether the process is reaction rate controlled (low ϕ) or diffusion controlled (high ϕ).

The relationship between enantiomeric excess, conversion, intrinsic stereospecificity, and Thiele modulus can be derived by noting that the time dependence of the extraparticle concentrations of the two isomers is proportional to their fluxes into the particle:

$$\frac{d[A_R]}{dt} \alpha - N_R \tag{24}$$

$$\frac{d[A_S]}{dt} \alpha - N_S \tag{25}$$

As before, the time variable can be eliminated by taking the ratio of the two equations,

$$\frac{d[A_R]}{d[A_S]} = \frac{N_R}{N_S} \tag{26}$$

and integrating the resulting expression. The final result for reactant A is given by the following expression:

$$\frac{\ln[(1 - c)(1 - ee)]}{\ln[(1 - c)(1 + ee)]} = E^{1/2} \frac{1/\tanh\phi - 1/\phi}{1/\tanh(\phi/E^{1/2}) - 1/(\phi/E^{1/2})} \tag{27}$$

By direct comparison of Equations 13 and 27 it can be seen that the "effective" E value for the system is given by

$$E_{eff} = E^{1/2} \frac{1/\tanh\phi - 1/\phi}{1/\tanh(\phi/E^{1/2}) - 1/(\phi/E^{1/2})} \tag{28}$$

It can be demonstrated that in the limit as $\phi \to 0$ (i.e., reaction rate control), the effective E value (E_{eff}) approaches the intrinsic value of E. However, in the limit as $\phi \to \infty$ (i.e., large diffusional resistance), the effective E value is equal to the square root of the intrinsic stereospecificity. Figure 8 shows the dependence of E_{eff} on the Thiele modulus for several values of E. The dependence of final enantiomeric purity as a function of Thiele modulus and conversion is illustrated in Figure 9.

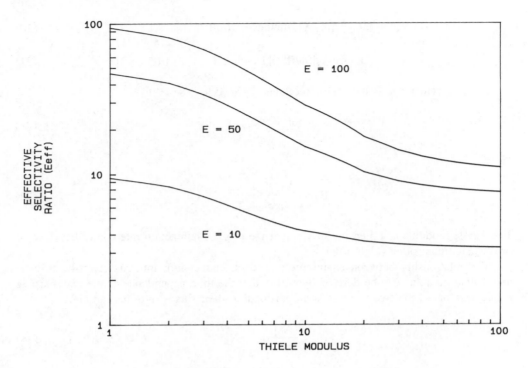

FIGURE 8. Effect of Thiele modulus on effective selectivity ratio.

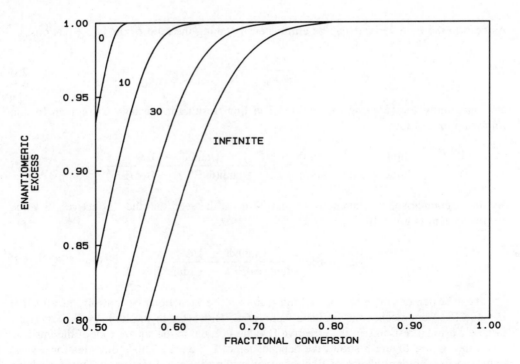

FIGURE 9. Enantiomeric excess of reactant A vs. conversion for several values of Thiele modulus.

As shown in Equations 19 and 22, the Thiele modulus increases with increasing immobilized enzyme concentration. It follows from this that the higher the enzyme concentration, the lower the effective enzyme stereospecificity and, thus, the higher the conversion required to achieve a desired enantiomeric purity of the unconverted reactant. By the same token, when the optical purity of the reaction product is of interest, a similar analysis shows a significant penalty associated with high enzyme loadings and appreciable diffusional resistance in the support.

Reduced stereospecificity in enzymatic resolutions is a penalty that must be paid for enzyme immobilization regardless of the geometry of the support, i.e., whether the enzyme is confined to a membrane or bound to a particulate support. Although the effect can never be eliminated, at least not with high-productivity bioreactors, an understanding of the effect of diffusional resistance on stereospecificity will facilitate the design of enzyme membrane reactors that are appropriately optimized for both minimum membrane area (i.e., high productivity) and production of a high-optical-purity product.

NOMENCLATURE

c	Reactant conversion
D_{eff}	Effective diffusivity of reactant
ee(A)	Enantiomeric excess for reactant A
ee(B)	Enantiomeric excess for product B
E	Enzyme stereospecificity (selectivity)
$[Ez^0]$	Total enzyme concentration in catalyst particle
[i]	Concentration of species i
[i]*	Concentration of species i inside catalyst particle
k_{2R}	Enzyme turnover number for isomer R
k_{2S}	Enzyme turnover number for isomer S
K_{mR}	Michaelis-Menten constant for isomer R
K_{mS}	Michaelis-Menten constant for isomer S
N_R	Flux of reactant R into catalyst particle
N_S	Flux of reactant S into catalyst particle
v_R	Reaction rate for isomer R
v_S	Reaction rate for isomer S
δ	Catalyst particle radius
ϕ	Thiele modulus (Equation 22)

REFERENCES

1. **Michaels, A. S. and Matson, S. L.,** Membranes in biotechnology: state of the art, *Desalination*, 53, 231, 1985.
2. **Matson, S. L.,** Membrane Reactors, Ph.D. dissertation, University of Pennsylvania, Philadelphia, 1979.
3. **Lopez, J. L.,** Carrier-Mediated Transport in Membrane Reactors: Deacylation of Benzylpenicillin, Ph.D. dissertation, University of Pennsylvania, Philadelphia, 1983.
4. **Matson, S. L. and Quinn, J. A.,** Membrane reactors in bioprocessing, *Ann. N.Y. Acad. Sci.*, 469, 152, 1986.
5. **Chen, C.-S., Fujimoto, Y., Girdaukas, G., and Sih, C. J.,** Quantitative analyses of biochemical kinetic resolutions of enantiomers, *J. Am. Chem. Soc.*, 104, 7294, 1982.

Chapter 34

PROTEIN ENGINEERING FOR THE DEVELOPMENT OF NOVEL AFFINITY SEPARATIONS

Kevin M. Ulmer

TABLE OF CONTENTS

I. INTRODUCTION: AFFINITY PURIFICATION

In a recent survey of modern protein purification techniques, Bonnerjea et al.[1] have noted that affinity separation is the second most commonly applied technique after ion exchange. Affinity methods are unique, however, in offering both the highest average purification factor (100-fold) and the highest maximum purification factor (3000-fold) of all the techniques surveyed. Indeed, these authors have found that affinity purification is at least an order of magnitude better in this respect than any other method. In addition, they have further noted that affinity techniques are also extremely versatile in that they can be utilized at essentially any stage of a purification scheme.

Affinity purification is not without its limitations, however. As noted in the survey described above, affinity ligands are relatively expensive. Affinity columns are also prone to fouling. In combination this often dictates that affinity methods not be utilized until the later stages of purification. However, maximum advantage of the tremendous purification factor afforded by affinity methods is obtained when they are employed as early in the purification scheme as possible. Furthermore, those affinity ligands which potentially offer the tightest binding and, therefore, the highest purification factor introduce a further complication for elution of the product from the column. Often extremes of pH and/or ionic strength are required which may denature or otherwise damage the protein product.

Thus, the ideal affinity column would have the following properties:

1. Low cost
2. General method for obtaining an affinity ligand to any protein of interest
3. Resistance to fouling and to inactivation by crude lysates or culture broths
4. High specificity and tight binding
5. Ability to elute with high efficiency under nondenaturing conditions
6. Long column recycle lifetime
7. Stable ligand-matrix bond (i.e., no leaching of ligand from column)

II. IMMUNOAFFINITY PURIFICATION

As noted in the purification method survey, about 20% of the affinity procedures used triazine dyes as affinity ligands. Such low-molecular-weight compounds are most commonly employed, but it is also possible to use monoclonal or polyclonal antibodies immobilized on suitable supporting matrices as affinity ligands.[2] Immunoaffinity columns are extremely attractive for protein purification. The availability of monoclonal antibody techniques now makes it generally possible to obtain high-specificity antibodies to essentially any protein or peptide of interest, thus satisfying criterion number 2 above. Usually it is possible to obtain a family of monoclonals with a range of binding constants for the protein antigen, thus providing some of the flexibility required to meet criteria 4 and 5.

Unfortunately, however, most immunoaffinity columns do not meet criteria 1, 3, and 6, but instead suffer from high cost, ligand leaching, sensitivity to fouling and inactivation, and short column lifetimes. As a consequence, immunoaffinity chromatography is generally restricted to laboratory use and is presently appropriate only for very high-unit-cost (i.e., therapeutic) proteins. At least one Food and Drug Administration (FDA)-approved biologic is manufactured using a murine monoclonal immunoaffinity step: Hoffmann-LaRoche's Roferon®-A (recombinant human interferon α-2a).[3] More general application of this powerful purification technique will require substantial reductions in cost and increases in column stability and lifetime.

III. PROTEIN ENGINEERING

The rapidly developing technology of protein engineering[4] may provide the key to the broader application of affinity purification techniques based on engineered antibodies. Through the use of site-directed mutagenesis or synthetic gene technology, it is now possible to make precise and facile changes to the amino acid sequence of any protein for which the gene has been cloned and expressed by recombinant DNA techniques. When such amino acid changes are selected on the basis of detailed knowledge of both the three-dimensional atomic structure and the biochemical/biophysical properties of the protein, it is possible to alter the structure and function of the protein in a rational, engineering fashion. Early success has already been reported in the engineering of a range of properties, including increasing stability, altering substrate specificity, modifying kinetic parameters, and shifting the pH profile (for recent reviews, see References 5 and 6).

Thermal stability has been increased by introducing novel disulfide bonds into T4 lysozyme,[7] subtilisin BPN',[8] and λ repressor,[9] while stabilization to denaturing agents has been similarly accomplished with dihydrofolate reductase.[10] Oxidation resistance in subtilisin BPN' has been achieved by replacing a sensitive methionine.[11] Other approaches to enhancing stability which are currently being explored include increased van der Waals and hydrogen bonding, additional salt bridges or ion binding sites, and the modification or replacement of loops. Further development of these methods is likely to lead to general techniques for engineering very robust proteins.

By substitution of amino acids which interact with the substrate at the active site, protein engineering has also been successfully employed to alter the substrate specificities, kinetic parameters, and pH profiles of subtilisin BPN',[12] trypsin,[13] and tyrosyl-tRNA synthetase,[14] among others.

IV. ENGINEERED ANTIBODIES

Several groups have now reported the successful production of recombinant antibodies by cloning and expressing the appropriate heavy- and light-chain genes in *Escherichia coli*,[15] yeast,[16] oocytes,[17] or mammalian cells.[18] In such systems it is now possible to consider applying protein engineering techniques in order to enhance the properties of the antibody for affinity separation.

The first target would be to dramatically reduce the cost of producing the antibody while simultaneously increasing the stability and extending the useful column lifetime. Most of the antibody molecule is, in fact, not required for antigen binding. All of the determinants are found in the two variable domains of the heavy and light chains. In limited cases it has been possible to isolate so-called F_v fragments by proteolytic cleavage of intact immunoglobulins. Such minimal antigen-binding fragments now can be produced more generally and simply by inserting a stop codon at the appropriate positions in the heavy- and light-chain genes in order to produce the truncated polypeptides. In such a case, each chain would contain only about 110 amino acids. Unfortunately, it is still difficult to obtain proper folding and assembly of these variable domains when expressed in microbial hosts, and a further simplification may be required in order to decrease the production cost for kilogram quantities of engineered antibody. It has recently been reported[19,20] that it is possible to fuse the heavy and light chains into a single polypeptide which would fold with monomolecular kinetics. Such ''single-chain'' antibodies are ultimately the goal for the general production of small, stable, relatively inexpensive antibody binding sites. Further stabilization to protease resistance or denaturation could then be achieved by employing additional protein engineering techniques such as described previously. Additionally, it should be possible to develop highly efficient immobilization procedures for such engineered antibodies by providing a unique, reactive amino acid in an optimal position on the antibody surface to avoid interference with the antigen binding site.[21]

Once a small, stable, and efficiently immobilized antibody binding site is achieved, it will then be possible to explore mechanisms for controlling the binding and release of antigen. This is likely to involve modification of electrostatic interactions in order to achieve very large changes in antigen binding, constant over very narrow pH and/or ionic strength ranges, which are still within the normal stability range of the target protein antigen. Such methods would essentially involve very precise control of the pKs of side chains by proper distribution and selection of ionizable or polar amino acids at the antibody surface.[22] An alternative approach would be to engineer the antibody to reversibly unfold and refold with high efficiency over a similarly narrow range of physical and chemical conditions[23] or to undergo a conformational change in a similar fashion.

Such an ultimate, engineered antibody should satisfy all of the criteria outlined above for the ideal affinity ligand. The research costs for developing such an engineered antibody may appear prohibitively high, and, indeed, they would be if it were necessary to repeat this entire process with every new protein to be purified. Fortunately, it is likely that much of this technology could be generalized.[20] Antibodies have a relatively invariant core structure, and the availability of many X-ray structures, including several recent ones of complexes with protein antigens,[24,25] should greatly facilitate protein engineering. Furthermore, it is likely that monoclonals will already have been isolated for analytical use in the purification process, and these would provide the appropriate starting point. With the availability of suitable probes, the cloning of the necessary portion of the required heavy and light chains is now very straightforward, and it may indeed be simpler to sequence the variable domains and simply alter the hypervariable loops in a previously cloned and already engineered single-chain antibody. It is the generality and versatility which makes antibodies so attractive for engineering as immunoaffinity ligands.

Engineered antibodies will be the obvious choice for the affinity purification of proteins and peptides, but their use may be further extended to include lower molecular weight compounds. By suitably conjugating such small molecules to higher molecular weight immunogens, it is possible to generate monoclonal antibodies which specifically bind the hapten. Such antibodies could extend the useful range of immunoaffinity techniques to higher value compounds of only a few hundred daltons molecular weight.

V. CATALYTICALLY INACTIVATED ENZYMES

An alternative strategy for developing suitable specific affinity ligands for smaller molecules would be to start with an enzyme which binds the compound of interest as either a natural substrate, product, or inhibitor. One of the easiest things to do by directed mutagenesis is to substitute a catalytically essential amino acid in the active site of the enzyme, thereby eliminating all enzymatic activity. If done properly, the inactive enzyme will still bind the target compound, but will not modify it catalytically in any way. The binding constant (and its dependence on pH, ionic strength, or the binding of an allosteric effector) can then be further engineered by modifying the amino acids involved in stabilizing the enzyme/substrate complex.[14]

In the future, such an approach may be particularly attractive for affinity purification of smaller molecules (e.g., antibiotics, hormones, growth factors, etc.), especially if the enzyme involved in production of the compound has already been subjected to protein engineering in order to enhance the biosynthesis of the compound through increased stability, activity, and immobilization efficiency. Much of the development cost for such a catalytically inactive enzyme affinity ligand could then be shared with the production development.

VI. CONCLUSION

Protein engineering will provide a new technology for the development of improved affinity purification schemes based on engineered antibodies and enzymes. By lowering the production costs, increasing the stability, and extending the useful range of such protein affinity ligands, it should be possible to greatly extend the power of affinity separations to provide lower-cost and higher-purity separations for lower unit cost products.

REFERENCES

1. **Bonnerjea, J., Oh, S., Hoare, M., and Dunnill, P.,** Protein purification: the right step at the right time, *Bio/Technology,* 4, 954, 1986.
2. **Solomon, B., Koppel, R., and Katchalski-Katzir, E.,** Use of a specific monoclonal antibody for the preparation of highly active immobilized carboxypeptidase A, *Bio/Technology,* 2, 709, 1984.
3. Roferon®-A package insert, Hoffman-LaRoche, Nutley, NJ, June 1986.
4. **Ulmer, K. M.,** Protein engineering, *Science,* 219, 666, 1983.
5. **Knowles, J. R.,** Tinkering with enzymes: what are we learning?, *Science,* 236, 1252, 1987.
6. **Oxender, D. L. and Fox, C. F.,** *Protein Engineering,* Alan R. Liss, New York, 1987.
7. **Perry, L. J. and Wetzel, R.,** Disulfide bond engineering into T4 lysozyme: stabilization of the protein toward thermal inactivation, *Science,* 226, 555, 1984.
8. **Pantoliano, M. W., Ladner, R. C., Bryan, P. N., Rollence, M. L., Wood, J. F., and Poulos, T. L.,** Protein engineering of subtilisin BPN': enhanced stabilization through the introduction of two cysteines to form a disulfide bond, *Biochemistry,* 26, 2077, 1987.
9. **Sauer, R. T., Hehir, K., Stearman, R. W., Weiss, M. A., Jeitler-Nilsson, A., Suchanek, E. G., and Pabo, C. O.,** An engineered intersubunit disulfide enhances the stability and DNA binding of the N-terminal domain of λ repressor, *Biochemistry,* 25, 5992, 1986.
10. **Villafranca, J. E., Howell, E. E., Oatley, S. J., Xoung, N., and Kraut, J.,** An engineered disulfide bond in dihydrofolate reductase, *Biochemistry,* 26, 2182, 1987.
11. **Estell, D. A., Graycar, T. P., and Wells, J. A.,** Engineering an enzyme by site-directed mutagenesis to be resistant to chemical oxidation, *J. Biol. Chem.,* 260, 6518, 1985.
12. **Estell, D. A., Graycar, T. P., Miller, J. V., Powers, D. B., Burnier, J. P., Ng, P. G., and Wells, J. A.,** Probing steric and hydrophobic effects on enzyme-substrate interactions by protein engineering, *Science,* 233, 659, 1986.
13. **Graf, L., Craik, C. S., Patthy, A., Rocziak, S., Fletterick, R. J., and Rutter, W. J.,** Selective alteration of substrate specificity by replacement of aspartic acid-189 with lysine in the binding pocket of trypsin, *Biochemistry,* 26, 2616, 1987.
14. **Fersht, A. R., Shi, J.-P., Knill-Jones, J., Lowe, D. M., Wilkinson, A. J., Blow, D. M., Brick, P., Carter, P., Waye, M. M. Y., and Winter, G.,** Hydrogen bonding and biological specificity analysed by protein engineering, *Nature (London),* 314, 235, 1985.
15. **Boss, M. A., Kenten, J. H., Wood, C. R., and Emtage, J. S.,** Assembly of functional antibodies from immunoglobulin heavy and light chains synthesized in *E. coli, Nucleic Acids Res.,* 12, 3791, 1984.
16. **Wood, C. R., Boss, M. A., Kenten, J. H., Calvert, J. E., Roberts, N. A., and Emtage, J. S.,** The synthesis and *in vivo* assembly of functional antibodies in yeast, *Nature (London),* 314, 446, 1985.
17. **Roberts, S. and Rees, A. R.,** The cloning and expression of an anti-peptide antibody: a system for rapid analysis of the binding properties of engineered antibodies, *Protein Eng.,* 1, 59, 1986.
18. **Deans, R. J., Denis, K. A., Taylor, A., and Wall, R.,** Expression of an immunglobin heavy chain gene transfected into lymphocytes, *Proc. Natl. Acad. Sci. U.S.A.,* 81, 1292, 1984.
19. **Klausner, A.,** 'Single-chain' antibodies become a reality, *Bio/Technology,* 4, 1041, 1986.
20. **Ladner, R. C.,** U.S. Patent 4,704,692, 1987.
21. **Dao-Pin, S., Alber, T., Bell, J. A., Weaver, L. H., and Matthews, B. W.,** Use of site-directed mutagenesis to obtain isomorphous heavy-atom derivatives for protein crystallography: cysteine-containing mutants of phage T4 lysozyme, *Protein Eng.,* 1, 115, 1987.
22. **Wells, J. A., Powers, D. B., Bott, R. R., Graycar, T. P., and Estell, D. A.,** Designing substrate specifically by protein engineering of electrostatic interactions, *Proc. Natl. Acad. Sci. U.S.A.,* 84, 1219, 1987.

23. **Goto, Y. and Hamguchi, K.,** Role of amino-terminal residues in the folding of the constant fragment of the immunoglobulin light chain, *Biochemistry,* 26, 1879, 1987.

24. **Colman, P. M., Laver, W. G., Varghese, J. N., Baker, A. T., Tulloch, P. A., Air, G. M., and Webster, R. G.,** Three-dimensional structure of a complex of antibody with influenza neuraminidase, *Nature (London),* 326, 358, 1987.

25. **Amit, A. G., Mariuzza, R. A., Phillips, S. E. V., and Poljak, R. J.,** Three dimensional structure of an antigen-antibody complex at 2.8 A resolution, *Science,* 233, 747, 1986.

Chapter 35

MICROBIAL METAL LEACHING AND RESOURCE RECOVERY PROCESSES

G. J. Olson and F. E. Brinckman

TABLE OF CONTENTS

I. INTRODUCTION

Microbial processing of metals is not a new technology; it has been used, at least in the case of ore leaching, for centuries. Specifically, copper was recovered by man as early as 1000 B.C. from metal-laden waters which passed through copper ore deposits and mines.[1] However, the microbial role in the process of copper sulfide ore dissolution was only discovered about 30 years ago. Microbially assisted leaching of low-grade copper ores and wastes at mines in the western U.S. has been practiced for decades.[2] Since the discovery in the 1950s of the participation of bacteria in metal ore leaching processes, a small group of international investigators has continued to study the bioleaching of ores and the physiology of ore leaching bacteria, most notably *Thiobacillus ferrooxidans*. Recently, the depletion of high-grade metal ore deposits, stricter environmental laws, and the high cost of conventional mining and smelting have stimulated interest in exploring alternative metal recovery processes for a variety of metal ores, concentrates, and metal-containing wastes.[3] In the past few years, a new in-place uranium bioleaching operation has begun in Canada.[4] Microbial preleaching of gold-containing host rock to assist in the cyanide process for gold recovery seems to be on the verge of commercialization.[5] Small companies have formed which are developing metal biorecovery techniques. Many scientists and engineers are studying bioleaching of strategic metals, bioprocessing of coal, and metal recovery from process or waste streams.[1,6-8]

This surge of interest in the recovery of metals by bioprocessing may stem, in part, from the increased awareness of and possibilities engendered by genetic engineering and the publicity surrounding the "new biotechnology".

However, we are still some distance from applying genetic engineering to improve metal recovery processes, especially ore bioleaching. Much basic work remains to be done on the physiology, ecology, and engineering aspects of microbial transformation of metals before we can make rational choices for desired characteristics to be enhanced in the organisms. In addition, there are severe technical obstacles to be overcome in developing genetic systems for the ore-leaching bacteria, since they grow at low pH and are difficult to grow on solid media. Nevertheless, progress has been made in recent years. Several groups have begun studying genetic systems for the thiobacilli.[9,10] Also, new organisms have been discovered in the past few years which may be applied to metal sulfide oxidation. Some of these are thermophilic bacteria[11] which may be important in metal sulfide oxidation in self-heating leach dumps. Much has been learned in recent years about metal resistance mechanisms,[12] which often involve metal transformations, that might find commercial application. Finally, engineering science is becoming more integrated into research, which is a necessary component of commercialization.

II. ENGINEERING CONSIDERATIONS IN METAL BIORECOVERY

Pure-culture, controlled bioreactor processes are associated with the production of pharmaceuticals, enzymes, and other organic compounds. However, the bioprocessing of inorganic materials and ores is characterized by extreme conditions of pH, temperature, heavy metal content, heterogeneous conditions, and mixed cultures. Reactions may occur on an immense scale, as in dump leaching operations, and have long processing times.

The extent and nature of bioprocess engineering for metal recovery, at present and in the future, depend on many factors. Especially important are the volume of material to be processed and the value of the products. For example, bioprocesses for the recovery of gold from ore concentrates might be made cost effective by employing grinding of the substrate for maximizing the rates and extent of the bioreaction. On the other hand, copper ore bioleaching or prospective coal desulfurization using microorganisms involves processing large volumes of material of relatively lower value. Simple engineering designs and lower processing costs will be necessary in these systems.

III. MICROBIAL LEACHING OF ORES AND WASTES FOR METAL RECOVERY

It is well known that the activities of acidophilic bacteria in the western U.S. copper leaching dumps play an important role in copper dissolution.[1,2] The activities of these bacteria also contribute to production of acidic coal mine drainage.[13] Microbial oxidation of pyrites (FeS_2) also assists in uranium recovery at the Denison mine in Canada.[4] The ferric sulfate solutions produced by this microbial activity oxidize the uranium to more soluble species which are recovered from solution. Better control of such bioleaching processes is sought, either to maximize metal dissolution or to minimize the leaching of acid and iron salts into environments surrounding active and abandoned coal mines. In addition, prospects for microbial leaching of strategic and precious-metal-containing ores or concentrates are being investigated worldwide.

Bioleaching of copper and uranium ores and mine wastes occurs due to the activity of certain unusual bacteria that oxidize metal sulfides to soluble metal sulfates in acidic solutions. *T. ferrooxidans* is the most extensively studied of these bacteria. It obtains energy from the oxidation of reduced sulfur species (e.g., S^0, $S_2O_3^{2-}$) and from certain reduced metal species as well, most importantly ferrous ions. This organism can be readily grown in 0.01 N H_2SO_4 containing an energy source such as pyrite or sphalerite (ZnS) and inorganic salts supplying P, N, Mg, and Ca. Carbon dioxide can serve as the sole carbon source (fixed by a mechanism very similar to that of green plants), and oxygen is required as an electron acceptor for the oxidation of the reduced iron or sulfur energy source. Thus, the organism prospers in an inorganic environment at low pH. *T. ferrooxidans* is found associated with metal sulfide weathering and mining environments around the world.

Pyrite is an important substrate for the organisms in many mining environments. It is the principal metal sulfide associated with coal deposits and is often found in commercial metal sulfide ore deposits as well. *T. ferrooxidans* accelerates pyrite oxidation, generating acidic ferric sulfate solutions which readily oxidize and solubilize other metal sulfides. In addition, this organism oxidizes many metal sulfides directly. A combination of these reactions probably occurs in copper dump leaching environments to release copper.

However, *T. ferrooxidans* is not the only microorganism found in low pH, metal-rich environments associated with metal sulfide weathering or mining. Other iron- and sulfur-oxidizing bacteria, some thermophilic, have been isolated from such environments, along with acidophilic heterotrophic bacteria.[14,15] We know relatively little about the microbial ecology of the mining environment and its effect on metal dissolution. Nonetheless, laboratory experiments employing mixed cultures of bacteria from mining environments often show enhanced metal sulfide leaching rates when compared to rates using pure cultures of *T. ferrooxidans*.[16,17]

In addition to metal ore deposits, ore concentrates, and mining wastes, certain other types of wastes may be amenable to bioleaching. For example, bacterial leaching of metal sulfides for metal recovery from sewage sludge has been discussed.[18] Commerical smelting and refining of lead produces furnace slags and mattes containing impurity metals present as sulfides.[19] These residues are chemically leached,[19] but may also be amenable to bioleaching. We have been evaluating the leaching of cobalt from samples of the furnace matte using strains of *T. ferrooxidans*. The residues were added to a mineral-salts medium at a pulp density of 4% and inoculated with pyrite-grown *T. ferrooxidans*. Soluble cobalt increased with time to a maximum of several hundred milligrams per liter after 6 to 7 weeks

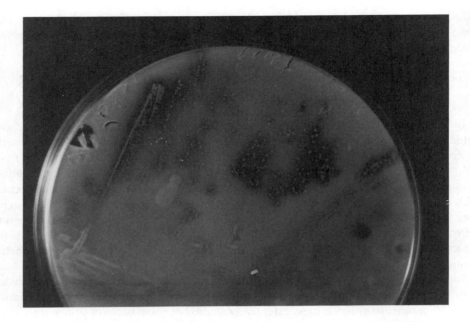

FIGURE 1.   Colonies of calcium phosphate-dissolving microorganisms surrounded by clear zones in agar. The defined medium contained a suspension of calcium phosphate as the sole source of phosphorus.

of bioleaching. Sterile controls showed little Co dissolution. We could maintain cultures of *T. ferrooxidans* on the metal sulfides in the furnace matte as a sole source of energy.

IV. MICROBIAL UPGRADING OF ORES AND FOSSIL FUELS

Undesirable components of ores and fossil fuels may be removed through the action of microorganisms. Examples include the removal of phosphorus from iron ores, desulfurization of coal, and the removal of pyrite from certain gold ores.

Some domestic deposits of iron ore contain elevated levels of insoluble phosphate minerals. Inexpensive methods for removal of at least part of this phorphorus are sought because phosphorus impurities can adversely affect the properties of finished steels. Removal of phosphorus is most easily accomplished at the raw materials stage.

It has been known for some time that extracellular products of certain soil microorganisms dissolve insoluble phosphates.[20,21] These products have in some cases been identified as organic or mineral acids. Phosphate-dissolving microoganisms are readily isolated from soils using a defined solid (agar) growth medium containing a finely suspended insoluble phosphate, e.g., $Ca_3(PO_4)_2$, as the sole source of phosphorus. Colonies of phosphorus-dissolving microorganisms are easily distinguishable on the agar surface (Figure 1). We have found that culture filtrates of a fungus isolated from Chesapeake Bay sediments dissolve insoluble phosphate minerals and remove phosphorus from a Michigan iron ore containing elevated levels (0.06%) of phosphorus. The culture filtrates had pH values of <3.0 (initial pH was 6.5 to 7.0), suggesting formation of organic acids from glucose, which was the sole carbon substrate in the medium. Liquid chromatography and Fourier transform infrared spectroscopy techniques can be used to identify metabolites in the culture medium and their effects on ore dephosphorization.

Several processes for bioleaching of phosphorus from iron ore may be possible, including application of purified metabolites, whole culture fluids from fermenters, or growth of the organisms in the ore or ore concentrate. Collaboration of microbiologists and chemists with chemical and mining engineers is necessary to evaluate the most cost-effective options.

Microbial processes for cleaning or upgrading fossil fuels have also recently received much attention, although the concept was first addressed in the late 1950s and early 1960s.[22,23] Substantial laboratory work worldwide over the past decade has investigated the microbial potential for desulfurization of coal. Most studies have employed *T. ferrooxidans* to remove pyritic sulfur from coal. Although some authors contend that the process is cost effective for certain coals,[24] there has yet to be an industrial demonstration project. Relatively slow rates of biodesulfurization of coal and the costs of incorporating a water-based desulfurization scheme into existing coal-burning processes at power generating plants have been the chief objections to the microbial process.

Reports of microbial removal of organic sulfur from coal have also appeared[25,26] and are of interest because organic sulfur removal from coal by other techniques is very difficult. However, until analytical difficulties associated with measuring organic sulfur in coal are overcome and until we better understand the speciation of sulfur in coal, it will be difficult to prove conclusively that organic sulfur is removed from coal by microbial processes.

Other research has been directed toward coal conversion, i.e., degradation of low-rank coals to liquid products. A variety of fungi have been found which convert certain low-rank coals into liquid products.[27,28] The product has been partially characterized; it is highly polar with a high molecular weight. It is not clear how useful these products may be. However, some investigators have proposed that anaerobic microbial consortia may be used on a large scale, perhaps in underground caverns, to convert lignite coal to methane gas.[29] The coal would first be liquified by heat and alkali, or perhaps by the above bioliquefaction route.

V. THE IMPORTANCE OF STANDARDS

Research into bioleaching of ores and removal of pyritic sulfur from coal is conducted in many countries, and many studies are published which report on the dissolution of metal sulfides by various strains of *T. ferrooxidans* and other organisms. It is still uncertain as to which properties of *T. ferrooxidans,* if enhanced, would lead to more rapid oxidation of pyrite or other metal sulfides. Obviously, the strains which most rapidly oxidize these substrates are sought. However, it is difficult to compare leaching rates among the many laboratories conducting such studies because of the diversity of incubation conditions, measurements, strains, substrates, and expressions of leaching rates. The availability of a well-characterized metal sulfide reference material and the use of standardized procedures to conduct bioleaching tests and report results would greatly assist in data comparison. Pyrite seems an ideal initial candidate for a standard reference material (SRM) given its importance in connection with microbial ore leaching and coal desulfurization. An interlaboratory comparison of pyrite bioleaching rates by a specified strain of *T. ferrooxidans* would show both the feasibility and the limits of data comparison.

As a first step in this direction, we have begun to evaluate the prospects for standardization of bioleaching tests and the potential for producing a pyrite SRM. These activities are being conducted with the American Society for Testing and Materials Committee E-48 on Biotechnology and with the National Institute of Standards and Technology (NIST) Office of Standard Reference Materials. Very few studies report variance in leaching rates obtained in replicate flasks. We evaluated leaching rate in triplicate flasks using a South Carolina pyrite and three different strains of *T. ferrooxidans.* Soluble total iron (after centrifugation of the sample) was determined by an *o*-phenanthroline colorimetric procedure,[30] and the rate of pyrite dissolution was calculated from the linear portion of the metal release curve as discussed by Torma and Sakaguchi.[31] The three strains showed pyrite leaching rates of 1.7 to 2.2 mg Fe per liter per hour (Figure 2). The relative standard deviations of the pyrite

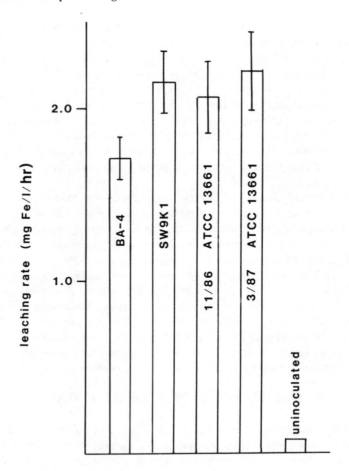

FIGURE 2.   South Carolina pyrite leaching rates (bars denote standard deviation; triplicate flasks) for three strains of *T. ferrooxidans* in shake flasks.

bioleaching rates ranged from 7 to 11%, suggesting that interlaboratory comparisons of bioleaching rates should be possible. The reproducibility of Fe leaching rates was tested again 4 months later with the ATCC strain and was nearly identical (Figure 2). After additional studies on other pyrite minerals, we will suggest standardized testing protocols, conduct interlaboratory testing, and perhaps produce a well-characterized pyrite SRM.

VI. METAL ACCUMULATION OR PRECIPITATION BY MICROORGANISMS

Microorganisms may selectively or nonselectively accumulate and precipitate metals from solution by a variety of mechanisms.

In general, the microbial cell envelope has a net negative charge and a number of negatively charged functional groups which can bind metal cations rather nonselectively.[32,33] Nonliving biomass preparations have been found to be effective sorbents for heavy metals[8,34] and may be used for selective metal recovery by manipulation of solution pH and complexing agents.[35] Less sophisticated applications of biomass for metal removal from waste streams are presently used commercially. For example, in Missouri, mining wastes are passed through meandering streams where algae, bacteria, and other organisms remove dissolved lead and zinc.[36] Microbial processes are used to treat metal- and cyanide-containing wastewaters emanating from gold mines. The Homestake Mine in South Dakota, the leading gold producer in the nation, has a 5.5 million gal/d wastewater treatment plant using rotating biological contractors with adsorbed microbial populations for adsorption of toxic metals and for degrading cyanides to ammonia and nitrates.[37]

Microorganisms also metabolically reduce and precipitate oxidized forms of metals and metalloids. For example, oxidized selenium species are reduced to elemental selenium by certain microorganisms.[38] In this case, the selenium species act as electron acceptors in respiration. Controlled treatment systems employing organisms which carry out this reaction may be useful in cleaning up selenium-contaminated waters in the western U.S. Magneto-tactic bacteria accumulate soluble iron species and intracellularly synthesize fine particles of magnetite.[39]

Metabolic products may also precipitate metals. Examples include hydrogen sulfide (produced on a large scale in nature by sulfate-reducing bacteria), which reacts with many kinds of metal ions in solution to form highly insoluble metal sulfides. Other biogenic metabolites, such as trimethyl tin cations, reduce certain metals to their elemental forms via a reductive demethylation mechanism.[40]

VII. SUMMARY AND FUTURE PROSPECTS

There is a rapidly expanding interest in applying microbial processing for metal dissolution and recovery from ores and wastes. We are still learning about mechanisms and routes of microbial metal transformations. New measurement methods and standards directed toward characterizing such heterogeneous systems are needed. Process optimization manipulations through bioengineering will range from simple and inexpensive to more complex, depending on product value. As we learn more about the selective biotransformation of metals in mixtures, it may be possible to engineer "intelligent" bioprocessing systems where the metal of interest could not only be removed, but recovered in a specified oxidation state, particle size, or molecular composition. This may require multistep or multiorganism configurations.

REFERENCES

1. **Brierley, C. L., Kelly, D. P., Seal, K. J., and Best, D. J.,** Materials and biotechnology, in *Biotechnology, Principles and Applications*, Higgins, I. J., Best, D. J., and Jones, J., Eds., Blackwell Scientific, Oxford, 1985, 163.
2. **Duncan, D. W., Walden, C. C., Russell, P. C., and Lowe, E. A.,** Recent advances in the microbiological leaching of sulfides, *Trans. Soc. Min. Eng. AIME*, 238, 122, 1967.
3. **Ritchey, G. M.,** Hydrometallurgy — 10 years later and a look to the future, *Hydrometallurgy*, 15, 1, 1985.
4. **Wadden, D. and Gallant, A.,** The in-place leaching of uranium at Denison Mines, *Can. Metall. Q.*, 24, 127, 1985.
5. **Lawrence, R. W. and Bruynesteyn, A.,** Biological pre-oxidation to enhance gold and silver recovery from refractory pyritic ores and concentrates, *CIM Bull.*, 76, 107, 1983.
6. **Olson, G. J. and Brinckman, F. E.,** Bioprocessing of coal, *Fuel*, 65, 1638, 1986.
7. **Olson, G. J. and Kelly, R. M.,** Microbiological metal transformations: biotechnical applications and potential, *Biotechnol. Prog.*, 2, 1, 1986.
8. **Hutchins, S. R., Davidson, M. S., Brierley, J. A., and Brierley, C. L.,** Microorganisms in the reclamation of metals, *Annu. Rev. Microbiol.*, 40, 311, 1986.
9. **Holmes, D. S. and Yates, J. R.,** Genetic engineering of biomining organisms, *World Biotechnol. Rep.*, 2, A67, 1984.

10. **Woods, D. and Rawlings, W.,** Molecular genetic studies on the thiobacilli and the development of improved biomining bacteria, *BioEssays,* 2, 8, 1985.
11. **Wood, A. P. and Kelly, D. P.,** Autotrophic and mixotrophic growth of three thermoacidophilic iron-oxidizing bacteria, *FEMS Microbiol. Lett.,* 20, 107, 1983.
12. **Williams, J. W. and Silver, S.,** Bacterial resistance and detoxification of heavy metals, *Enzyme Microb. Technol.,* 6, 530, 1984.
13. **Lundgren, D. C., Vestal, J. R., and Tabita, F. R.,** The microbiology of mine drainage pollution, in *Water Pollution Microbiology,* Mitchell, R., Ed., John Wiley & Sons, New York, 1972, 69.
14. **Marsh, R. M. and Norris, P. R.,** Mineral sulfide oxidation by moderately thermophilic acidophilic bacteria, *Biotechnol. Lett.,* 5, 585, 1983.
15. **Harrison, A. P., Jr.,** *Acidiphilium cryptum* gen. nov., sp. nov., hetero-trophic bacterium from acidic mineral environments, *Int. J. Syst. Bacteriol.,* 31, 327, 1981.
16. **Dugan, P. R. and Apel, W. A.,** Microbiological desulfurization of coal, in *Metallurgical Applications of Bacterial Leaching and Related Microbiological Phenomena,* Murr, L. E., Torma, A. E., and Brierley, J. A., Eds., Academic Press, New York, 1978, 223.
17. **Groudev, S. N.,** Oxidation of arsenopyrite by pure and mixed cultures of *Thiobacillus ferrooxidans* and *Thiobacillus thiooxidans, C. R. Acad. Bulg. Sci.* 34, 1139, 1981.
18. **Schönborn, W. and Hartmann, H.,** Bacterial leaching of metals from sewage sludge, *Eur. J. Appl. Microbiol. Biotechnol.,* 5, 305, 1978.
19. **Kennedy, D. C., Becker, A. P., and Worcester, A. A.,** Development of an ion exchange process to recover cobalt and nickel from primary lead smelter residues, in Proc. Conf. Metal Speciation, Separation and Recovery, Industrial Waste Elimination Research Center, Illinois Institute of Technology, Chicago, July 1986, VII-1.
20. **Agnihotri, V. P.,** Solubilization of insoluble phosphate by some soil fungi isolated from nursery seedbeds, *Can. J. Microbiol.,* 16, 877, 1970.
21. **Duff, R. B., Webley, D. M., and Scott, R. O.,** Solubilization of minerals and related materials by 2-ketogluconic acid producing bacteria, *Soil Sci.,* 95, 105, 1963.
22. **Silverman, M. P., Rogoff, M. H., and Wender, I.,** Removal of pyritic sulphur from coal by bacterial action, *Fuel,* 42, 113, 1963.
23. **Zarubina, Z. M., Lyalikova, N. N., and Shmuk, Y. I.,** Investigation of the microbiological oxidation of the pyrite in coal, *Izv. Akad. Nauk SSSR Otd. Tekh. Nauk Metall. Topl.,* 117, 1959; as cited in **Schönborn, W. and Hartmann, H.,** *Eur. J. Appl. Microbiol. Biotechnol.,* 5, 305, 1978.
24. **Dugan, P. R.,** Microbiological desulfurization of coal and its increased monetary value, *Biotechnol. Bioeng. Symp.,* 16, 185, 1986.
25. **Kargi, F. and Robinson, J. M.,** Removal of organic sulfur from coal using the thermophilic organism *Sulfolobus acidocaldarius, Fuel,* 65, 397, 1986.
26. **Isbister, J. B. and Kobylinski, E. A.,** Microbial desulfurization of coal, in *Processing and Utilization of High Sulfur Coals,* Attia, Y. A., Ed., Elsevier, Amsterdam, 1985, 627.
27. **Wilson, B. W., Bean, R. M., Franz, J. A., Thomas, B. L., Cohen, M. S., Aaronson, H., and Gray, E. T., Jr.,** Microbial conversion of low-rank coal: characterization of biodegraded product, *Energy Fuels,* 1, 80, 1987.
28. **Cohen, M. S. and Gabriele, P. D.,** Degradation of coal by the fungi *Polyporus versicolor* and *Poria monticola, Appl. Environ. Microbiol.,* 44, 23, 1982.
29. **Wise, D. A.,** Methane production from coal-derived materials, in *Proc. Department of Energy Biological Treatment of Coals Workshop,* Herndon, VA, June 23 to 25, 1986, 226.
30. **American Society for Testing and Materials,** Standard test methods for iron in water, in *Annual Book of ASTM Standards,* Vol. 11.01, American Society for Testing and Materials, Philadelphia, 1986, 523.
31. **Torma, A. E. and Sakaguchi, H.,** Relation between the solubility product and the rate of metal sulfide oxidation by *Thiobacillus ferrooxidans, J. Ferment. Technol.,* 56, 173, 1978.
32. **Hoyle, B. and Beveridge, T. J.,** Binding of metallic ions to the outer membrane of *Escherichia coli, Appl. Environ. Microbiol.,* 46, 749, 1983.
33. **Beveridge, T. J. and Murray, R. G. E.,** Uptake and retention of metals by cell walls of *Bacillus subtilis, J. Bacteriol.,* 127, 1502, 1976.
34. **Darnall, D. W., Greene, B., Henzl, M. T., Hosea, J. M., McPherson, R. A., Sneddon, J., and Alexander, M. D.,** Selective recovery of gold and other metals from an algal biomass, *Environ. Sci. Technol.,* 20, 206, 1986.
35. **Greene, B., Hosea, M., McPherson, R., Henzl, M., Alexander, M. D., and Darnall, D. W.,** Interaction of gold(I) and gold(III) complexes with algal biomass, *Environ. Sci. Technol.,* 20, 627, 1986.
36. **Gale, N. L. and Wixson, B. G.,** Removal of heavy metals from industrial effluents by algae, *Dev. Ind. Microbiol.,* 20, 259, 1978.

37. **Whitlock, J. L. and Mudder, T. I.,** The Homestake wastewater treatment process: biological removal of toxic parameters from cyanidation wastewaters and bioassay effluent evaluation, in *Fundamental and Applied Biohydrometallurgy,* Lawrence, R. W., Branion, R. M. R., and Ebner, H. G., Eds., Elsevier, Amsterdam, 1986, 327.
38. **Ehrlich, H. L.,** *Geomicrobiology,* Marcel Dekker, New York, 1981, 301.
39. **Blakemore, R. P.,** Magnetotactic bacteria, *Annu. Rev. Microbiol.,* 36, 217, 1982.
40. **Brinckman, F. E. and Olson, G. J.,** Chemical principles underlying bioleaching of metals from ores and solid wastes, and bioaccumulation of metals from solution, *Biotechnol. Bioeng. Symp.,* 16, 35, 1986.

Chapter 36

COAL BIOSOLUBILIZATION

Linda L. Henk, Loni Gibbs, M. N. Karim, and James C. Linden

TABLE OF CONTENTS

I. INTRODUCTION

Several laboratories have examined biosolubilization of low-ranked coals by streptomycetes[1-3] and fungi.[3,4] The objective of our research at Colorado State University was to examine the kinetics of coal biosolubilization by several microorganisms known to degrade lignin or by fungal cultures that had been isolated from coal. The scope of our research includes the following parameters: (1) pretreatment of coal, (2) screening for biosolubilization, (3) shake-flask experiments, and (4) coal liquefaction by pyridine and biosolubilization.

A low-ranked Texas lignite was subjected to nitric acid oxidation[2,4] and to ammonia freeze explosion (AFEX).[5] These pretreated coals, along with untreated lignite, were used either in surface culture biosolubility tests or shake-flask experiments.

Preliminary results indicate that nitric acid pretreatment of Texas lignite is an effective pretreatment. Liquefaction of untreated and AFEX-treated coal is much slower than nitric acid-pretreated coal. Coal liquefaction by pyridine is examined as a possible assay method for the amount of coal in biosolubilization broths.

II. MATERIALS AND METHODS

A. PRETREATMENT OF COAL

1. Nitric Acid-Oxidized Coals

Texas lignite oxidized with nitric acid was received from Idaho National Engineering Laboratory (INEL), Idaho Falls. Coal was first sieved through -20-mesh screens, oxidized by 8 M nitric acid for 48 h, and then extracted with water.[6]

2. Ammonia Freeze-Explosion Coals

Texas lignite, -20 mesh, was subjected to AFEX pretreatment.[5] AFEX pretreatment, developed by Dale at Colorado State University, has been used successfully as a pretreatment of lignocellulosics prior to cellulase hydrolysis. Conditions of the pretreatment follow: 125 g Texas lignite, 250 g ammonia, and 30% water (w/w), at 250 psig for 60 min. Immediately after pretreatment, the coal was removed from the reactor, sealed tightly, and frozen for later use.

B. SCREENING FOR BIOSOLUBILIZATION

The following microorganisms were screened for ability to liquefy samples of control, nitric acid-pretreated, and AFEX-pretreated coals: *Streptomyces viridosporus* T7a, *S. flavovirens* #28, *S. setonii* #75, *Penicillium* sp. RWL-5, fungus ACL-13, and fungus YML-21.

Cultures were grown on yeast-malt agar (YMA, Difco Laboratories, Detroit) for 8 d before an arbitrary amount of sterile coal was sprinkled on the culture surface.

All coal types were sterilized at 121°C for 30 min prior to use. Cultures were observed visually for any sign of liquefaction at 3 and 24 h after coal addition.

C. SHAKE-FLASK EXPERIMENTS

Parameters for shake-flask experiments follow: no pH control, 5% inoculum, 105 rpm, and ambient temperature. Organisms studied were *Penicillium* sp. RWL-5 and *S. setonii*. The cultures were grown on yeast-malt broth (YMB, Difco). Samples were taken from day zero with RWL-5 on a timely basis in order to examine biosolubilization in cell-free filtrates vs. time. In addition, a 48-h culture received 1.0 g of sterile nitric acid-pretreated coal. This coal was added directly to the cells in suspension and was sampled after an additional 48 h of contact.

S. setonii 7-day cultures received coal directly to examine biosolubilization in the presence of cells. At timely intervals, flasks were filtered and the filtrates were examined at 500 nm for evidence of liquefaction of coal.

D. COAL LIQUEFACTION
1. Pyridine Solubilization
The pyridine liquefaction of coal is used to create standardized curves of concentration vs. absorbance. Each coal type should be subjected individually to liquefaction by pyridine in order to establish a standard curve for that particular kind of coal. These curves can then be used to determine the amount of coal solubilized by various microorganisms from the measured absorbance of the culture medium.

The procedure is as follows:

1. Weigh out a known amount of coal into a glass scintillation vial or test tube. Establish a series of concentrations to be used for the standard curve.
2. Add a known amount of pyridine to the coal.
3. Allow the coal to solubilize in the pyridine for a set time period, from 1 to 3 h. Prior to taking a sample, mix gently. Allow the coal to settle before a sample is taken.
4. Dilute each sample in the medium used for fermentation experiments (e.g., YMB).
5. Scan the pyridine:coal dilutions from 750 to 200 nm with the appropriate light source (either visible or UV) to establish the maximum absorbances. Use the medium as a reference. (Pyridine has a maximum absorbance of 260 nm.)
6. Determine the amount of residual coal which has not been dissolved. This establishes the actual concentration of coal in the solution. Filter the residual coal through preweighed Gooch crucibles, wash thoroughly with distilled water, and dry overnight in an oven at 100°C. Reweigh and calculate the actual concentration of coal in solution by difference.
7. Once the maximum absorbance has been determined, establish a standard curve at that particular wavelength for absorbance vs. actual concentration. Determine the concentration of dissolved coal in the fermentation broths by comparison to the standard curves for each coal type.

2. Biosolubilization
Nitric acid-pretreated coal and control coal (1.0-g amounts) were added to 10 ml of cell-free culture filtrate. The coal was allowed to solubilize for 24 h before it was filtered, and the filtrate was analyzed spectrophotometrically for coal liquefaction.

III. RESULTS

A. SCREENING FOR BIOSOLUBILIZATION
In Table 1, the results of the screening experiments are presented. After 3 h, some liquefaction of AFEX coal was observed with all six test cultures. However, *Penicillium* sp. RWL-5 was the only organism to sufficiently liquefy AFEX material in 24 h. Liquefaction of untreated and AFEX coal is much slower than that of nitric acid coal. The three streptomycetes and *Penicillium* sp. RWL-5 appeared to be the organisms of choice. All were capable of fairly complete liquefaction of nitric acid coal. *S. flavovirens* #28 and the *Penicillium* sp. also demonstrated an ability to liquefy either AFEX or control coals.

B. SHAKE-FLASK EXPERIMENTS AND COAL LIQUEFACTION
Preliminary comparisons were made of cell-free filtrates of RWL-5 fermentations, which were used to liquefy control coal and nitric acid-pretreated coal, with pyridine-liquefied

TABLE 1
Results: Screening for Biosolubilization

Culture	Coal type	3 h	24 h
YML-21, fungus	Control	None	None
	AFEX	Slight (2 pieces)	Slight $+/-$
	HNO$_3$	None	Slight $+/-$
ACL-13, fungus	Control	None	None
	AFEX	Fair $+$	Fair $+$
	HNO$_3$	None	Good $+++$
Penicillium sp. RWL-5	Control	None	None
	AFEX	Moderate $++$	Good $+++$
	HNO$_3$	None	Excellent $++++$
Streptomyces flavovirens	Control	Slight $+/-$	Slight $+/-$
#28	AFEX	Slight	Slight $+/-$
	HNO$_3$	Moderate $++$	Excellent $++++$
S. setonii	Control	None	None
#75	AFEX	Slight $+/-$	None
	HNO$_3$	Moderate $++$	Excellent $+++$
S. viridosporus	Control	Slight $+/-$	None
T7A	AFEX	Slight $+/-$	None
	HNO$_3$	Moderate $++$	Excellent $++++$

Note: Visual observation of coal biosolubilization at 3 and 24 h after coal came in contact with fungal mat surface. None: no liquifaction; slight $+/-$: 2—3 droplets; fair $+$: 10% liquifaction; moderate $++$: 25% liquifaction; good $+++$: 50% liquifaction; excellent $++++$: 70% liquifaction.

coal. Very low concentrations of coal were present in the medium after a 24-h period. The concentration of control coal was maximally 2.5×10^{-3}, while the concentration of nitric acid-pretreated coal was 1.0×10^{-5} mg/ml.

Nitric acid coal was added directly to RWL 5 cultures which were 48 h old, and the coal was allowed to solubilize for an additional 48 h. The concentration of nitric acid-pretreated coal in solution corresponded to 0.0245 mg/ml, or roughly a 1000-fold increase when compared to that solubilized in the cell-free filtrate.

S. setonii was unable to liquefy either AFEX or the control coal after 48 h of contact time. This result was consistent with the preliminary screening results obtained with agar plating.

IV. CONCLUSIONS

These preliminary results indicate that nitric acid pretreatment of Texas lignite is an effective treatment prior to biosolubilization. Surface cultures of the streptomycetes and the *Penicillium* sp. were capable of producing greater concentrations of solubilized products when compared to biosolubilization by cell-free filtrates. This suggests that direct contact with cells is necessary for liquefaction to occur, perhaps due to an oxygen dependency in the reaction.

ACKNOWLEDGMENTS

This work was partially supported by EG & G Idaho Company, Idaho National Engineering Laboratory, which provided the coal and the microorganisms for this work, and by the Colorado State University Experiment Station under Series No. 383..

REFERENCES

1. **Strandberg, G. W. and Lewis, S. N.,** Solubilization of coal by an extracellular product from *Streptomyces setonii* 75Vi2, *J. Ind. Microbiol.,* 1, 371, 1987.
2. **Crawford, D. L., Gupta, R. K., and Spiker, J. K.,** Biotransformation of Coal by Ligninolytic *Streptomyces,* paper submitted to the Idaho Agricultural Experiment Station regarding U.S. Department of Energy grant DE-FG07-86ER13586, 1986.
3. **Quigley, D. R., Wey, J. E., Breckenridge, C. R., and Stoner, D. L.,** The influence of pH on biological solubilization of oxidized, low rank coal, *Resour. Recycling,* 1, 163, 1988.
4. **Strandberg, G. W. and Lewis, S. N.,** A method to enhance the microbial liquefaction of lignite coals, *Biotechnol. Bioeng. Symp.,* 17, 153, 1986.
5. **Dale, B. E. and Moriera, M. J.,** A freeze-explosion technique for increasing cellulose hydrolysis, *Biotechnol. Bioeng.,* 12, 31, 1982.
6. **Wey, J. E.,** Standard Conditions for Nitric Acid Oxidation of Coal, personal communication, Idaho National Engineering Laboratory, Idaho Falls, ID, 1987.

Index

INDEX

P